Surface Preparation and Cleaning Conference (SPCC 2018)

Cambridge, Massachusetts, USA
10-11 April 2018

ISBN: 978-1-5108-6214-2

Printed from e-media with permission by:

Curran Associates, Inc.
57 Morehouse Lane
Red Hook, NY 12571

Some format issues inherent in the e-media version may also appear in this print version.

Printed by Curran Associates, Inc. (2018)

Surface Preparation and Cleaning Conference (SPCC)
c/o Mark Thirsk
Linx Consulting-PO Box 384
Mendon, MA 01756-0384
USA

Phone: 617-273-8837

mthirsk@linx-consulting.com

Additional copies of this publication are available from:

Curran Associates, Inc.
57 Morehouse Lane
Red Hook, NY 12571 USA
Phone: 845-758-0400
Fax: 845-758-2633
Email: curran@proceedings.com
Web: www.proceedings.com

SPCC 2018 Agenda

Linx Cleaning Conferences

Session 1: Session Chair – Joel Barnett

Session 2: Session Chairs – Akshey Sehgal, Evelyn Kennedy

Linx Cleaning Conferences

2:15 PM BEOL pre-metallization wet clean: post-etch residue removal and metal compatibility – QuocToan Le, imec 96

2:35 PM Selective Removal of Post-Etch Residues Formed by Patterning of High-K Materials Through Precise Control of Water during Cleaning – Jerome Daviot, Technic France 108

Session 2 (Cont.): Session Chairs – Matthew Thorum, Jagdish Prasad

3:15 PM Dry removal of a surface functionalization chemistry used for pattern collapse prevention – Guy Vereecke, imec 119

3:35 PM Cleaning Surfaces from Nanoparticles with Polymer Film: Impact of the Polymer Stripping – Adeline Lallart, STM 128

Session 3: Panel Discussion **Moderator: Mike Corbett**

4:05 PM Panel Discussion

5:20 PM: Wrap Up/Additional Questions/Adjourn Organizers

5:30 PM: Day 1 End

SPCC 2018 Agenda

Linx Cleaning Conferences

Session 4: Session Chairs – Rick Chen, Chris Sparks

9:15 AM **Invited: Process challenges associated with Nano-imprint masks – Nobuyoshi Sato, Toshiba 140**

9:40 AM Etching of Silicon Nitride with High Temperature Water and Deuterium Oxide – Joshua Barclay, Univ. of North Texas 151

10:00 AM Selective wet removal of the SiN ESL prior to contact formation – Antoine Pacco, imec 159

10:20 AM Novel EHS-Friendly Ru Select Etch and SPM Alternatives for 5nm Applications – Chien-Pin Sherman Hsu, Avantor 169

10:40 AM Development of Wet-etch Chemistries for Tungsten Word-line Recess – CK Ge, Versum 178

Session 4 (Cont.): Session Chairs – Shariq Siddiqui, Thomas Phely-Bobin

11:25 AM Study of cobalt etch control by pH and oxidizer – Yuuichi Ogawa, Kurita 186

11:45 AM Developing Wet Cleans for a Cobalt Contact Integration Scheme 200 – Akshey Sehgal, GLOBALFOUNDRIES

Linx Cleaning Conferences

12:05 PM	Deposition of Volatile Chlorohydric Acid on Copper Wafer Depending on Humidity and HCl Airborne Concentration – Minh-Phuong Tran, CEA-Leti 202
12:25 PM	Ammonia cross contamination: FOUP to wafer evaluation and their volatile acids comparison – Paola Gonzalez, Entegris 210
2:00 PM:	**INVITED: The Business Cycle Upswing Impact on Semiconductors — Duncan Meldrum Ph.D., Chief Economist, Hilltop Economics LLC** N/A
2:25 PM	Acidic Cleaning Solutions for Post InGaAs CMP Cleaning – Jin-Goo Park, Hanyang University 219
2:45 PM	Fundamentals of Post-CMP Cleaning of Dielectric Surface Contaminated with Ceria (Nano-to Micro) Particles – Atanu Das, Entegris 230
3:05 PM	Characterization of incoming PVA brush for 10 nm below post CMP cleaning Process – Jung-Hwan Lee, Hanyang University 240
3:25 PM	Post Cleaning Chemical for Tungsten Chemical Mechanical Planarization – Jun-Yong Kim, Samsung 254

Linx Cleaning Conferences

Posters Skyline Suites

![imec logo]

SELECTIVE ETCH REQUIREMENTS FOR THE NEXT GENERATION OF SEMICONDUCTOR DEVICES

FRANK HOLSTEYNS

ON BEHALF OF THE SURFACE AND INTERFACE PREPARATION GROUP OF THE UNIT PROCESS DEPARTMENT

10TH OF APRIL 2018

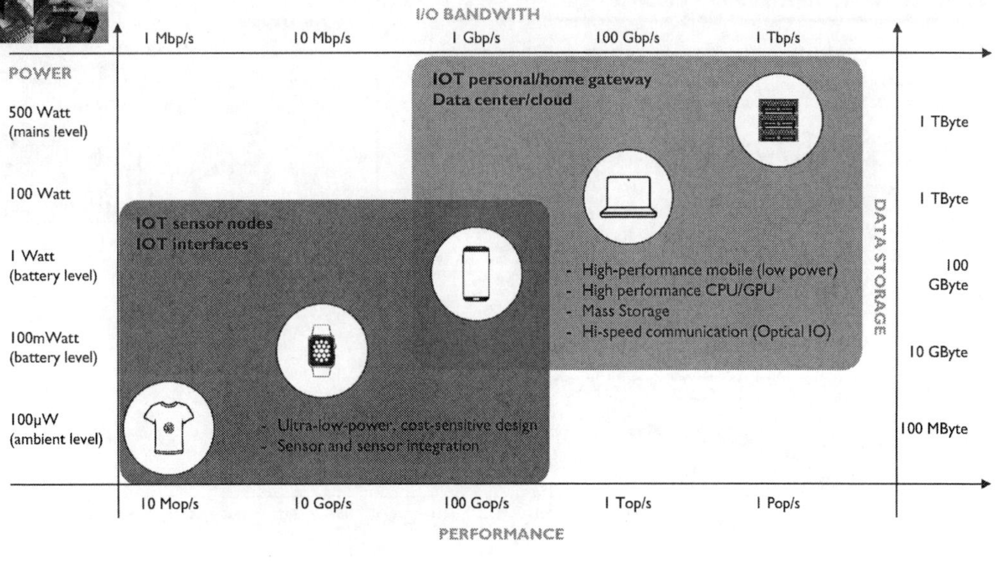

imec

DIMENSIONAL SCALING CHALLENGES, DEVICE ARCHITECTURE & MATERIAL INNOVATION

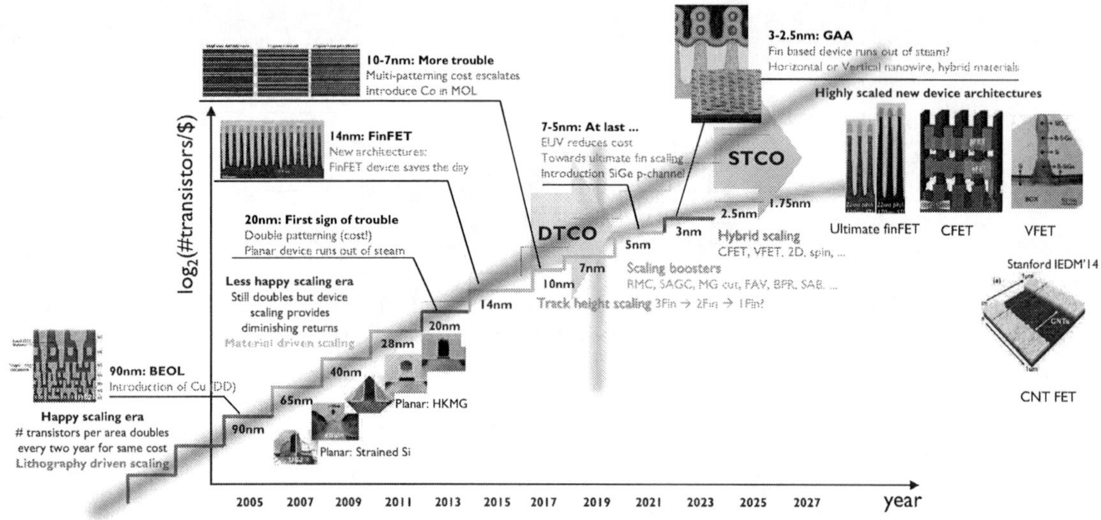

imec

3

DIMENSIONAL SCALING CHALLENGES, DEVICE ARCHITECTURE & MATERIAL INNOVATION

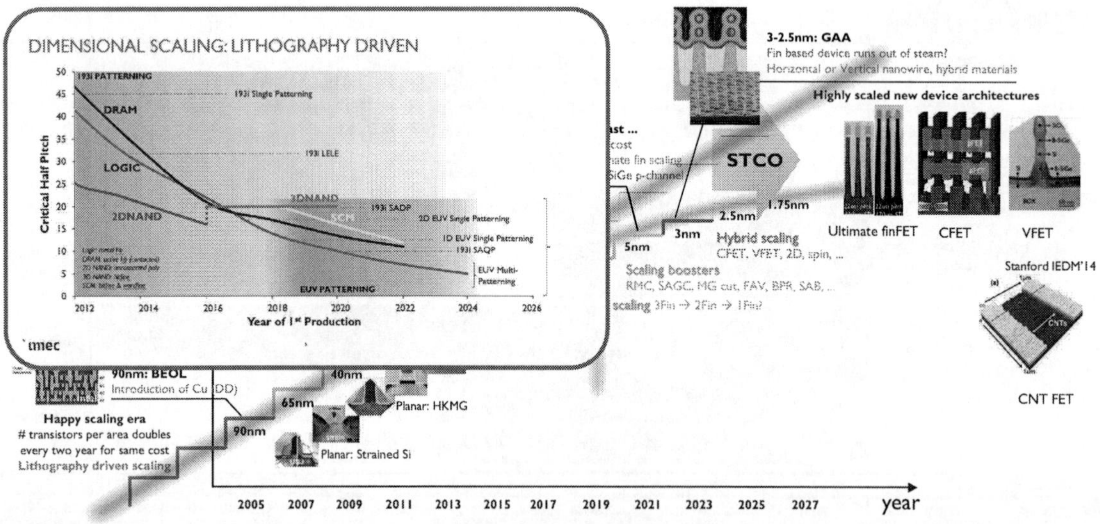

imec

4

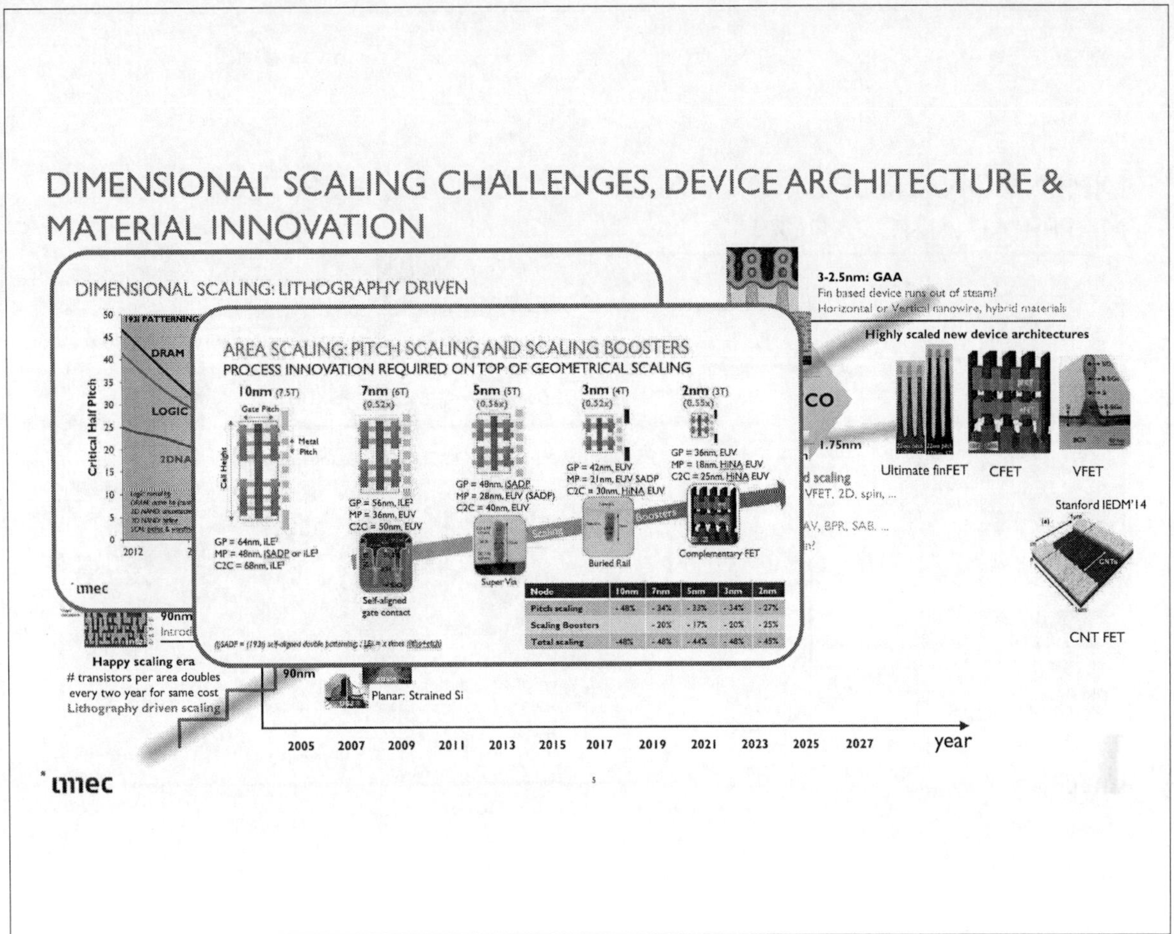
DIMENSIONAL SCALING CHALLENGES, DEVICE ARCHITECTURE & MATERIAL INNOVATION
DIMENSIONAL SCALING: LITHOGRAPHY DRIVEN
Critical Half Pitch
DRAM
LOGIC
2DNA
AREA SCALING: PITCH SCALING AND SCALING BOOSTERS
PROCESS INNOVATION REQUIRED ON TOP OF GEOMETRICAL SCALING
10nm (7.5T)
7nm (6T)
5nm (5T)
3nm (4T)
2nm (3T)
Gate Pitch
Metal Pitch
Cell Height
GP = 64nm, iLE
MP = 48nm, iSADP or iLE
C2C = 68nm, iLE
GP = 56nm, iLE
MP = 36nm, EUV
C2C = 50nm, EUV
GP = 48nm, iSADP
MP = 28nm, EUV (SADP)
C2C = 40nm, EUV
GP = 42nm, EUV
MP = 21nm, EUV SADP
C2C = 30nm, HiNA EUV
GP = 36nm, EUV
MP = 18nm, HiNA EUV
C2C = 25nm, HiNA EUV
Self-aligned gate contact
Super Via
Buried Rail
Complementary FET
Scaling
Boosters
Node | 10nm | 7nm | 5nm | 3nm | 2nm
Pitch scaling | -48% | -34% | -33% | -34% | -27%
Scaling Boosters | | -20% | -17% | -20% | -25%
Total scaling | -48% | -48% | -44% | -48% | -45%
3-2.5nm: GAA
Fin based device runs out of steam?
Horizontal or Vertical nanowire, hybrid materials
Highly scaled new device architectures
Ultimate finFET
CFET
VFET
Stanford IEDM'14
CNT FET
90nm
Happy scaling era
transistors per area doubles every two year for same cost
Lithography driven scaling
Planar: Strained Si
2005 2007 2009 2011 2013 2015 2017 2019 2021 2023 2025 2027
year
imec
5

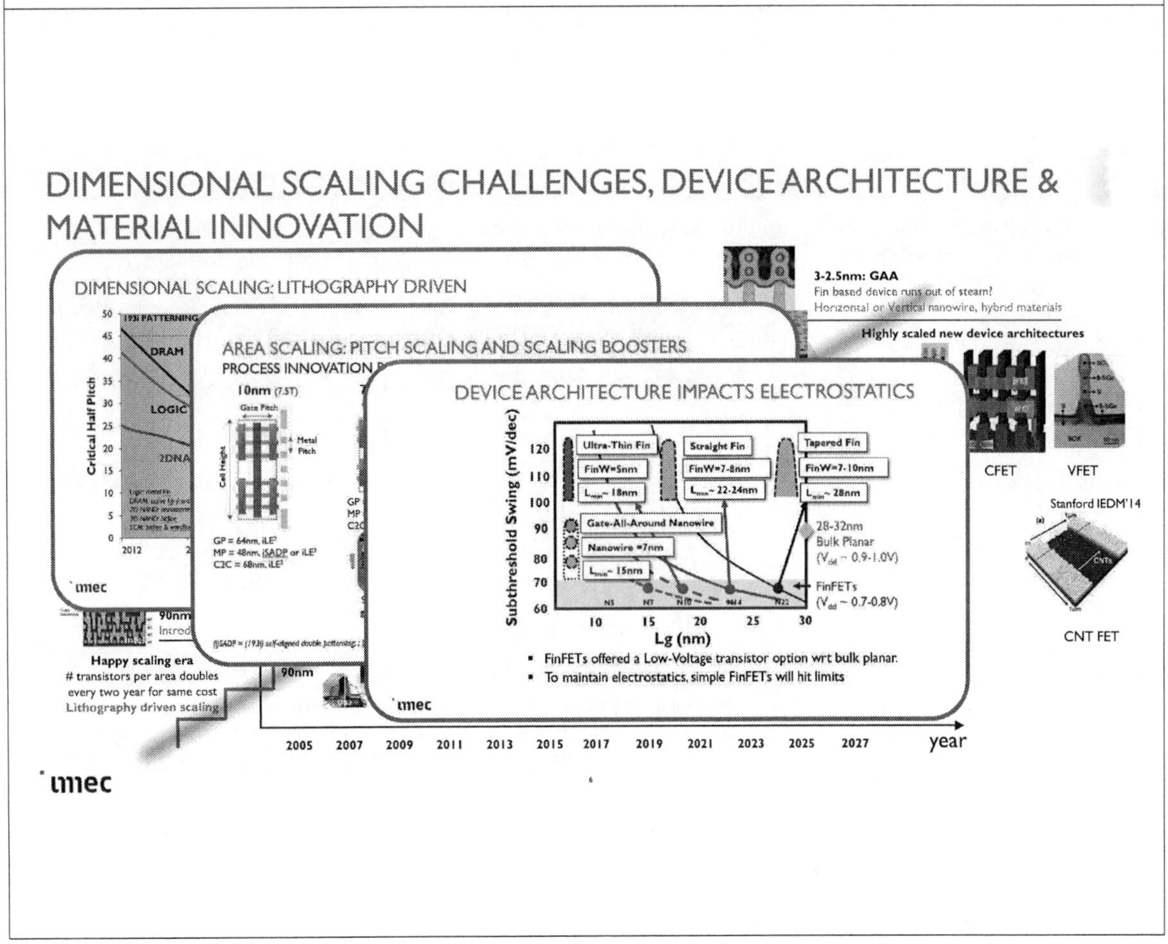
DIMENSIONAL SCALING CHALLENGES, DEVICE ARCHITECTURE & MATERIAL INNOVATION
DIMENSIONAL SCALING: LITHOGRAPHY DRIVEN
AREA SCALING: PITCH SCALING AND SCALING BOOSTERS
DEVICE ARCHITECTURE IMPACTS ELECTROSTATICS
Subthreshold Swing (mV/dec)
Ultra-Thin Fin
FinW=5nm
Lmin~ 18nm
Straight Fin
FinW=7-8nm
Lmin~ 22-24nm
Tapered Fin
FinW=7-10nm
Lmin~ 28nm
Gate-All-Around Nanowire
Nanowire =7nm
Lmin~ 15nm
28-32nm Bulk Planar (Vdd ~ 0.9-1.0V)
FinFETs (Vdd ~ 0.7-0.8V)
Lg (nm)
FinFETs offered a Low-Voltage transistor option wrt bulk planar.
To maintain electrostatics, simple FinFETs will hit limits
3-2.5nm: GAA
Fin based device runs out of steam?
Horizontal or Vertical nanowire, hybrid materials
Highly scaled new device architectures
CFET
VFET
Stanford IEDM'14
CNT FET
Happy scaling era
transistors per area doubles every two year for same cost
Lithography driven scaling
2005 2007 2009 2011 2013 2015 2017 2019 2021 2023 2025 2027
year
imec
6

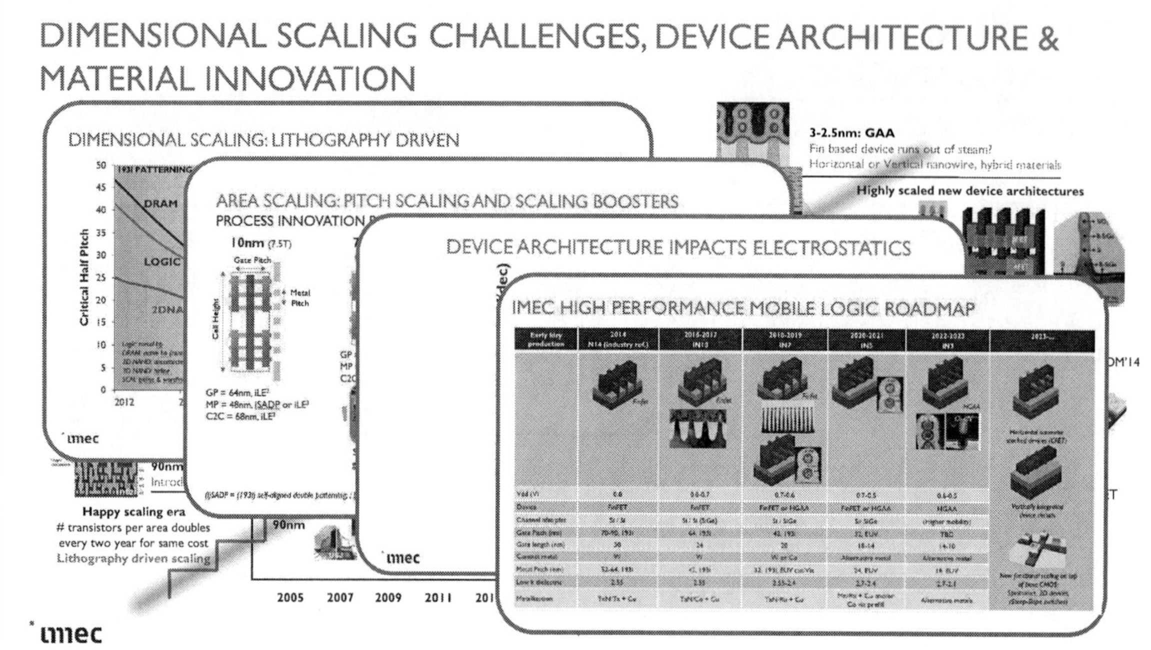
DIMENSIONAL SCALING CHALLENGES, DEVICE ARCHITECTURE & MATERIAL INNOVATION
DIMENSIONAL SCALING: LITHOGRAPHY DRIVEN
Critical Half Pitch
DRAM
LOGIC
AREA SCALING: PITCH SCALING AND SCALING BOOSTERS
PROCESS INNOVATION
10nm (7.5T)
Gate Pitch
Metal Pitch
Cell Height
GP = 64nm, iLE³
MP = 48nm, (SADP or iLE³
C2C = 68nm, iLE³
DEVICE ARCHITECTURE IMPACTS ELECTROSTATICS
IMEC HIGH PERFORMANCE MOBILE LOGIC ROADMAP
3-2.5nm: GAA
Fin based device runs out of steam?
Horizontal or Vertical nanowire, hybrid materials
Highly scaled new device architectures
Happy scaling era
transistors per area doubles every two year for same cost
Lithography driven scaling
90nm
2005 2007 2009 2011
imec

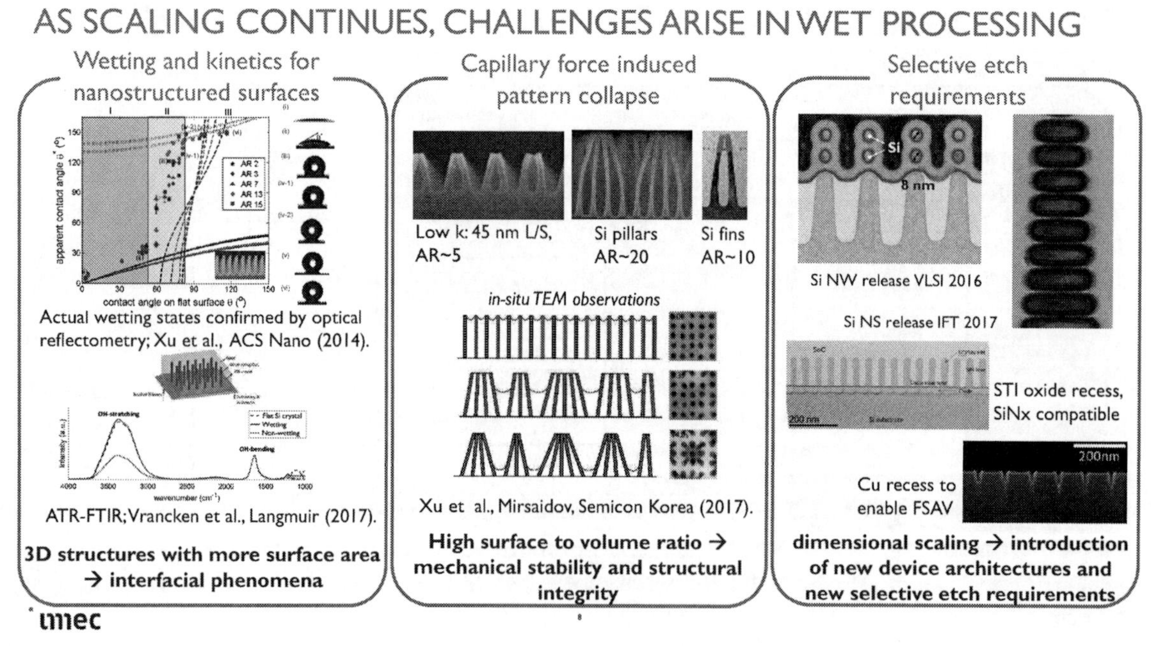
AS SCALING CONTINUES, CHALLENGES ARISE IN WET PROCESSING
Wetting and kinetics for nanostructured surfaces
apparent contact angle θ* (°)
contact angle on flat surface θ (°)
Actual wetting states confirmed by optical reflectometry; Xu et al., ACS Nano (2014).
ATR-FTIR; Vrancken et al., Langmuir (2017).
3D structures with more surface area → interfacial phenomena
Capillary force induced pattern collapse
Low k: 45 nm L/S, AR~5
Si pillars AR~20
Si fins AR~10
in-situ TEM observations
Xu et al., Mirsaidov, Semicon Korea (2017).
High surface to volume ratio → mechanical stability and structural integrity
Selective etch requirements
Si
8 nm
Si NW release VLSI 2016
Si NS release IFT 2017
STI oxide recess, SiNx compatible
200 nm
Cu recess to enable FSAV
200nm
dimensional scaling → introduction of new device architectures and new selective etch requirements
imec

KINETICS MAY VARY IN NANO-CONFINEMENT

Wetting	electrical double layers	water structuring
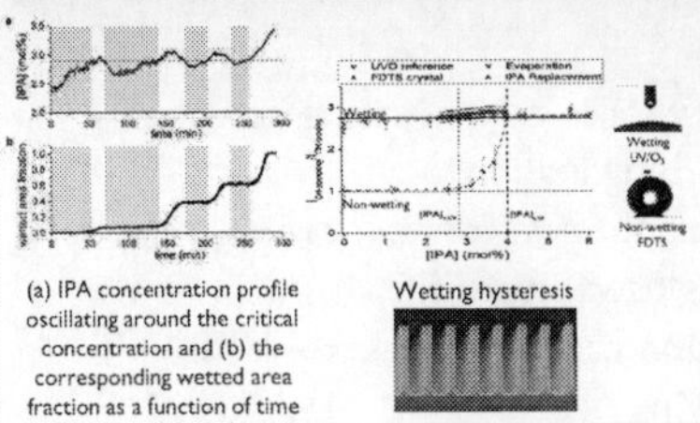 (a) IPA concentration profile oscillating around the critical concentration and (b) the corresponding wetted area fraction as a function of time; Wetting hysteresis	microchannel / nanochannel 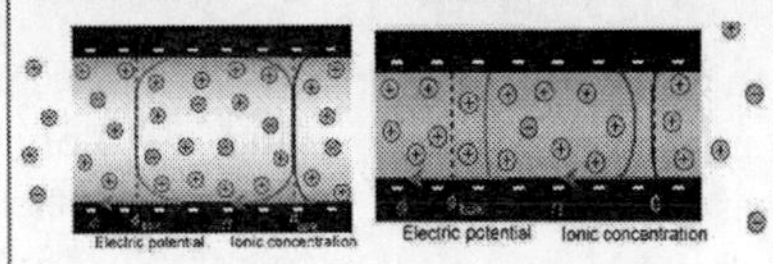 H. Daiguji, *Chem. Soc. Rev.* (2010).	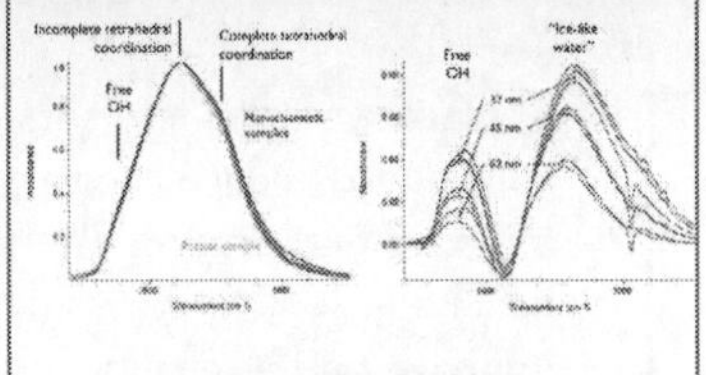 Vereecke, *ECIS* (2017).
Differentiation between Wenzel / Cassie-Baxter / Mixed wetting states	Overlap of electrical double layers (EDL) in nanochannel	Water confinement
In-situ study of wetting stability and hysteresis on initially non-wetting substrate	Depletion of ions with same charge as surface in channel: no electroneutrality	Formation of ice-like water in nanoconfined volumes
▪ Wetting hysteresis observed (Vrancken et al., Langmuir 2017).	▪ pH shift expected (D.Bottenus et al., Lab on Chip, 2009, 9, 219.) ▪ Depletion etchants (A. Okuyama et al., Solid State Phenom. 2015, 219, 115.)	▪ effect of water structuring on diffusivity of chemical species in nano-confined volumes expected

imec

CAPILLARY FORCE INDUCED PATTERN COLLAPSE

IN SITU CHARACTERIZATION OF DEWETTING AT NANOSCALE

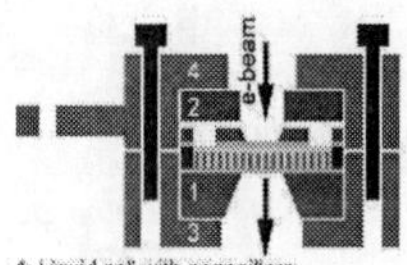

- Real-time visualization of pattern collapse with TEM in liquid cell.
- Polycrystalline Si nanopillars, height ≈ 450 nm.
- Formation of clusters due to capillary instabilities.

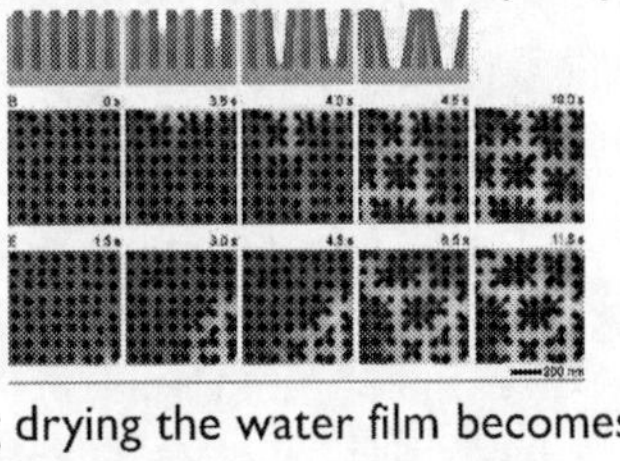

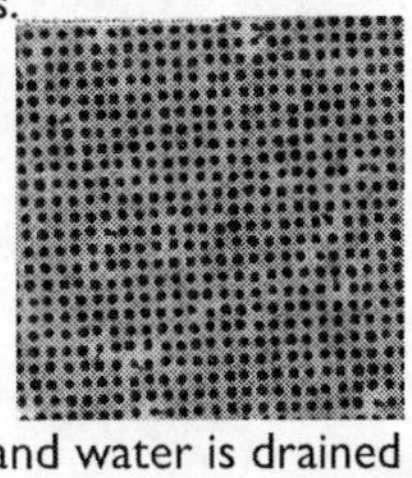

- During drying the water film becomes unstable, and water is drained gradually towards bended nanopillars islands.

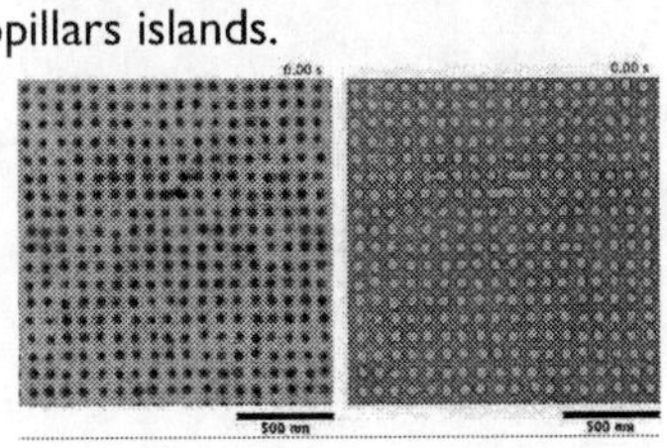

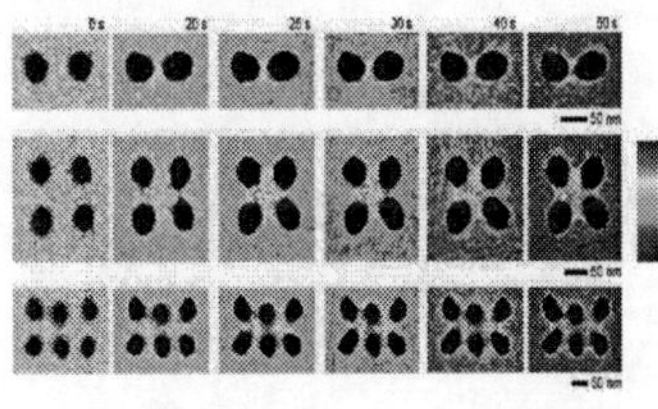

imec

10

PATTERN COLLAPSE/STICTION FREE DRYING

IPA dry

- Replace water by low surface tension (γ) solvents to reduce capillary force;
- Improve evaporation rate (gas flow and heat)
- Improve on IPA quality

Si nanopillars with native oxide

DIW (γ=0.072 N/m) IPA (γ=0.022 N/m)

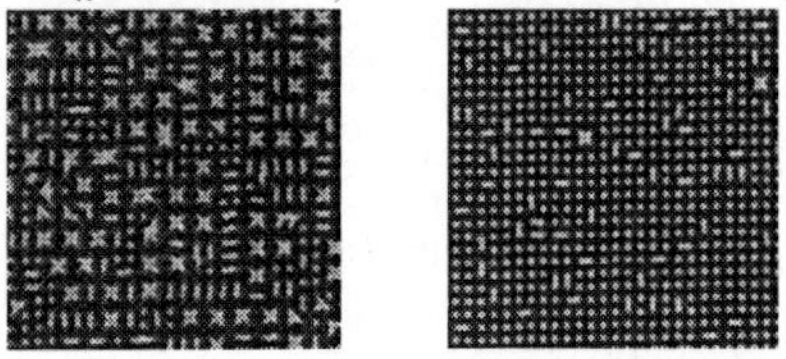

Surface functionalizing chemistry (SFC)

- Reduce capillary force by increasing contact angles ($\boldsymbol{\theta}$) of rinsing liquids;
- Reduce surface adhesion force, more relevant when IPA dry is used after SFC;
- Further reduction capillary force: towards sublimation drying

Increase surface hydrophobicity reduces pattern collapse in **water** (not necessarily for other solvents!)

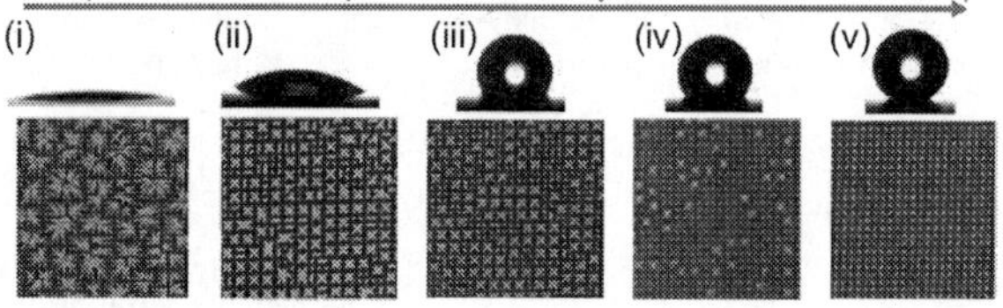

imec

SELECTIVE ETCH REQUIREMENTS

SELECTIVE/ISOTROPIC ETCH OPPORTUNITIES (WET/DRY)

FINFET → GAA → CFET

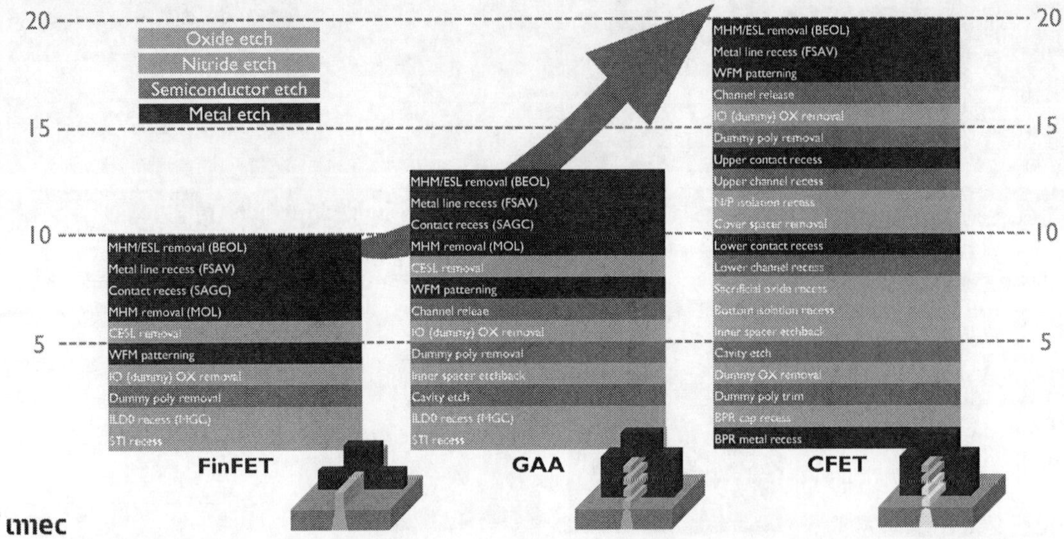

imec

SEMICONDUCTOR ETCH: GAA SELECTIVE ETCH REQUIREMENTS

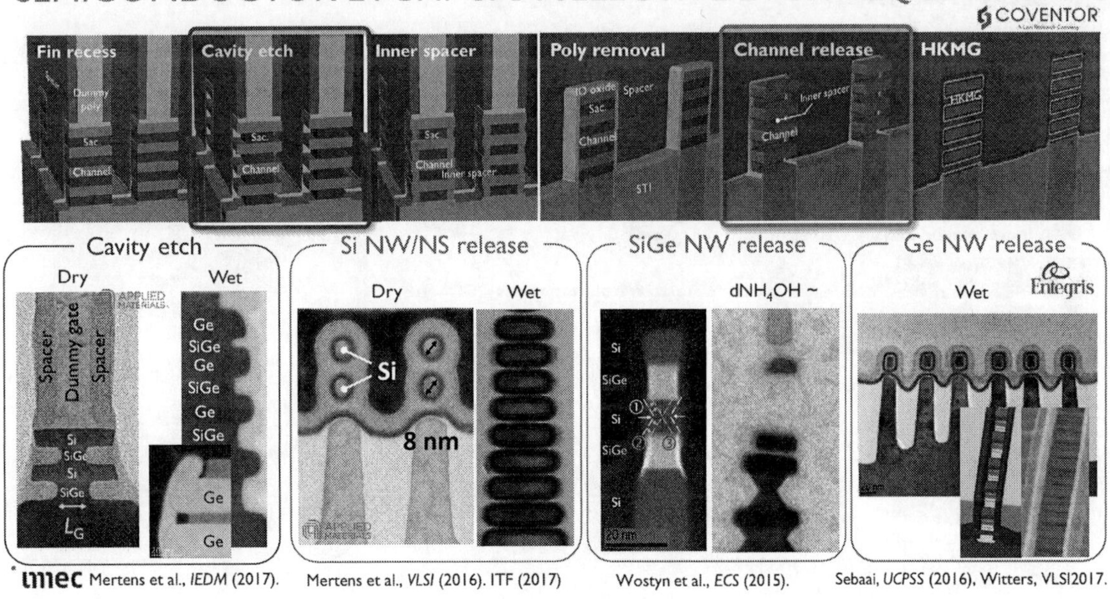

imec Mertens et al., *IEDM* (2017). Mertens et al., *VLSI* (2016). ITF (2017) Wostyn et al., *ECS* (2015). Sebaai, *UCPSS* (2016), Witters, VLSI2017.

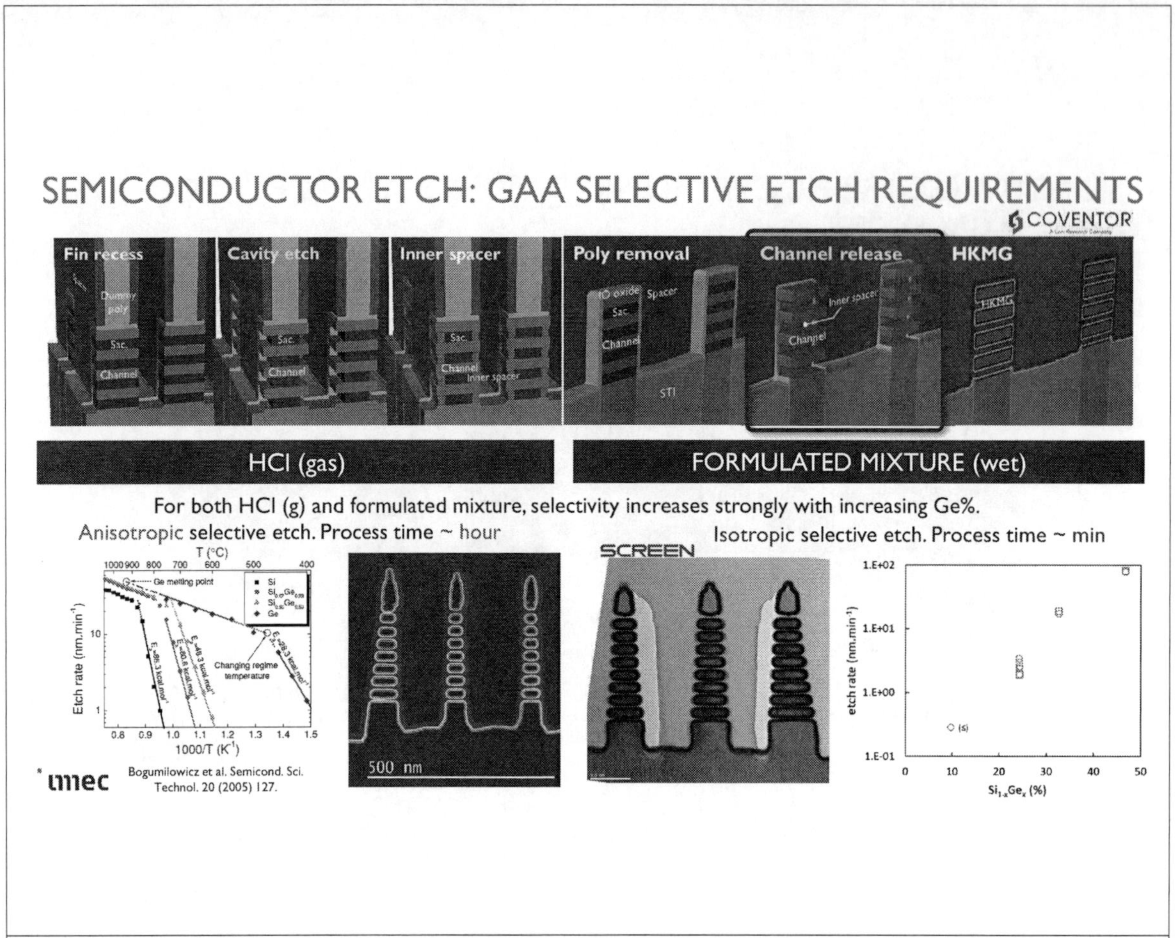
SEMICONDUCTOR ETCH: GAA SELECTIVE ETCH REQUIREMENTS
COVENTOR
Fin recess
Cavity etch
Inner spacer
Poly removal
Channel release
HKMG
Dummy poly
Sac.
Channel
Inner spacer
Spacer
STI
HKMG
HCl (gas)
FORMULATED MIXTURE (wet)
For both HCl (g) and formulated mixture, selectivity increases strongly with increasing Ge%.
Anisotropic selective etch. Process time ~ hour
Isotropic selective etch. Process time ~ min
T (°C)
Ge melting point
Changing regime temperature
Etch rate (nm.min⁻¹)
1000/T (K⁻¹)
500 nm
SCREEN
etch rate (nm.min⁻¹)
Si1-xGex (%)
imec
Bogumilowicz et al. Semicond. Sci. Technol. 20 (2005) 127.

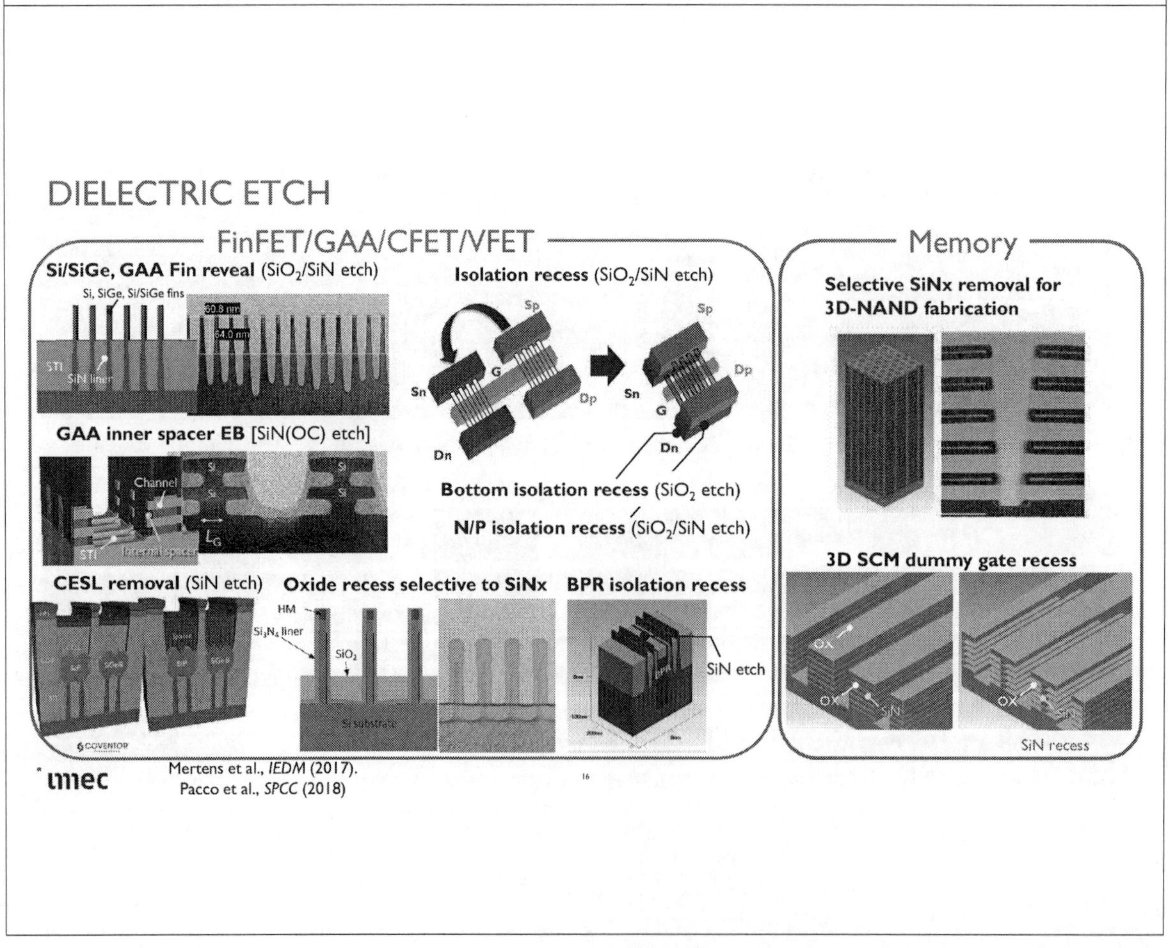
DIELECTRIC ETCH
FinFET/GAA/CFET/VFET
Memory
Si/SiGe, GAA Fin reveal (SiO2/SiN etch)
Si, SiGe, Si/SiGe fins
STI
SiN liner
60.8 nm
64.0 nm
Isolation recess (SiO2/SiN etch)
Sp
Sn
G
Dp
Dn
GAA inner spacer EB [SiN(OC) etch]
Channel
Internal spacer
Si
LG
Bottom isolation recess (SiO2 etch)
N/P isolation recess (SiO2/SiN etch)
Selective SiNx removal for 3D-NAND fabrication
3D SCM dummy gate recess
OX
SiN
SiN recess
CESL removal (SiN etch)
Oxide recess selective to SiNx
BPR isolation recess
HM
Si3N4 liner
SiO2
Si substrate
SiN etch
COVENTOR
imec
Mertens et al., IEDM (2017).
Pacco et al., SPCC (2018)

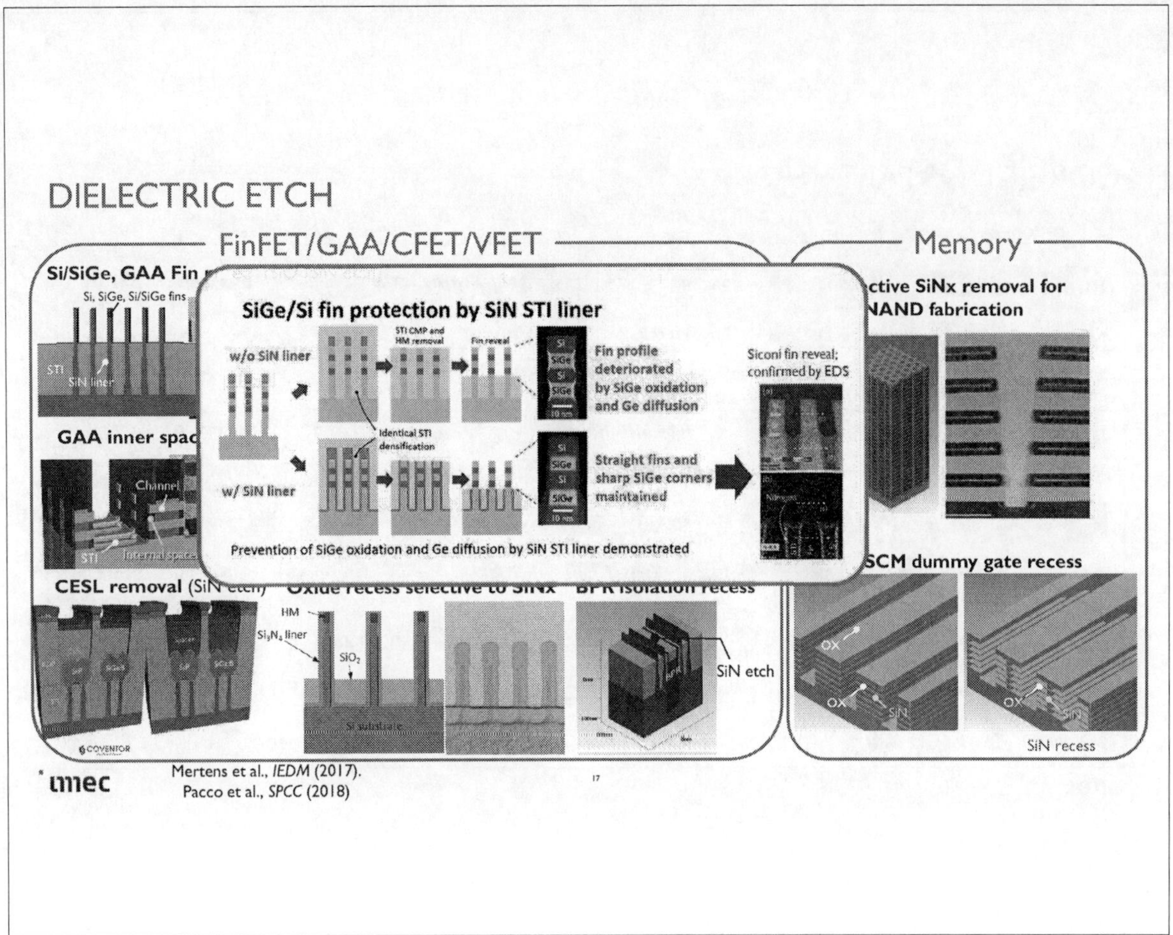
DIELECTRIC ETCH
FinFET/GAA/CFET/VFET
Memory
Si/SiGe, GAA Fin
Si, SiGe, Si/SiGe fins
STI
SiN liner
GAA inner spac
Channel
Internal spacer
CESL removal (SiN etch)
Oxide recess selective to SiNx
BPR isolation recess
HM
Si3N4 liner
SiO2
Si substrate
SiN etch
SiGe/Si fin protection by SiN STI liner
w/o SiN liner
w/ SiN liner
STI CMP and HM removal
Fin reveal
Identical STI densification
Fin profile deteriorated by SiGe oxidation and Ge diffusion
Straight fins and sharp SiGe corners maintained
Siconi fin reveal; confirmed by EDS
Prevention of SiGe oxidation and Ge diffusion by SiN STI liner demonstrated
ctive SiNx removal for NAND fabrication
SCM dummy gate recess
OX
SiN
SiN recess
Mertens et al., IEDM (2017).
Pacco et al., SPCC (2018).
17

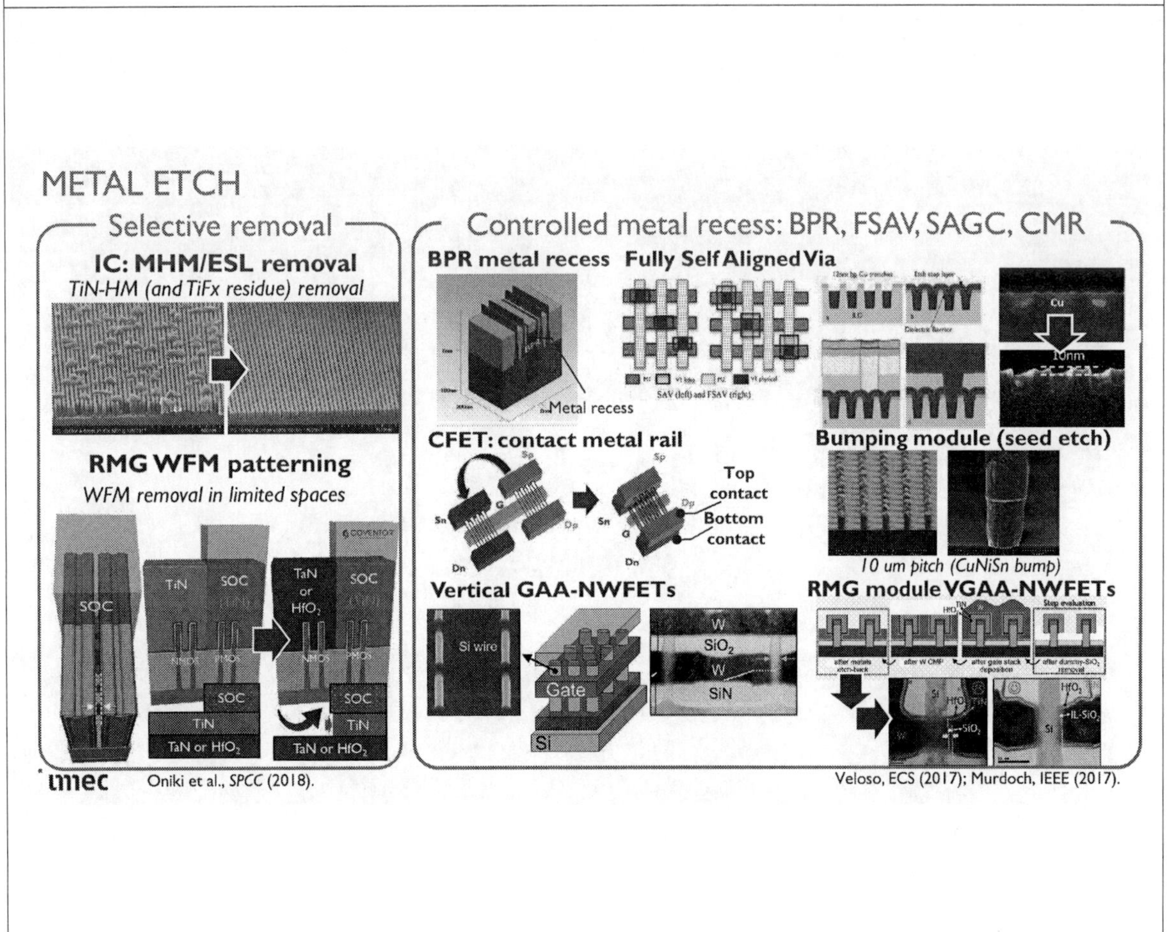
METAL ETCH
Selective removal
IC: MHM/ESL removal
TiN-HM (and TiFx residue) removal
RMG WFM patterning
WFM removal in limited spaces
SOC
TiN
TaN or HfO2
NMOS
PMOS
Controlled metal recess: BPR, FSAV, SAGC, CMR
BPR metal recess
Metal recess
Fully Self Aligned Via
SAV (left) and FSAV (right)
Cu
10nm
CFET: contact metal rail
Top contact
Bottom contact
Sp
Sn
Dp
Dn
G
Bumping module (seed etch)
10 um pitch (CuNiSn bump)
Vertical GAA-NWFETs
Si wire
Gate
Si
W
SiO2
SiN
RMG module VGAA-NWFETs
HfO2
IL-SiO2
Oniki et al., SPCC (2018).
Veloso, ECS (2017); Murdoch, IEEE (2017).

CORE CMOS PARTNERS

LOGIC / MEMORY IDM & FOUNDRIES

intel, SAMSUNG, tsmc, GLOBALFOUNDRIES, Micron, SK hynix, SanDisk, TOSHIBA

FABLESS & FABLITE

HUAWEI, HISILICON, ALTERA (now part of Intel), Panasonic, ARM, XILINX, socionext, SONY, QUALCOMM

EQUIPMENT & MATERIAL SUPPLIERS / OSAT / EDA / JDP PARTNERS

ASML, ASM, APPLIED MATERIALS, Hitachi High-Tech, KLA Tencor, Lam Research, SCREEN, TEL, CYMER, Canon (Canon Anelva Corporation), AIXTRON, NOVA, EVG, FEI, DISCO, Ultratech, SUSS MicroTec, OXFORD, ZEISS, POONGSAN, Park Systems, RIBER, TMEiC, NEOCERA, PVA TePla, Kurita, EBARA, Solmates, BRUKER, JORDAN VALLEY, CASCADE, Besi, Camtek, SEMILAB, NuFlare, Atlas Copco, EDWARDS, SMIC, Kaneka, JSR Micro, JX Nippon Mining & Metals, Entegris, MERCK, TOPPAN, inpria, siltronic, Soitec, NISSAN CHEMICAL INDUSTRIES, LTD., DOW, Shin-Etsu, SUMITOMO CHEMICAL, DONGJIN, tok, BASF, FUJIFILM, brewer science, SHINKO, Amkor Technology, UTAC, Quantum Wise, Panoramic Technology, cadence, Mentor (A Siemens Business), SYNOPSYS, global TCAD solutions, COVENTOR

19

imec

imec

embracing a better life

imec

CONFIDENTIAL – INTERNAL USE

RMG WET PROCESS CHALLENGES AND THE PATTERNING KNOBS TOWARDS N5 AND BEYOND LOGIC DEVICES

YUSUKE ONIKI *, LARS-ÅKE RAGNARSSON, GUY VEREECKE, FARID SEBAAI, HAROLD DEKKERS, EUGENIO DENTONI LITTA, TOM SCHRAM, FRANK HOLSTEYNS, AND NAOTO HORIGUCHI

imec, Kapeldreef 75, 3001 Leuven, Belgium, * Oniki.Yusuke@imec.be

SPCC 2018, 10-11st Apr. 2018, Cambridge, MA, US

PUBLIC

OUTLINE

- Introduction
  - imec view of logic roadmap
  - RMG scaling challenges
- RMG patterning development
  - Tri-layer patterning scheme
  - Digital wet etching of polycrystalline metal films
- RMG scaling and multi-Vt patterning knobs
  - D&GR
  - SiH_4 soak
  - "blocked" patterning
- Summary

imec

2

INTRODUCTION

IMEC VIEW OF LOGIC ROADMAP

STANDARD CELL & REPLACEMENT METAL GATE (RMG) SCALING

"i"N: imec node	iN10	iN7	iN5	iN3
Device	FF	FF	FF/GAA	FF/GAA/CFET
Channel	Si	Si	Si/SiGe	Si/SiGe
PP (poly pitch)	56	42	42	36-42
Track height	7.5/6.5	7.5/6.5	5.5/4.5	4.X/3.X
FP (fin pitch)	32	24	21	16-21
Lg (gate length)	20	16	12	10-12
MP (metal pitch)	40	32	21	16-21

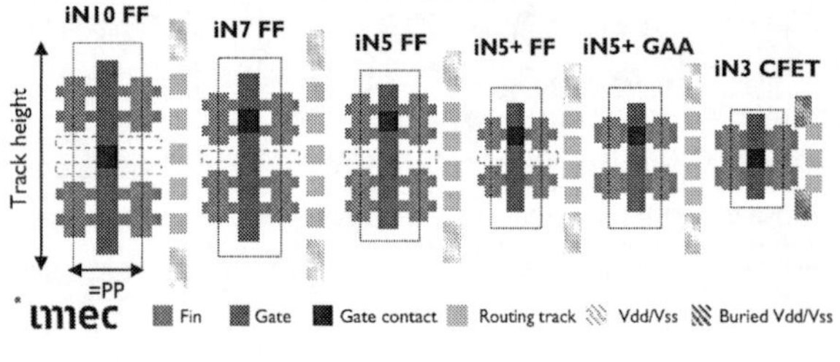

- Aggressive FP/Lg reduction needed
  - Continuous std. cell scaling
  - Device performance boost

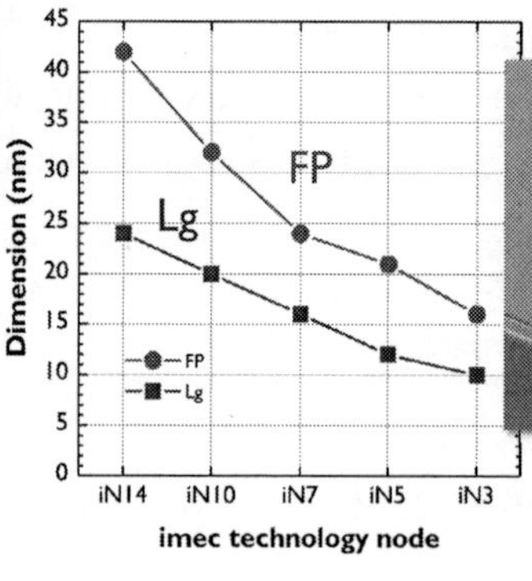

TYPICAL RMG PATTERNING SCHEME
BARC PATTERNING (DRY) + METAL REMOVAL (WET)

- Dummy poly removal
- Core/IO oxide patterning (IL formation)
- HK deposition (HfO_2)
- HK cap dep (TiN)
- Si-cap anneal (SSA)
- Barrier dep (TaN)
- **pWFM patterning**
- nWFM patterning
- W CMP

- Major challenges:
  - High aspect ratio BARC open etch (dry)
  - WFM selective removal in limited spaces (wet)

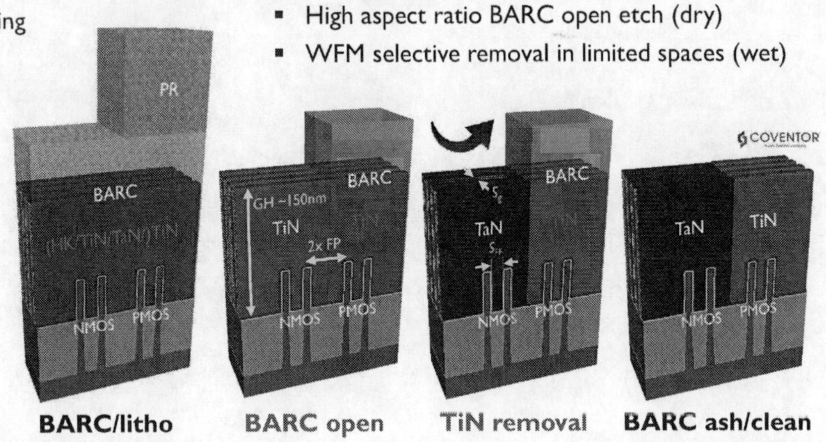

imec

RMG PATTERNING CHALLENGES
PATTERNING SPACE

- Limited spaces for filling and dry/wet etching, cleaning, rinsing and drying
  - MG thickness reduction, eWF tuning knobs, novel patterning and optimized unit process needed

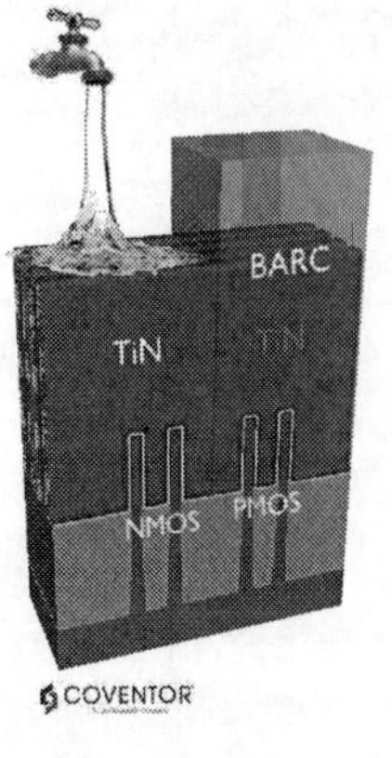

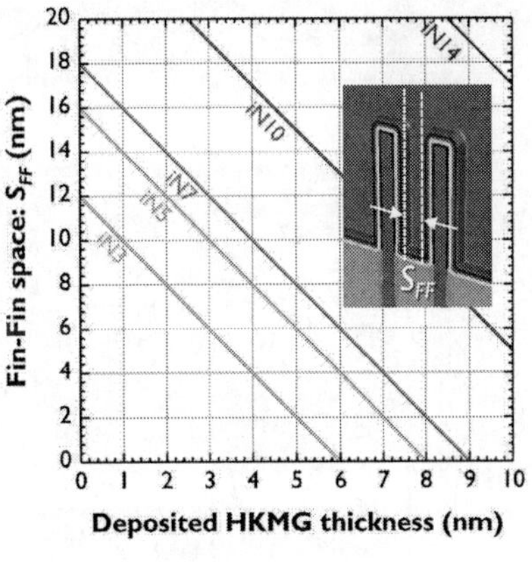

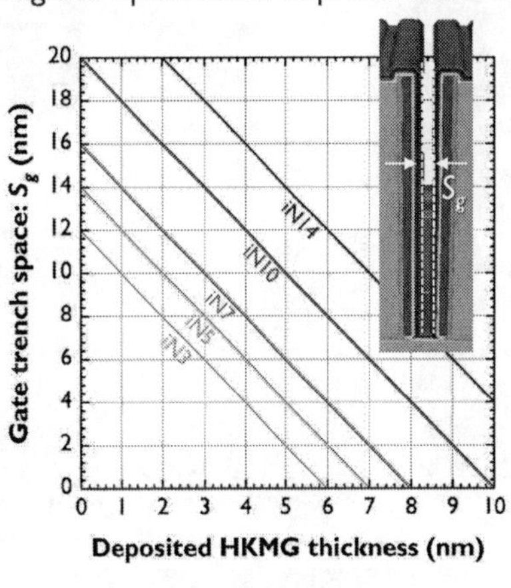

imec

RMG PATTERNING DEVELOPMENT

* T. Hopf (imec), SPIE 2017

RMG TRI-LAYER PATTERNING
N/P BOUNDARY LATERAL ETCH

- Tri-layer e.g. PR/SOG/SOC* as the patterning mask
  - Better gap-fill capability
  - To minimize N/P boundary lateral etch

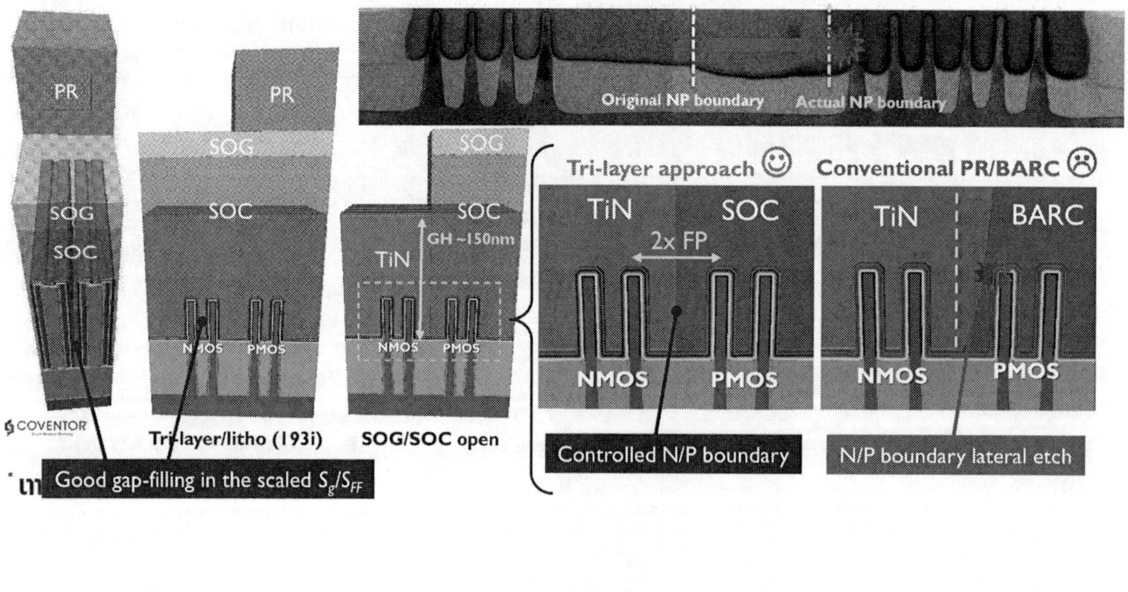

RMG TRI-LAYER PATTERNING
POST SOC ASH CLEAN

- Need to optimize SOC ash/clean

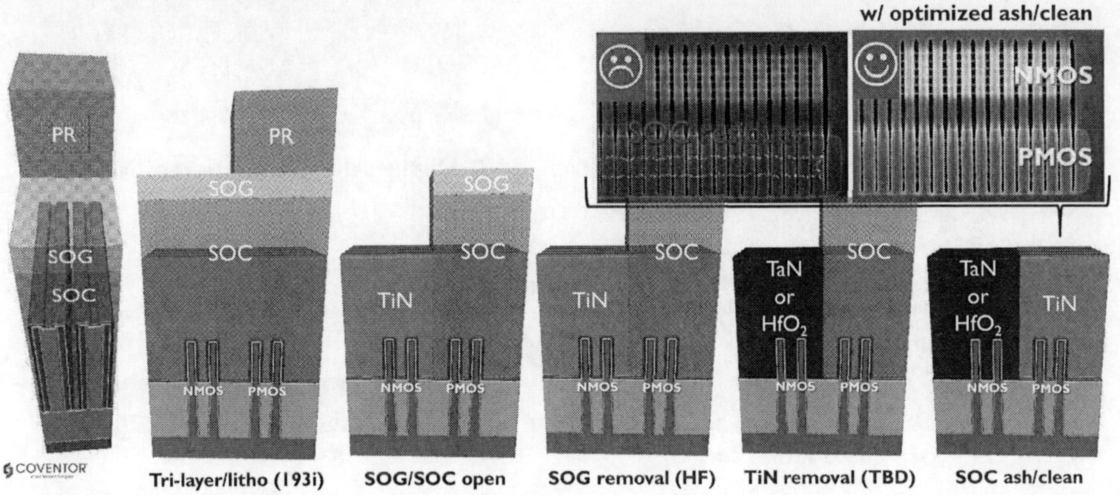

imec

WFM WET SELECTIVE ETCH
N/P BOUNDARY UNDERCUT

- Wet etch through the limited spaces (S_{FF}/S_g) may need prolonged process
  - Prolonged wet etch will induce N/P boundary undercut and damage HKMG material underneath

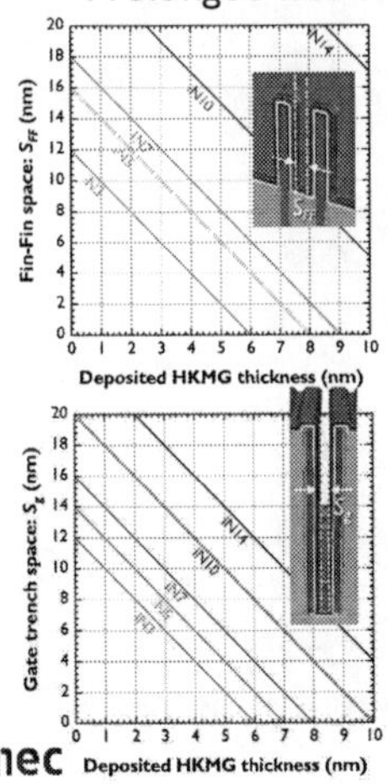

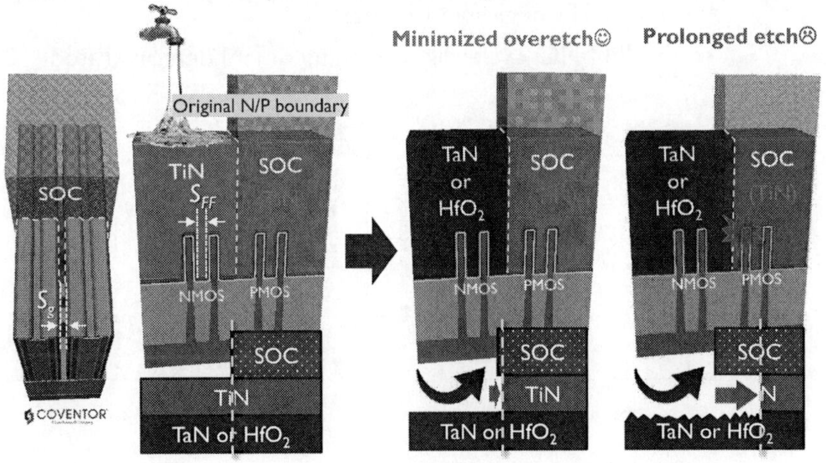

imec

DIGITAL WET ETCHING OF POLY CRYSTALLINE METAL FILMS
CONCEPT

- Two-step self-limiting process
  - 1. self-saturating metal oxide growth
  - 2. highly selective oxide removal

Ref. digital etch of semiconductors:
III-V G. C. DeSalvo, JES 1996
Si T. Hattori, JES 1998
III-V J. Lin, IEEE EDL 2014
III-V D. H. van Dorp, ECS JSSST 2015

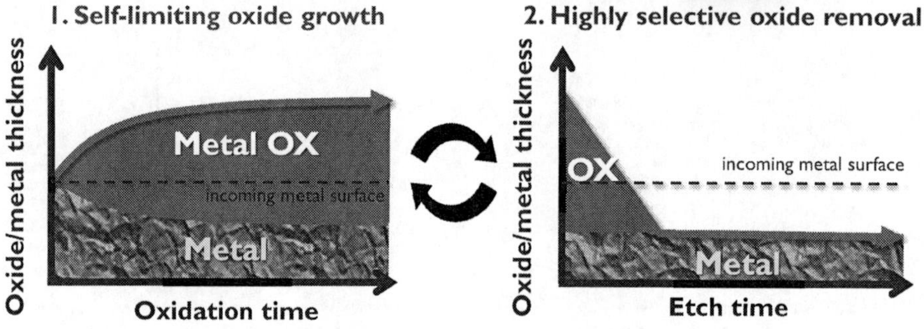

- The ER determined by the formed OX thickness and etch cycle#

imec

DIGITAL ETCHING OF TIN
FEASIBILITY

- dHPM (RT): self-limiting TiN etch (= self-limiting oxide growth)
- dHCl (HT): negligible TiN loss
- → HPM-HCl cycle: digital etching of TiN demonstrated (~0.66nm/cy)

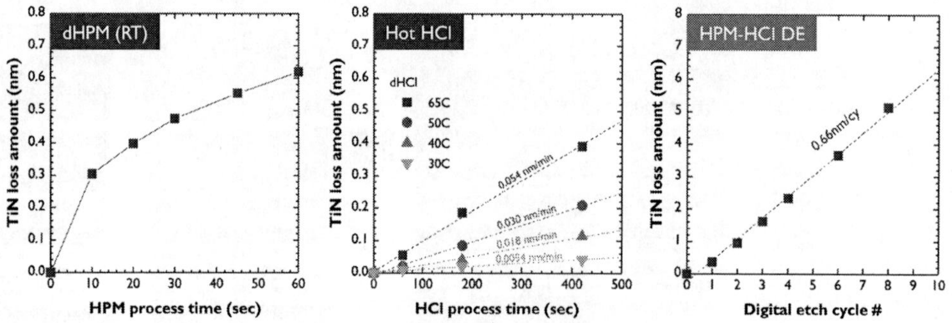

imec

12

DIGITAL ETCHING OF TIN
TIN TRENCH STRUCTURE

- Digital etch (HPM-HCl)
  - Conformal and uniform TiN etch demonstrated
  - Less sensitive to the TiN grain boundaries
- Ref. one-step etch (APM)
  - TiN surface roughening seen

Digital etch (HPM-HCl cycle)

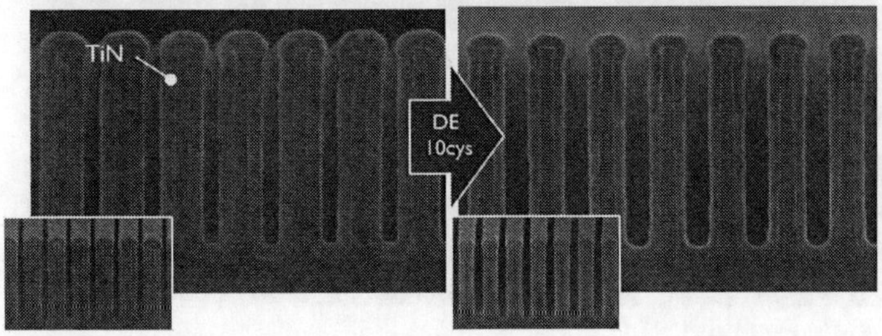

One-step etch (APM)

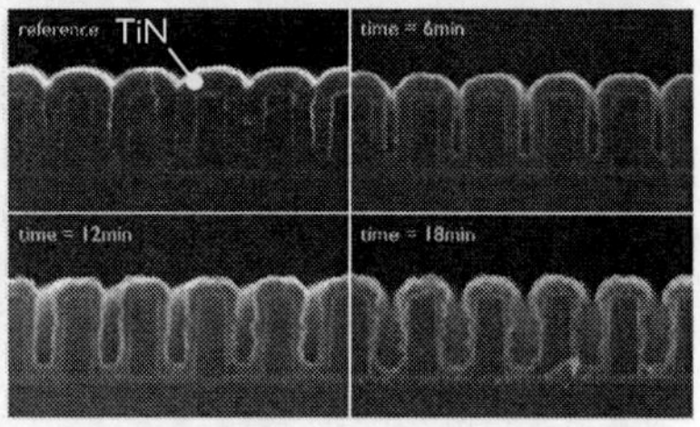

G. Vereecke (imec), SPCC 2017

imec

IMPACT ON ELECTRICAL PROPERTIES (MOSCAP)
DIGITAL ETCH (HPM-HCL CYCLE) OF TIN ON HK AND TAN

TiN DE: 5cys HPM-HCl DE on HfO_2

- STI patterning / channel doping
- ChemOX-IL (imec clean)
- ALD-HfO2 (1.5nm)
- ALD-TiN (2nm)
- TiN removal: 5cys HPM-HCl DE
- ALD-TiN(1nm)/TiAl(3nm)/TiN(3nm)
- CVD-W
- Gate etch

Ref.

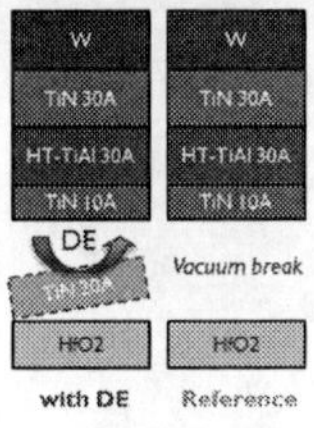

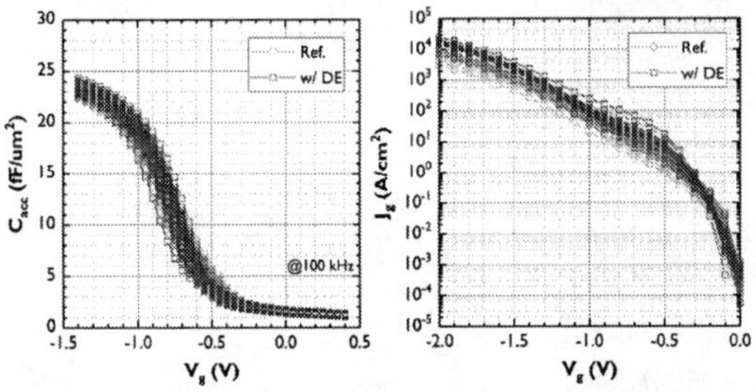

TiN DE: 5cys HPM-HCl DE on TaN

- STI patterning / channel doping
- ChemOX-IL (imec clean)
- ALD-HfO2 (1.5nm)
- ALD-TaN (1nm)
- ALD-TiN (2nm)
- TiN removal: 5cys HPM-HCl DE
- ALD-TiN(1nm)/TiAl(3nm)/TiN(3nm)
- CVD-W
- Gate etch

Ref.

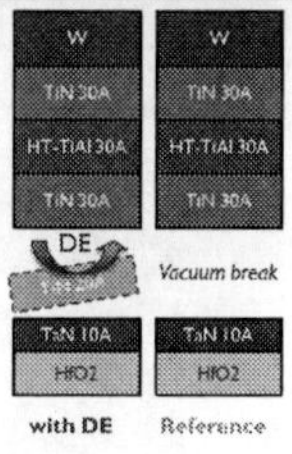

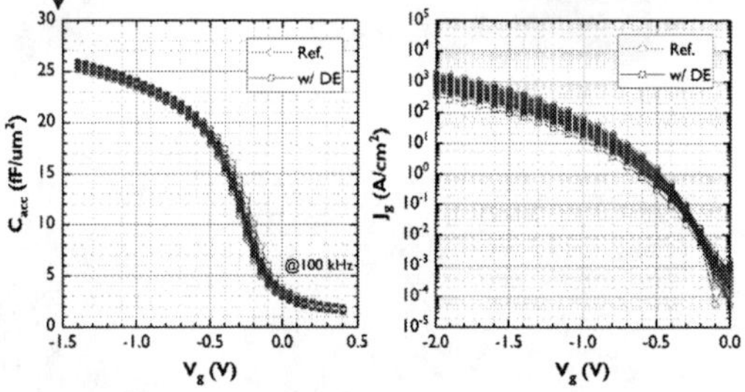

imec

- No negative impact on electrical properties (CET, Vfb, Jg, ...) observed

TIN DIGITAL ETCHING
CONFORMAL VS. PLUG ETCH

- Digital wet etching seems to be less sensitive to the grain boundaries and metal seam
  - Open trench → conformal etching
  - Closed trench → plug etching
  - enables metal recess etching

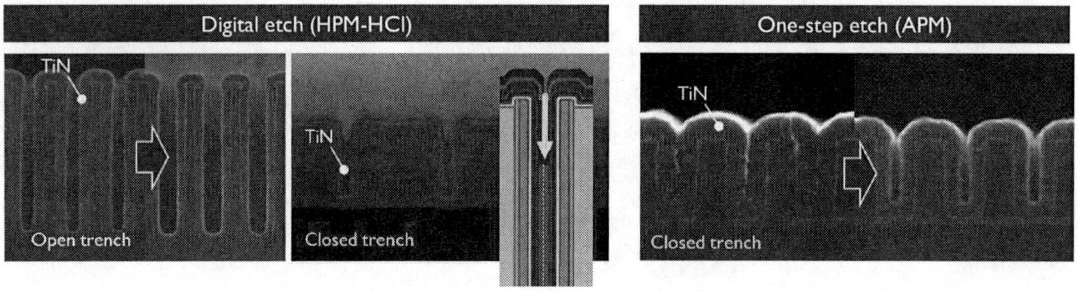

imec

15

APPLICATIONS
DIGITAL WET ETCHING METAL FILMS

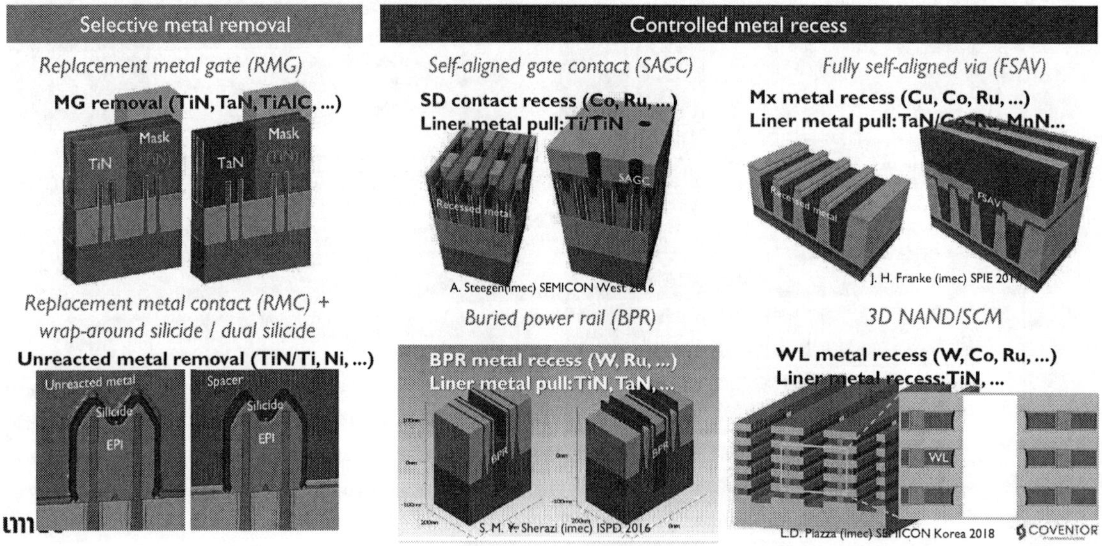

imec

RMG SCALING & MULTI-V_T PATTERNING KNOBS

RMG SCALING & MULTI-V_T PATTERNING KNOBS

	① D&GR	② SiH_4 soak	TiN patterning on HK	③ nWFM blocked patterning	Fin CD trim in RMG	HK 1st or HKMG 1st
Knobs	Dipole	O scavenging	Skip TaN barrier	CMP assisted RMG patterning	Improve electrostatic w/o R_{ext} penalty	Go back to old tech.
	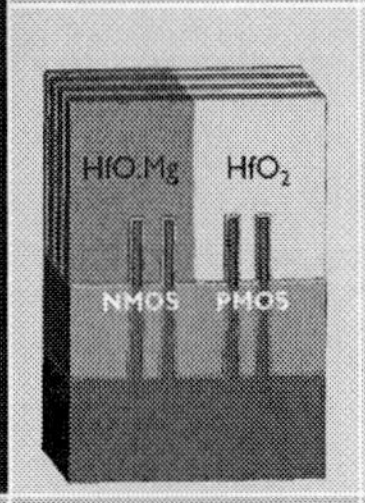	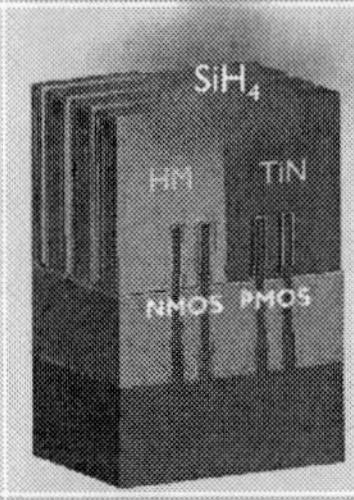	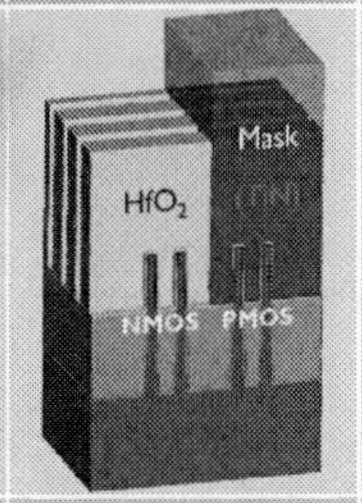			
Pros	Volumeless V_t tuning	Zero-thick V_t tuning	~2nm S_{FF}/S_g gain	Allow merged nWFM	~1nm S_{FF} gain	HKF: ~3nm S_g gain HKMGF: no S_g limit
Cons	Corrosive materials Patterning difficulty HM thickness scaling Post dry etch residue	HM thickness scaling SC/LC loading?	HK exposure/damage IL regrowth nWFM Al diffusion	Process complexity	I/O compatibility No S_g gain	FEOL metal contam Low thermal budget HKMG rem. bf. SD GAA incompatibility

imec

D&GR

DIFFUSION AND GATE REPLACEMENT (D&GR) FLOW

VOLUME-LESS EWF TUNING

- D&GR* flow with n-type WF shifter e.g. MgO, LaO, ...

* R. Ritzenthaler (imec), IEDM 2014

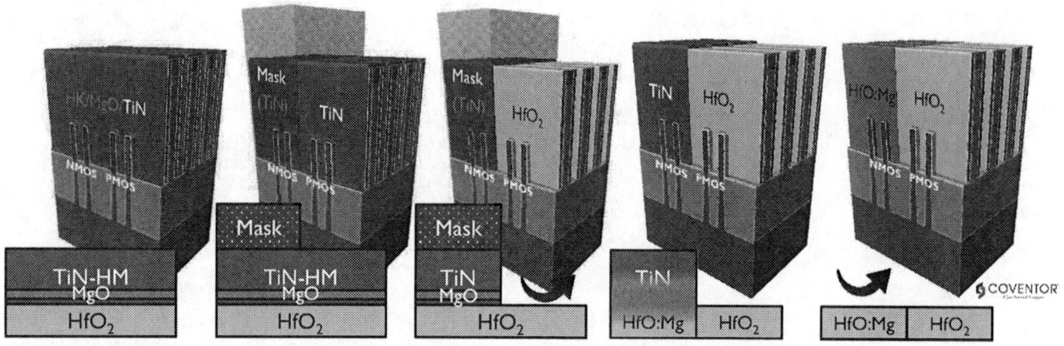

NMOS D&GR stack **Mask patterning** **TiN/MgO removal** **Drive-in anneal** **TiN removal**

imec

D&GR

D&GR RISK ASSESSMENT – 1/2

CORROSION

- Corrosive and hygroscopic MgO, LaO
  - Well optimized wet etching and rinsing solutions needed

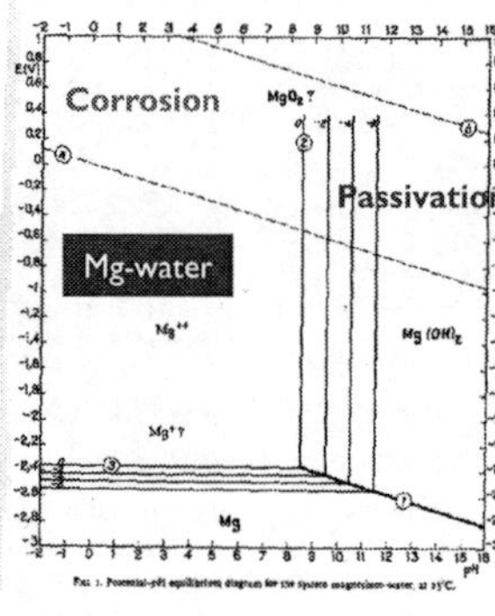

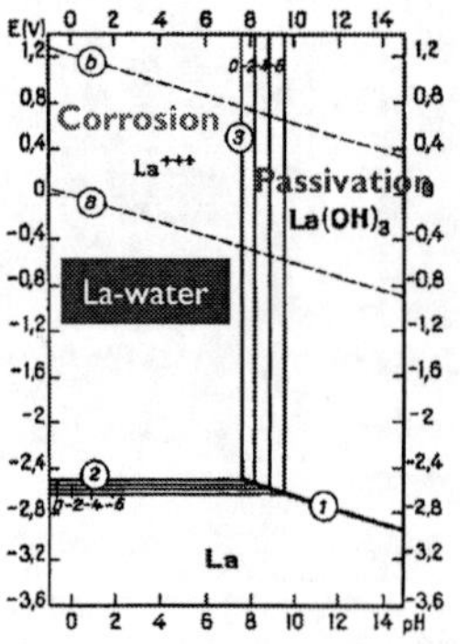

M. Pourbaix: "Atlas of Electrochemical Equilibria in Aqueous Solutions" (1974)
Mg-water: pp. 141, La-water: pp. 193.

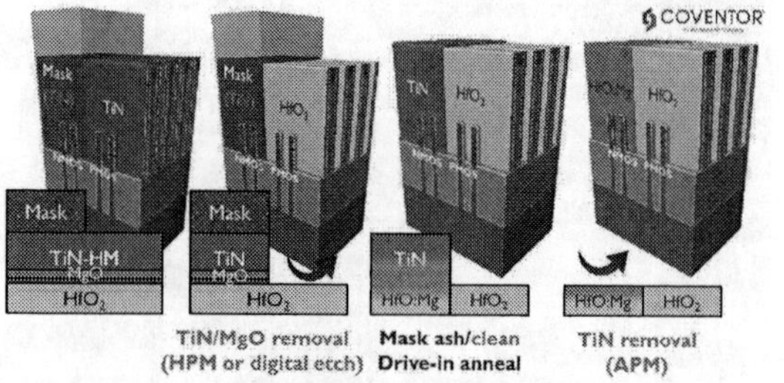

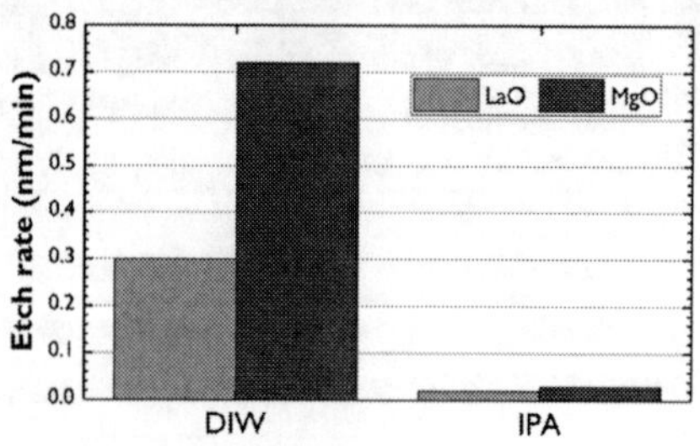

imec

❶ D&GR

D&GR RISK ASSESSMENT – 2/2

NON-SOLUBLE FLUORIDE RESIDUES

- Non-soluble fluoride residues to be generated if HfO:Mg, La exposed to fluorine containing chemistry (dry/wet)
  - LaFx, MgFx difficult to remove
  - Dry etch & post etch wet clean → fluorine free chemistries preferred
- Challenge/concern on PERR:
  - MGEB, MGC, x-couple, gate contact etches

Gate 1st planar:
Gate etch + HK removal (HF/HCl)

Gate 1st FinFET:
D&GR gate removal (HF + APM)

imec

21

❷ SiH_4 soak

SiH_4 SOAK

ZERO-THICKNESS EWF TUNING

- SiH_4 soak: O scavenging from the TiN/HfO_2 interface
  - provides p-type Vt shift
  - zero-thickness eWF tuning
- NP patterning done by a thin liner HM + SiH_4 soak
  - e.g. SiH_4 soak after SSA (sacrificial Si-cap anneal):

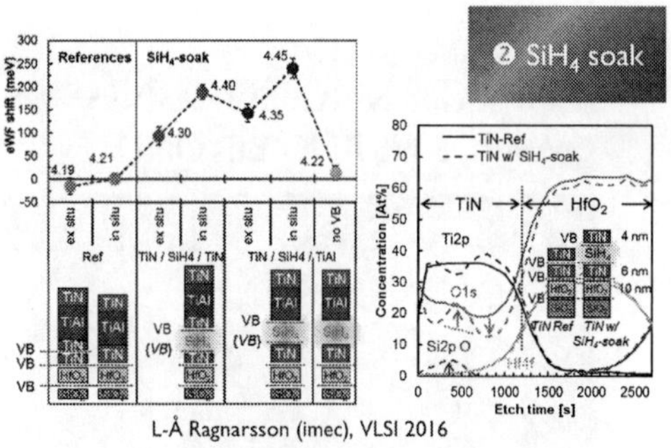

L-Å Ragnarsson (imec), VLSI 2016

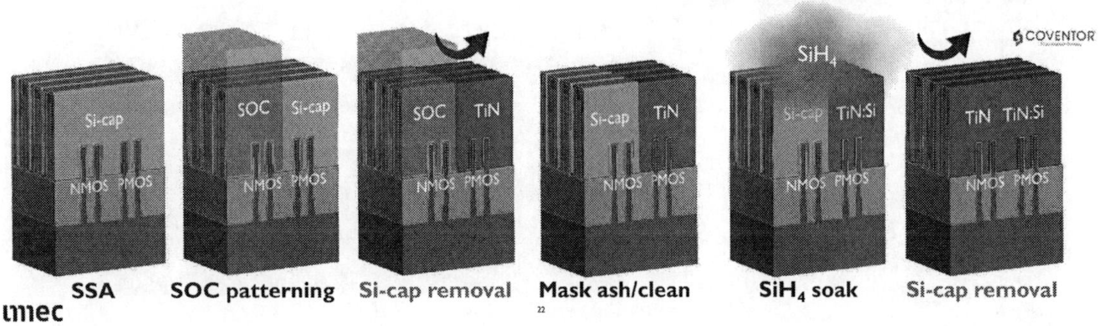

imec

22

❸ Blocked patterning

NWFM BLOCKED PATTERNING

HT-SOC & CMP ASSISTED "BLOCKED" PATTERNING

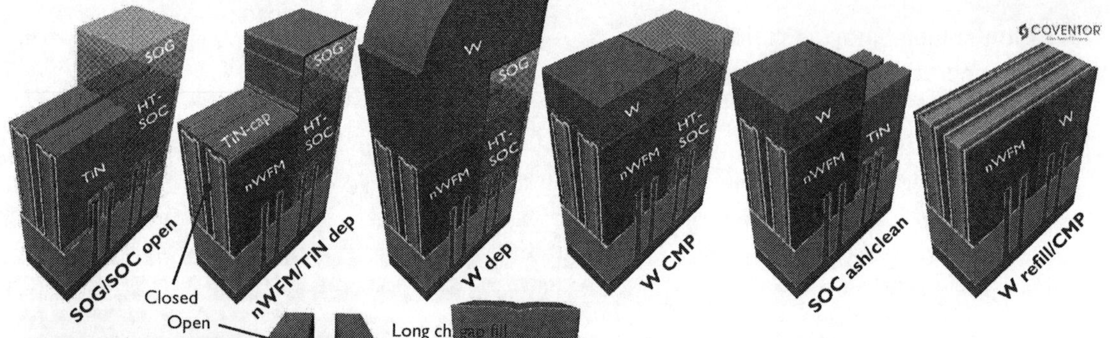

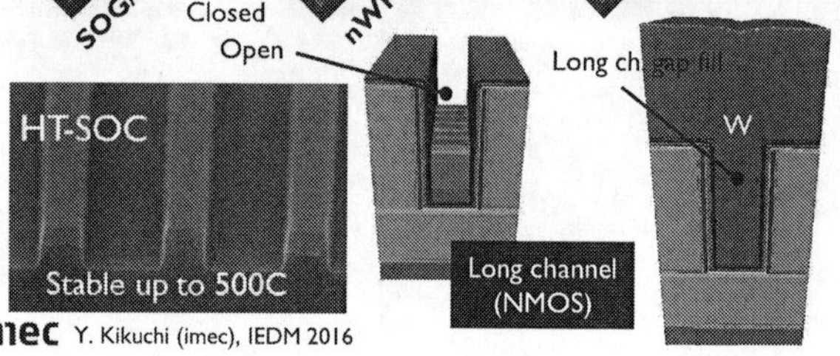

- CMP assisted "blocked" patterning
  - Enables closed gate patterning
  - Need thermally stable mask e.g. HT-SOC
  - No need for a wet metal etch!

imec Y. Kikuchi (imec), IEDM 2016

ULTIMATE RMG SCALING

TOWARDS N5 AND BEYOND

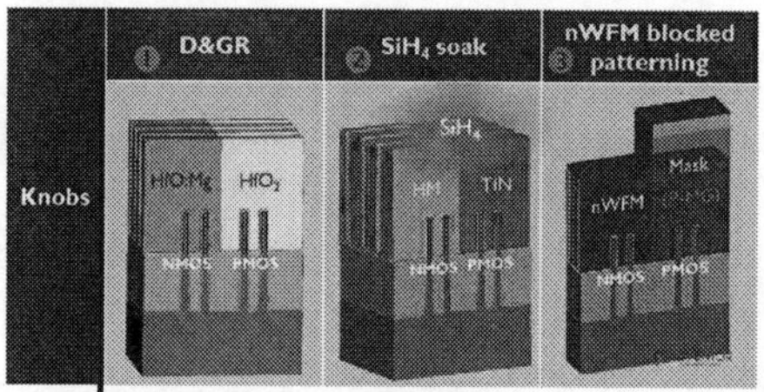

- Dummy poly removal
- Core/IO oxide patterning
- HK deposition
- D&GR ❶
- HK cap dep
- SSA/Masked SiH_4 soak ❷
- Barrier dep (can skip)
- nWFM blocked patterning ❸
- W CMP

imec

- Ultimate RMG scaling and multi-Vt enabled by combining several patterning knobs
  - e.g. D&GR + SiH_4 soak + blocked patterning

	N-LVT	*N-SVT*	*N-HVT*	*P-HVT*	*P-SVT*	*P-LVT*	
	W	W	W	W	W	W	
	TiN	TiN	TiN	TiN	TiN	TiN	
❸	nWFM	nWFM	nWFM				2-2.5nm
	TaN	TaN	TaN	TaN	TaN	TaN	0-1nm
❷	TiN	TiN	TiN:Si	TiN	TiN	TiN:Si	~1nm
❶	HfO:Mg	HfO	HfO	HfO:Mg	HfO	HfO	~1.5nm
	IL	IL	IL	IL	IL	IL	
	N-ch.	N-ch.	N-ch.	P-ch.	P-ch.	P-ch.	

~4.5 -6nm

24

SUMMARY

SUMMARY
RMG WET PROCESS CHALLENGES AND THE PATTERNING KNOBS

- RMG challenge
  - Not much space!
- RMG patterning development
  - Implementing tri-layer patterning (PR/SOG/SOC)
  - Digital wet etching of TiN successfully demonstrated
- RMG scaling & multi-V_t patterning knobs
  - ➊ **D&GR** to enable zero-thickness eWF tuning
  - ➋ **SiH_4 soak** as volumeless eWF tuning
  - ➌ **Blocked patterning** to pattern closed nWFM
  - Ultimate RMG scaling and multi-Vt patterning to be enabled by ➊+➋+➌

imec 26

imec

embracing a better life

imec

PUBLIC

cea tech

FROM RESEARCH TO INDUSTRY

CNRS | UNIVERSITÉ Grenoble Alpes | leti

WET AND SICONI ®

CLEANING SEQUENCES FOR SiGe EPITAXIAL REGROWTH

P.E. Raynal, V.Loup, L.Vallier, J.M. Hartmann, P. Besson

leti

2018 SPCC| Raynal Pierre-Edouard|

---

cea tech

RAISED SOURCE/DRAIN EPITAXY FOR SUB 22 NM FD-SOI

leti

FD-SOI architecture:

MOS bulk evolution:
- Low power consumption
- Good electrical performances

SOI substrate: Low thermal budget (Si Top layer)
SiGe channel: High air reactivity (re-oxidation)

Raised Source/Drain epitaxy

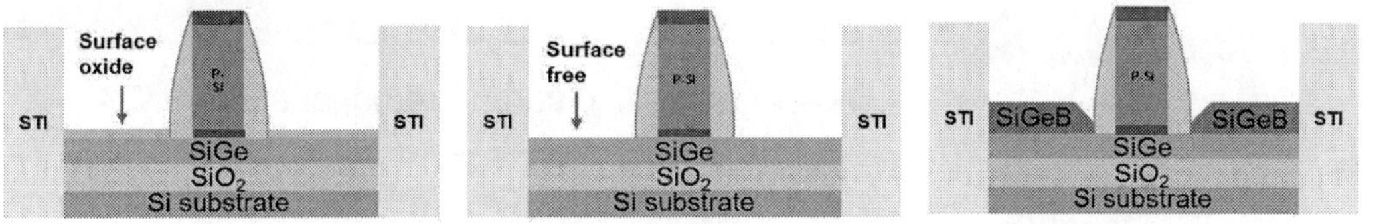

WET + H_2 Bake T°C <700°C **Epitaxial regrowth**

Efficient and low thermal budget pre-epitaxial treatment mandatory to avoid dewetting especially for SiGe starting layers

WET surface preparation
- Diluted HF/HCl

DRY surface preparation
- Remote plasma (Siconi ®)

| 2

OUTLINE

WET Surface Preparation	**HF oxide and carbon removal efficiency**
DRY Surface Preparation	Siconi® oxide and carbon removal efficiency
WET and DRY Surface Preparation	Alternative WET and DRY combination for surface preparation
SiGe Epitaxial regrowth	Validate the efficiency of the alternative WET and DRY combination

SPCC 2018 | Raynal Pierre-Edouard| | 3

OUTLINE

WET Surface Preparation	**HF oxide and carbon removal efficiency**
DRY Surface Preparation	Siconi® oxide and carbon removal efficiency
WET and DRY Surface Preparation	Alternative WET and DRY combination for surface preparation
SiGe Epitaxial regrowth	Validate the efficiency of the alternative WET and DRY combination

SPCC 2018 | Raynal Pierre-Edouard| | 4

SIGE 40% WET SURFACE PREPARATION

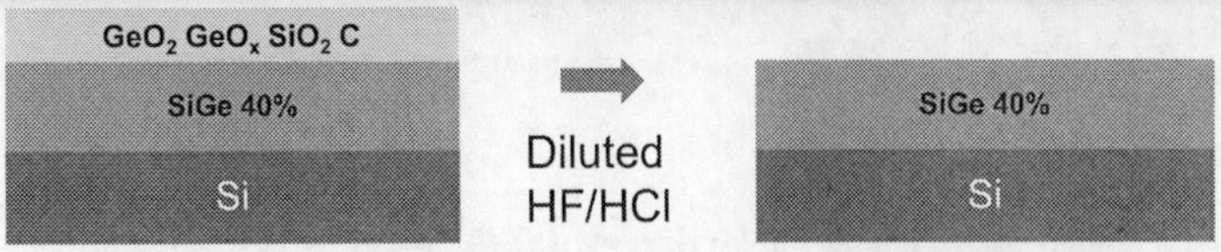

SiGe 40% grazing angle XPS quantification

	Native oxide	WET clean
Air exposure		4 minutes
C % (C1s)	15,5	5
GeO_x % (Ge3d)	1,26	N.D.
GeO_2 % (Ge3d)	5,41	0
SiO_2 % (Si2p)	14,83	0

HF/HCl surface preparation of SiGe 40%:
- **efficient SiO_2, GeO_2 and Carbon removal**
- **Subject to air break**

| 5

SiGe 40% SURFACE REACTIVITY

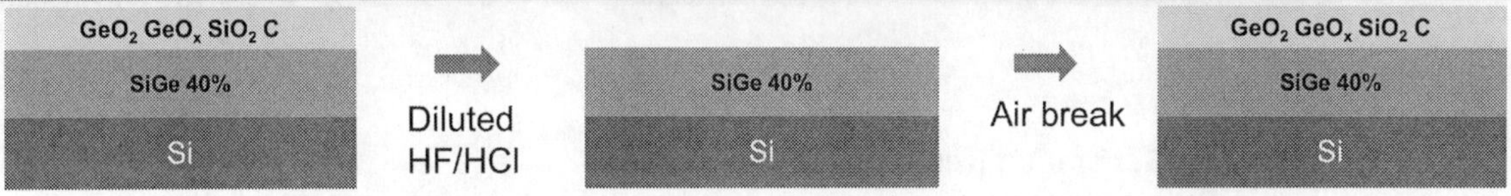

Evolution of **IV-H bond** in atmosphere after HF clean (FTIR-MIR)

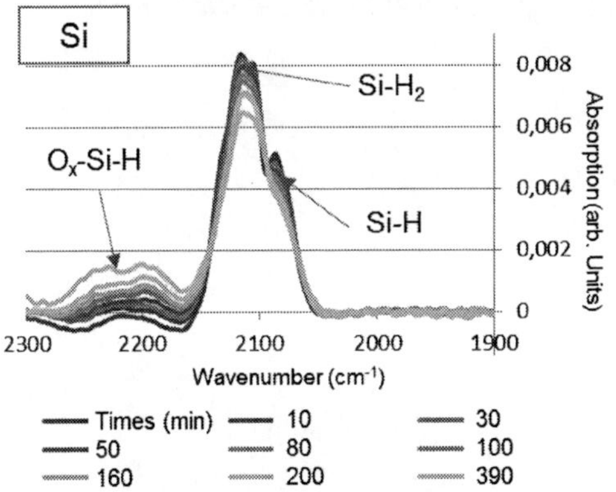

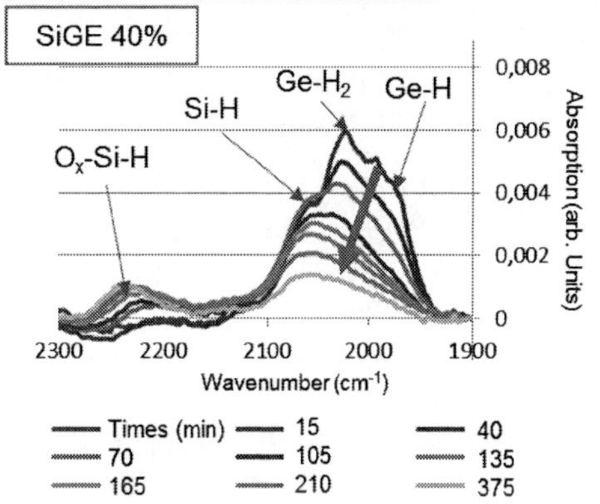

- IV-H bonds decrease after WET clean → formation of Si-O bonds
- Half of the Si-H and Ge-H loss after 2 hours for SiGe 40%

- SiGe 40% Air break can not be managed
- In-situ de-oxidation mandatory

| 6

AIR BREAK ISSUE

WET surface preparation:

Ex-situ WET CLEANING CHAMBER → **Epitaxial TOOL**

Air break unavoidable

DRY surface preparation:

In-situ DRY PLASMA CHAMBER → **Epitaxial TOOL**

No Air Break

DRY surface preparation useful for air sensitive material
- **Siconi®** → **remote plasma already used for SiO_2 removal and low thermal budget**

| 7

SICONI PRINCIPLE

1-Etchant Generation (remote plasma)

$$NF_3 + NH_3 \xrightarrow{RF} NH_4F + NH_4F.\,HF$$

2-Etch Process

$$SiO_2 + 4\,HF + 2\,NH_4F \rightarrow (NH_4)_2SiF_6\ (solid) + H_2O$$
$$GeO_2 + 4\,HF + 2\,NH_4F \rightarrow (NH_4)_2GeF_6\ (solid) + H_2O$$

3- Anneal Process (>100 °C)

$$(NH_4)_2\,SiF_6\ (solid) \rightarrow SiF_4(g) + NH_3\ (g) + HF(g)$$
$$(NH_4)_2\,GeF_6\ (solid) \rightarrow GeF_4(g) + NH_3\ (g) + HF(g)$$

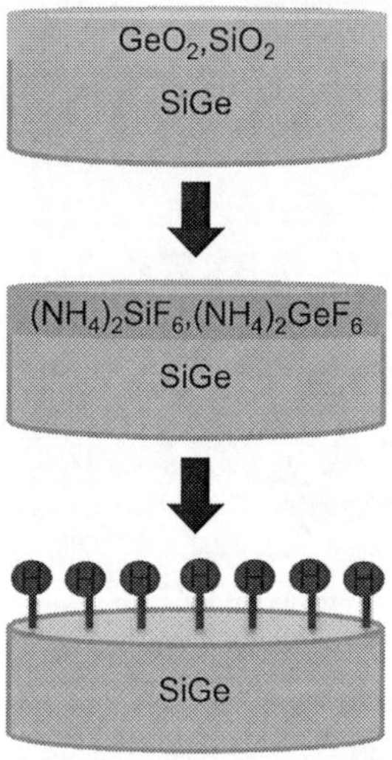

| 8

OUTLINE

WET Surface Preparation — HF oxide and carbon removal efficiency

DRY Surface Preparation — **Siconi® oxide and carbon removal efficiency**

WET and DRY Surface Preparation — Alternative WET and DRY combination for surface preparation

SiGe Epitaxial regrowth — Validate the efficiency of the alternative WET and DRY combination

IN-SITU SICONI® SURFACE PREPARATION OF SiGe 40%

SiGe 40% grazing angle XPS quantification

	Native oxide	Siconi®
Air exposure		4 minutes
GeO_2 % (Ge3d)	5,41	0
SiO_2 % (Si2p)	14,83	0
GeO_x % (Ge3d)	1,26	< LLD
C % (C1s)	15,5	12,6

Siconi® :
- **Efficient SiO_2 et GeO_2 removal**
- **Less efficient on Carbon and GeO_x**

Efficiency of Siconi® and HF-Last on GeOx removal

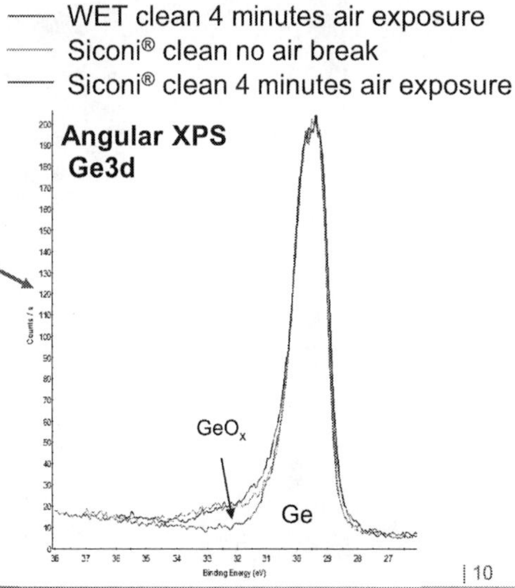

SUMMARY

WET Surface Preparation	HF oxide and carbon removal efficiency
DRY Surface Preparation	Siconi® oxide and carbon removal efficiency
WET and DRY Surface Preparation	**Alternative WET and DRY combination for surface preparation**
SiGe Epitaxial regrowth	Validate the efficiency of the alternative WET and DRY combination

ALTERNATIVE STRATEGY FOR SiGe 40% SURFACE PREPARATION

GeO_2 GeO_x SiO_2 C
SiGe 40%
Si

Native oxide

SiGe 40%
Si

Goal : Epi ready surface (without Oxygen, Carbon…)

ALTERNATIVE STRATEGY FOR SiGe 40% SURFACE PREPARATION

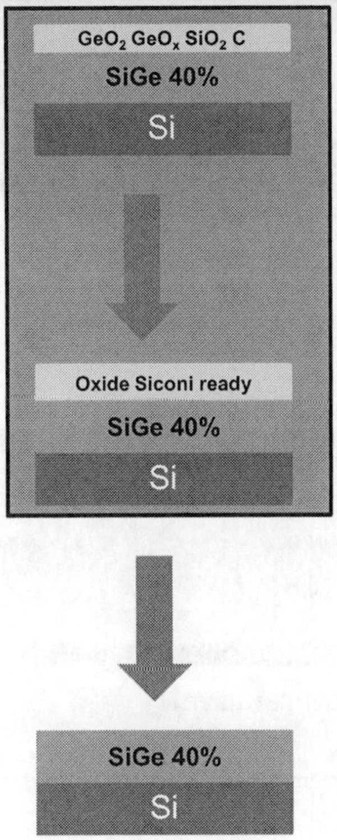

Native oxide

Can an oxidant solution:
- Change the oxide chemical composition?
- Remove the Carbon contamination?

Siconi ® less efficient on:
- GeO_2, GeO_x and Carbon

No air break ➡ Siconi®

Goal : Epi ready surface (without Oxygen, Carbon…)

| 13

ALTERNATIVE STRATEGY FOR SiGe 40% SURFACE PREPARATION

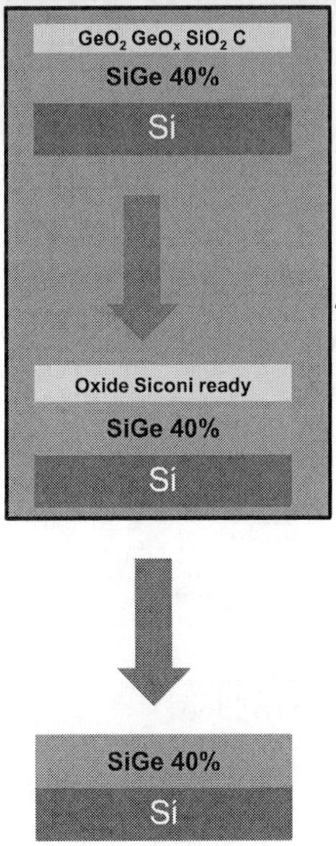

Native oxide

SiGe 40% chemical oxide

	Native oxide	HF-Ozone	HF-COLD SC1
SiO_2% (Si2p)	14,83	18,5	18
GeO_2% (Ge3d)	6,7	1,8	N.D.
C% (C1s)	15,5	3,5	N.D.

- Chemical oxide mostly SiO_2
- No carbon contamination

Does chemical oxide presence induce an efficient oxygen and carbon removal with Siconi?

No air break ➡ siconi®

Goal : Epi ready surface (without Oxygen, Carbon…)

| 14

SiGe 40%SURFACE STATE AFTER VARIOUS PREPARATIONS

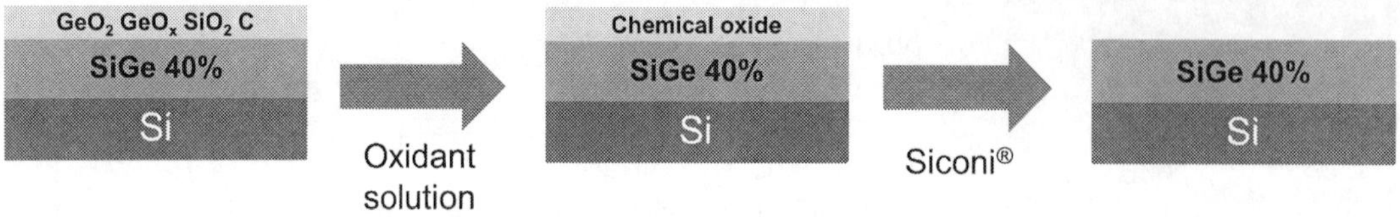

Angular XPS Quantification SiGe 40%

	Siconi	HF	HF-Siconi®	HF-COLD SC1-Siconi®	HF-HOT SC1-Siconi®	HF-Ozone-Siconi
O % (O1s)	10.41	8,54	6.78	N.D.	N.D.	N.D.
C % (C1s)	10.87	3.26	4.23	N.D.	N.D.	N.D.

Efficiency on Oxygen and Carbon removal →

HF-Chemical oxide-Siconi® → Lowest Oxygen and Carbon contamination

SUMMARY

WET Surface Preparation — HF oxide and carbon removal efficiency

DRY Surface Preparation — Siconi® oxide and carbon removal efficiency

WET and DRY Surface Preparation — Alternative WET and DRY combination for surface preparation

SiGe Epitaxial regrowth — **Validate the efficiency of the alternative WET and DRY combination**

STUDY OF EPITAXIAL REGROWTH

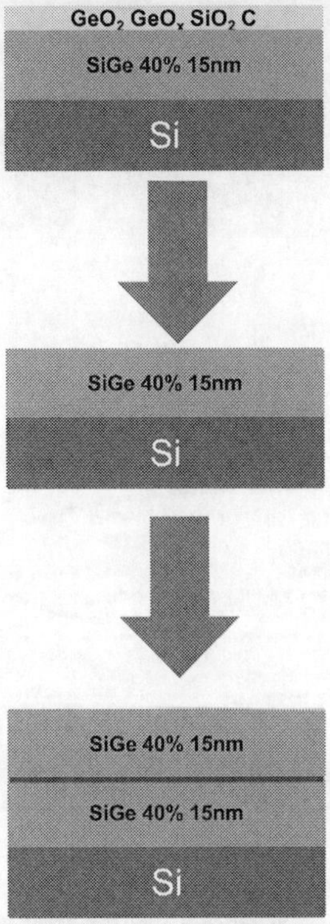

Various surface preparations for remove the oxide

Low temperature H_2 Bake and SiGe 40% epitaxial regrowth

Thickness of the layer with XRR
SIMS analysis of the interfacial contamination

SIGE 40% EPITAXIAL THICKNESS

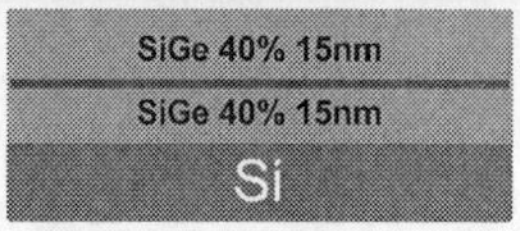

XRR SiGe 40% measurements

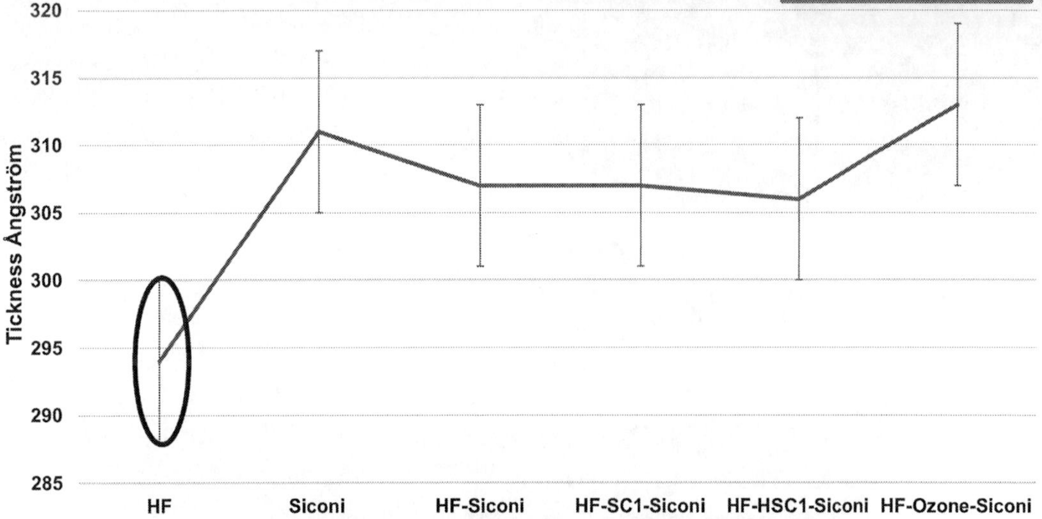

Growth delay with HF-Last surface preparation
Suggest high Oxygen interfacial contamination

OXYGEN INTERFACIAL CONTAMINATION (SIMS ANALYSIS)

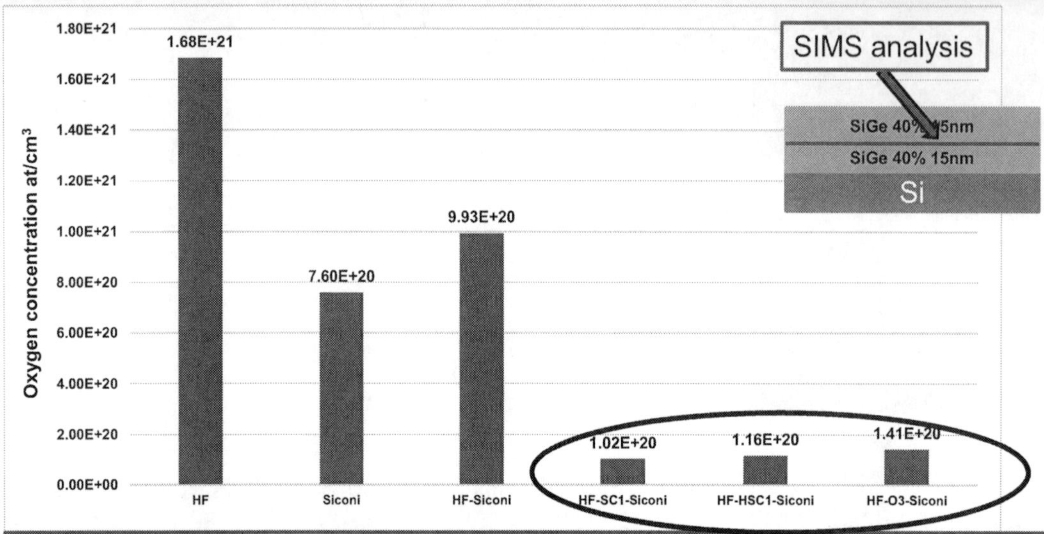

- HF-last: highest Oxygen concentration

- HF-Chemical Oxide-Siconi® yields the lowest oxygen concentration
- SIMS in line with XPS results

- No Carbon contamination detected for the various treatments
- Siconi® only: efficient removal of the C contamination by the H_2 bake

CONCLUSIONS

ex-situ HF-last :
- Surface sensitive to re-oxidation
- HF treatment suffers from regrowth delay
- highest Oxygen interfacial contamination

"In-situ" Siconi®:
- No regrowth delay
- Oxide interfacial contamination still present

HF-Chemical oxide-Siconi®:
- No regrowth delay
- Lowest Oxygen interfacial contamination
- The chemical oxide generated yields an efficient removal of the oxide with the Siconi process

| 20

ACKNOWLEDGMENT

Thanks to:

LTM: PHD supervisor — L. Vallier

WET: PHD supervisor — V. Loup, P. Besson

Epitaxy regrowth — J.M. Hartmann

This work was partially supported by:
- The LabEx Minos ANR-10-LABX-55-01
- ANR 10-EQPX-0030 (EQUIPEX FDSOI 11) and by Nano2017 project

THANKS YOU FOR YOUR ATTENTION

SPCC 2018 | Raynal Pierre-Edouard| | 21

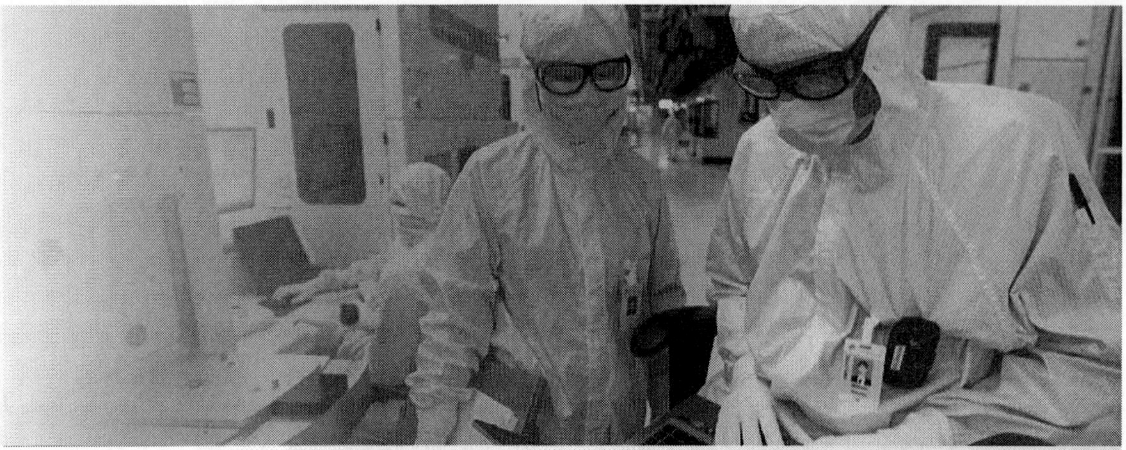

Pre SiGe Wet Cleans Development for sub 1x nm Technology Node

Akshey Sehgal, Anand Kadiyala, Michael DeVre and, Norberto Oliveria

April 10th, 2018

GLOBALFOUNDRIES

Background

- Due to higher aspect ratio features observed in advanced technology nodes (1x nm and smaller), the epi growth uniformity suffers across wafer
- Carbon, fluorine and oxygen residues incoming from post etch and ashing steps degrade the surface cleanliness and inhibit the epitaxial growth = missing epi defects
- Need C, O and F at low levels simultaneously going into epi deposition

Objective

- Improve pre-Epi wet cleans to ensure a pristine surface suitable for uniform epitaxial growth
- Ensure wet cleans equipment is not contributing to WiW non-uniformity

April 2, 2018 2

1 Achieving Pristine Wafer Surface with Pre-Epi Wet Clean

2 Improving Cavity Size WiW Uniformity

Sub 1x nm HVM Requirements for Pre-EPI Wet Cleans

- POR wet cleans are two-step cleans (back to back)
- HVM Requirements:
  - Eliminate carbon, fluorine and oxygen residues from cavity and SPCR Ox
  - SPCR Ox loss is minimal (= limited dHF usage in wet clean steps)
  - Retain cavity shape and size
  - Zero WiW non-uniformity

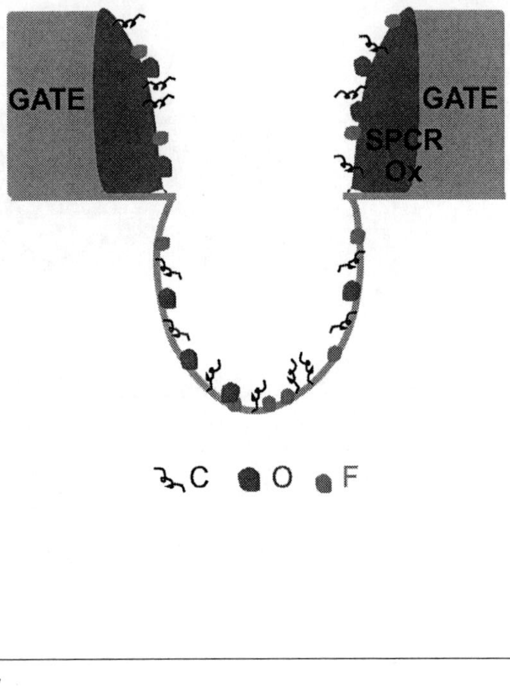

April 2, 2018

4

Pre-EPI Wet Clean Splits and XPS Results

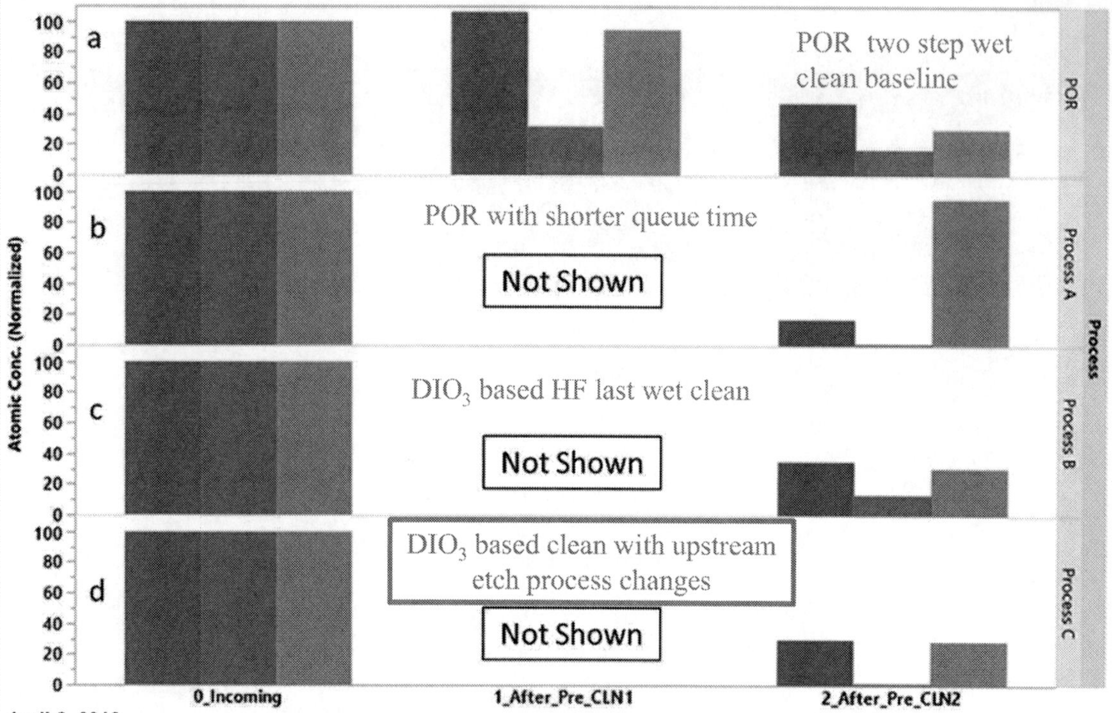

April 2, 2018

5

Sub 1x nm HVM Requirements for Pre-EPI Wet Cleans

- DIO_3 based wet clean did not lower missing epi defectivity to required levels

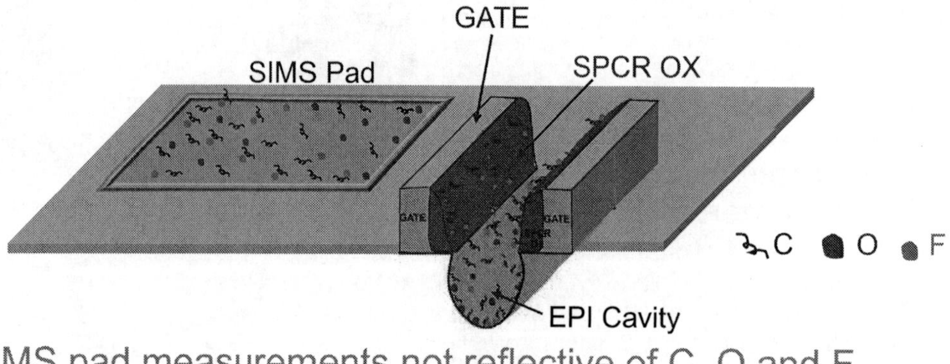

- SIMS pad measurements not reflective of C, O and F removal from SPCR Ox and epi cavity

☞ Therefore, we had to develop a new pre-epi wet clean

April 2, 2018

6

Results from 1st Split Lot

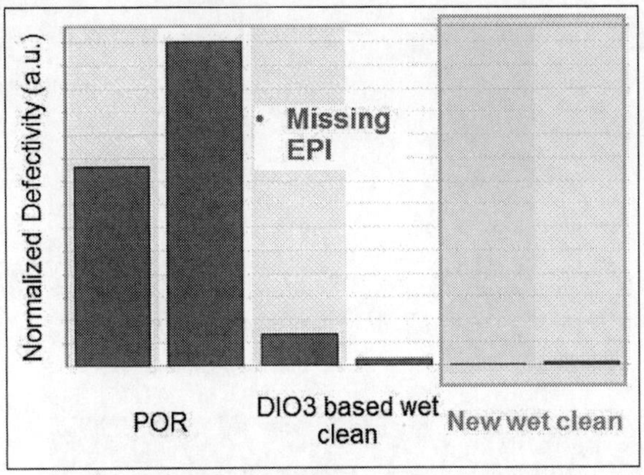

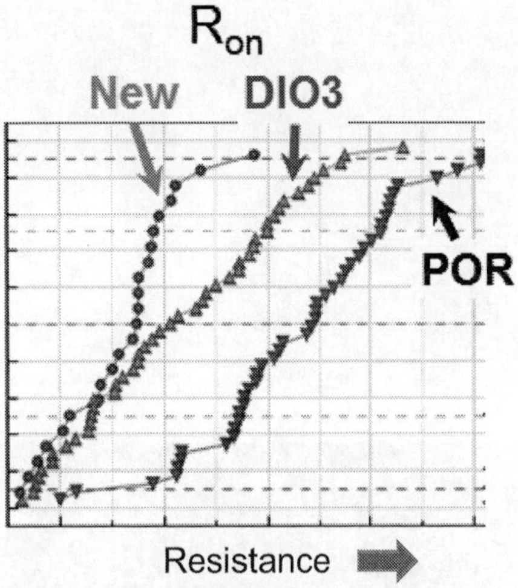

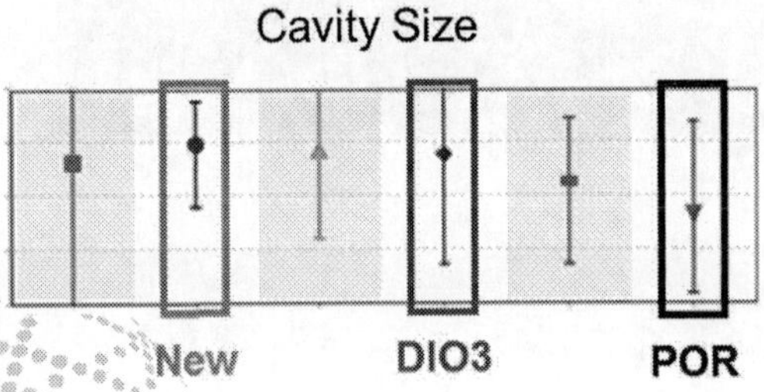

- ❑ **New pre-epi wet clean**
  - ❑ R_{on} ⇩ to largest extent with New wet clean
  - ❑ Cavity size is unchanged
  - ❑ ~ 50x ⇩ in missing epi defects

April 2, 2018 7

Post Implementation Results

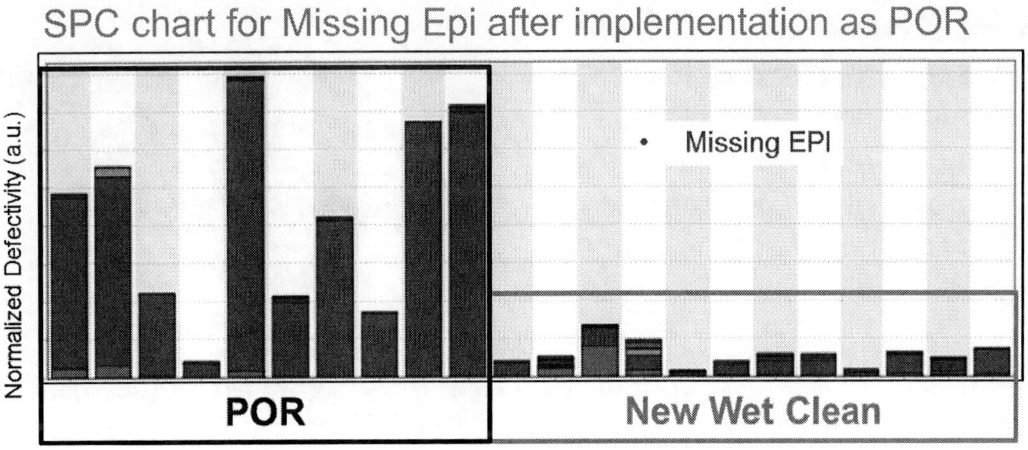

- ❑ **New Pre-Epi Wet Clean**
  - ❑ Accomplished with no change in SPCR dimension from old POR
  - ❑ Pristine wafer surface required additional changes in upstream processes
  - ❑ Further optimization in progress

April 2, 2018 8

1 Achieving Pristine Wafer Surface with Pre-Epi Wet Clean

2 Improving Cavity Size WiW Uniformity

Need for WiW Uniformity Improvement

- ❑ In sub 1x nm wet cleans, the pre-epi wet clean requires complete post etch residue removal and precise partial removal of SPCR Ox while leaving the rest of the exposed oxide and other materials intact

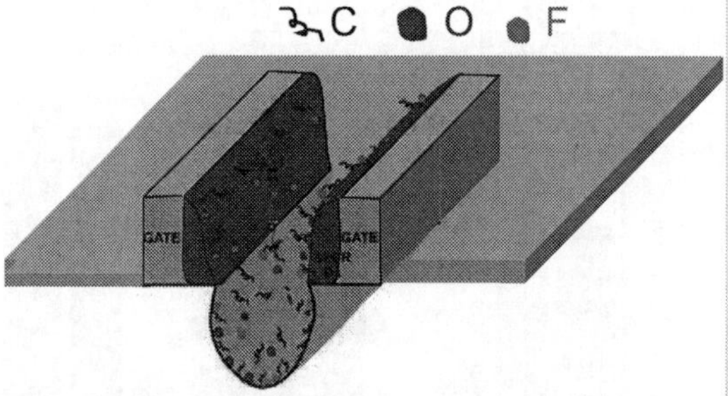

- ❑ Pre-epi wet clean step done with dHF /IPA drying on single wafer clean SNK does not meet cavity size WIW uniformity process specifications

- ❑ Hence, a decision was made to transfer this critical clean step to a different wet clean vendor toolset, from Vendor A to Vendor B

Improving WiW Uniformity: Adding BS N_2

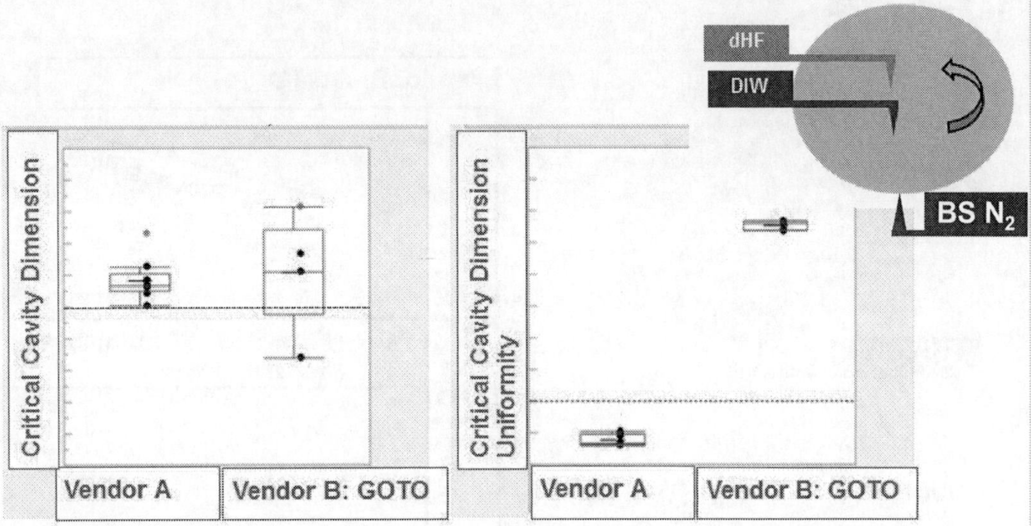

- ❑ Adding N_2 to wafer backside in vendor B tool gave worse performance

☞ Need to change process/ hardware/ software settings in GO TO tool to ⇩ WiW non-uniformity

April 2, 2018 11

Improving WiW Uniformity: Adding dHF on Wafer BS

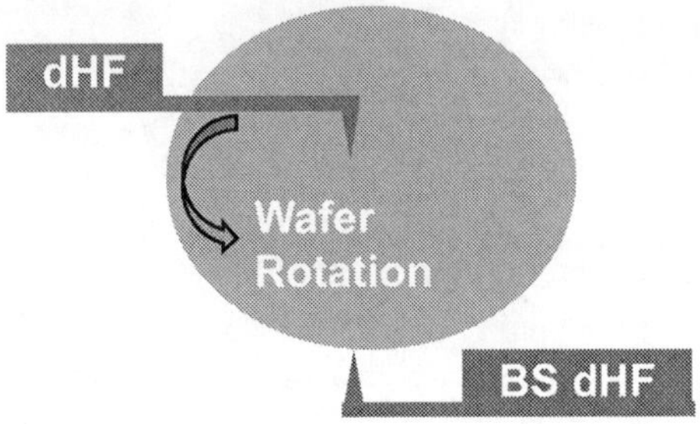

- ❑ Had to increase FS dHF time to match removal in Vendor A tool
- ❑ *Generating many BS particles with dHF only on wafer BS*

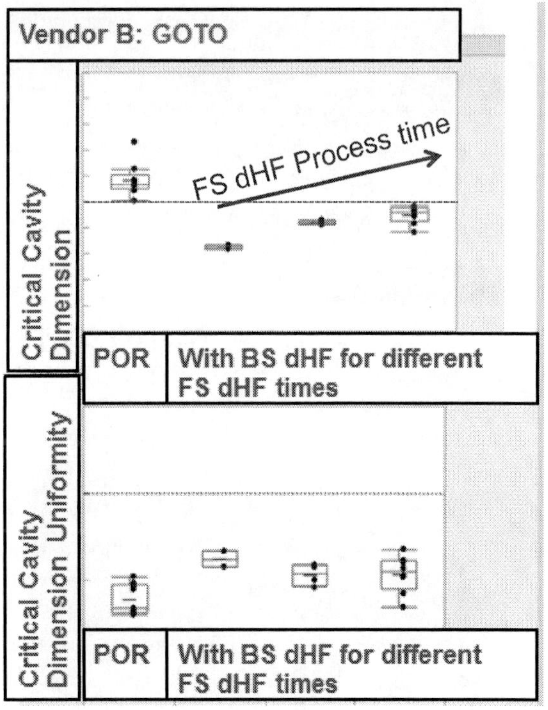

April 2, 2018 12

Improving WiW Uniformity: Adding DIW on Wafer BS

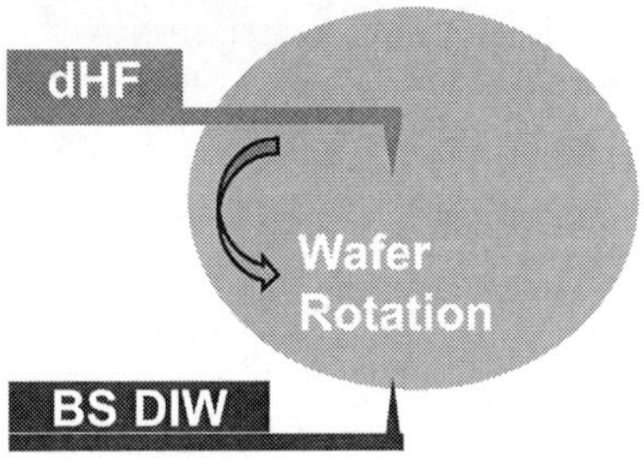

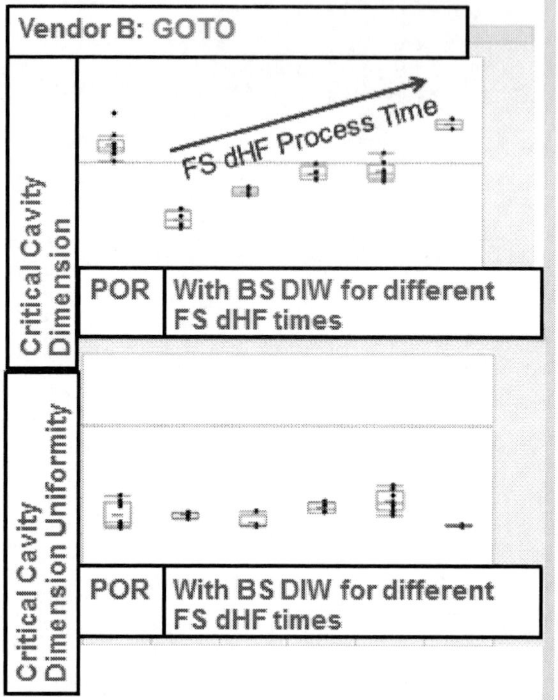

- ❑ With FS dHF and BS DIW process, only process time needs to be dialed in to achieve target critical cavity dimension

Improving WiW Uniformity: Effect of Flow on Wafer BS

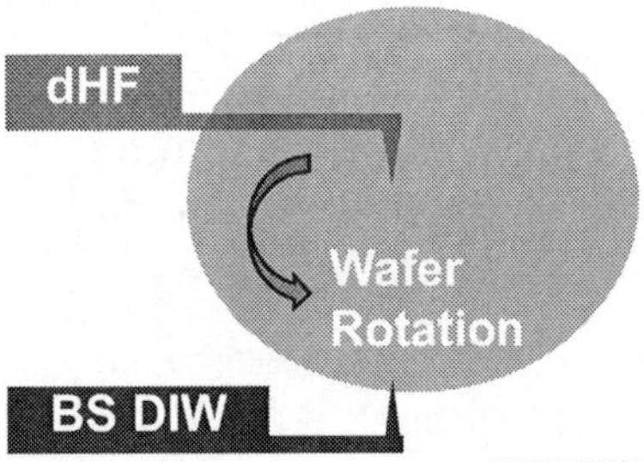

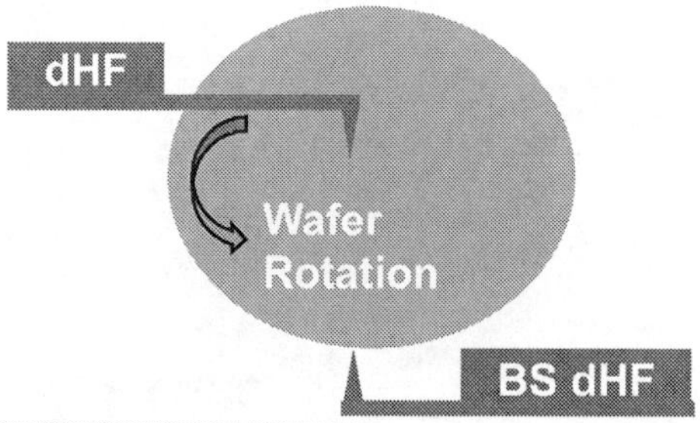

Blanket OX Test Wafer Data Shown

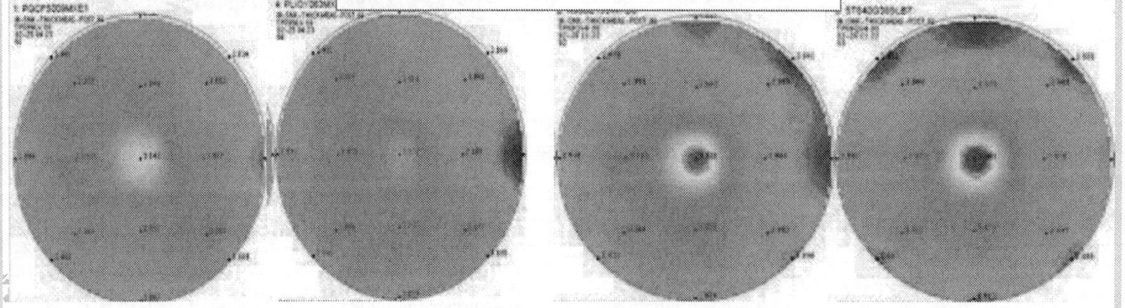

- ❑ BS dHF is generating excessive removal on wafer FS!

Improving WiW Uniformity: Adding DIW on Wafer BS + FS dHF scan

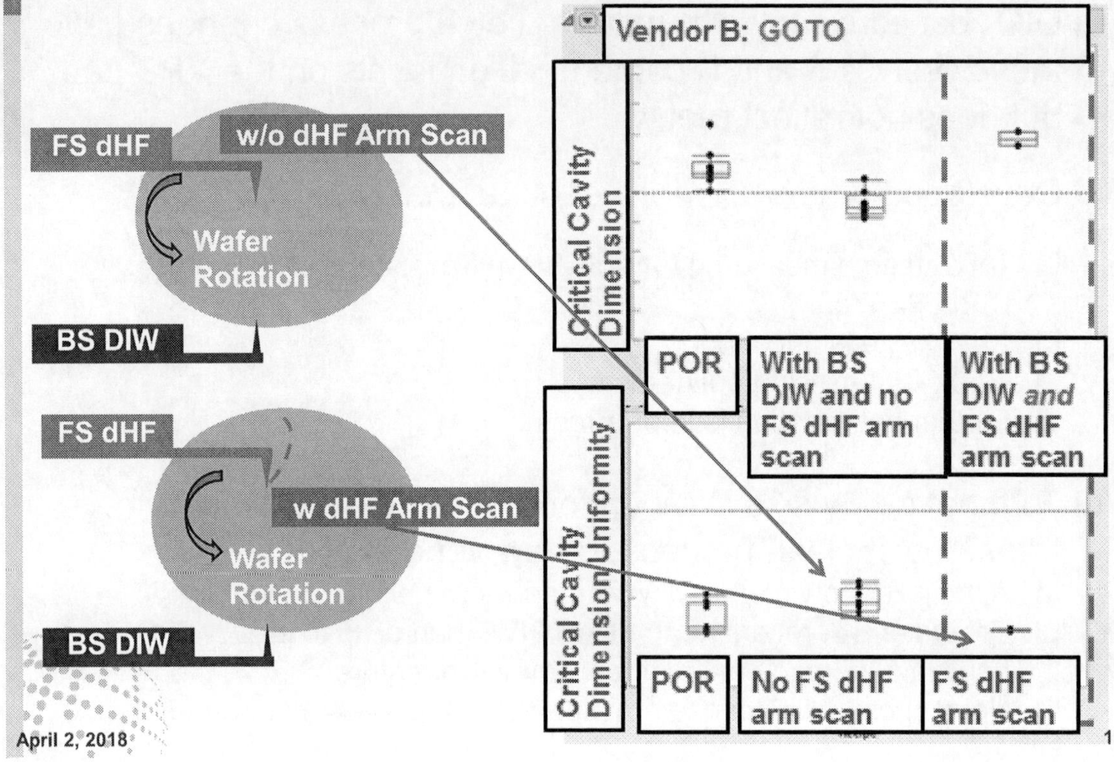

Final HVM Setting on Vendor B Tool

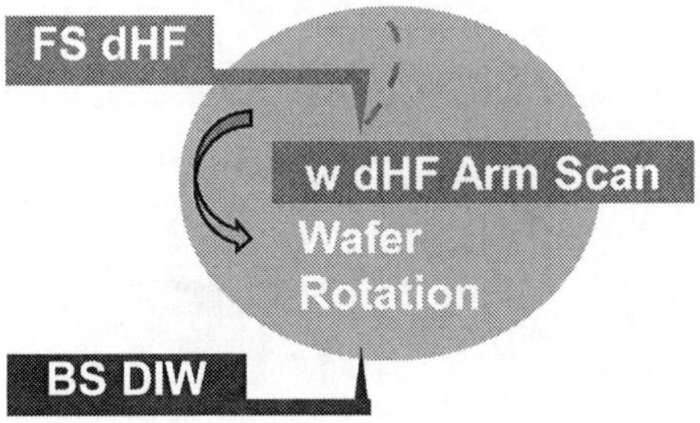

HVM LEARNINGS:

- ❑ Need to change process time to match critical cavity dimension achieved on Vendor A tool
- ❑ Adding BS flow makes WiW temperature uniform on wafer FS
- ❑ For wafer BS, recommend using DIW, than dHF, to prevent BS particle generation and center removal signature
- ❑ Need chemical dispense arm scan to minimize WiW non-uniformity

Summary

- DIO$_3$ based clean with upstream etch process changes gave the lowest C, O and F contamination levels on the XPS pad but does not reflect reality
- Developed a new pre-epi wet clean that:
  - ⇩ R_{on} (same measured for not shown transistor metrics such as DIBL, I_{eff} etc.)
  - Cavity size is unchanged
  - ~ 50x ⇩ in missing epi defects
  - Further optimization is in progress
- To meet cavity size WiW uniformity requirements, had to transfer wet clean process to new vendor SNK
  - Adding BS flow makes WiW temperature uniform on wafer FS
  - For wafer BS, recommend using DIW, than dHF, to prevent BS particle generation and center removal signature
  - Need chemical dispense arm scan to minimize WiW non-uniformity

April 2, 2018 17

Acknowledgements

Grateful thanks to Fab 8 colleagues in

- Wet Cleans, Metrology and Defect Inspection
- Integration, Device and, TCAD

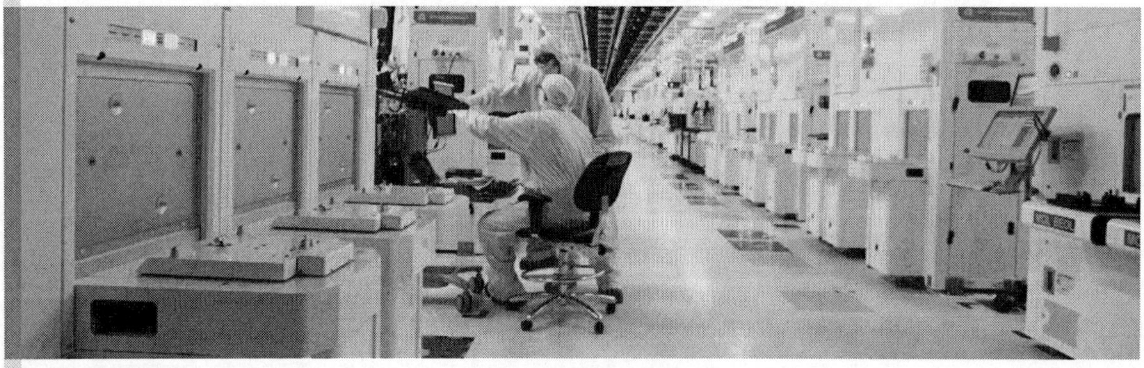

Thank You

GLOBALFOUNDRIES®

Trademark Attribution

GLOBALFOUNDRIES®, the GLOBALFOUNDRIES logo and combinations thereof, and GLOBALFOUNDRIES' other trademarks and service marks are owned by GLOBALFOUNDRIES Inc. in the United States and/or other jurisdictions. All other brand names, product names, or trademarks belong to their respective owners and are used herein solely to identify the products and/or services offered by those trademark owners.

FROM RESEARCH TO INDUSTRY

cea tech

leti

CNRS | UNIVERSITÉ Grenoble Alpes | LTM LABORATOIRE DES TECHNOLOGIES DE LA MICROÉLECTRONIQUE | leti

GeSn SURFACE PREPARATION BY WET CLEANING AND IN-SITU PLASMA TREATMENTS PRIOR TO METALLIZATION

SPCC 2018 | Raynal Pierre-Edouard|

P.E. Raynal, A. Quintero, V. Loup, Ph. Rodriguez, L. Vallier, J. Aubin, J.M. Hartmann, N. Chevalier and P. Besson

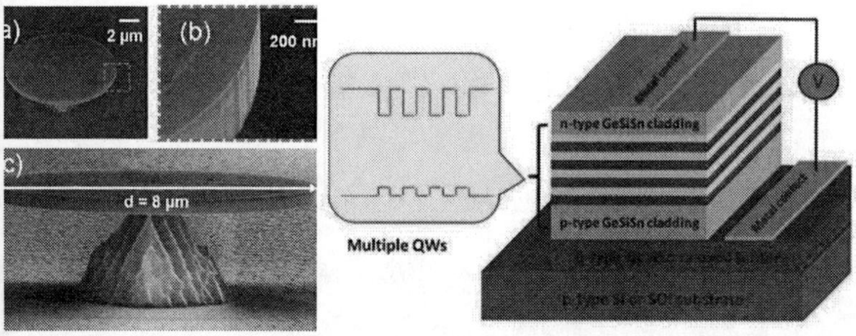

---

cea tech

GESN INTRODUCTION

leti

- Source/Drain compressive stressor in Ge channel MOS transistors (hole mobility increase)

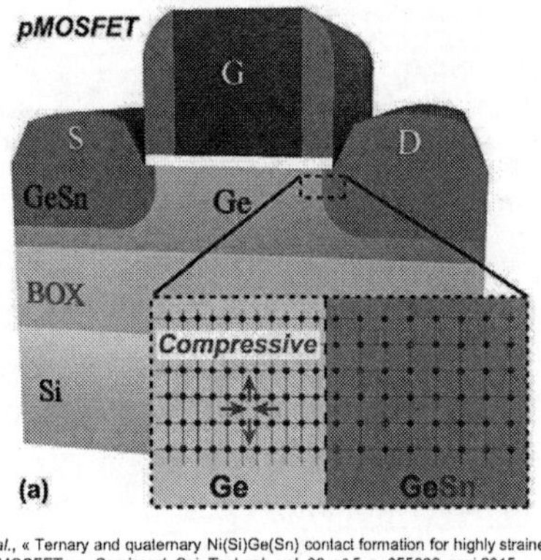

S. Wirths *et al.*, « Ternary and quaternary Ni(Si)Ge(Sn) contact formation for highly strained Ge p- and n-MOSFETs », *Semicond. Sci. Technol.*, vol. 30, n° 5, p. 055003, mai 2015.

- Direct-band gap materials (photonics applications) between 2,5 and 3,1 um wavelengths with GeSn 10% to 16%

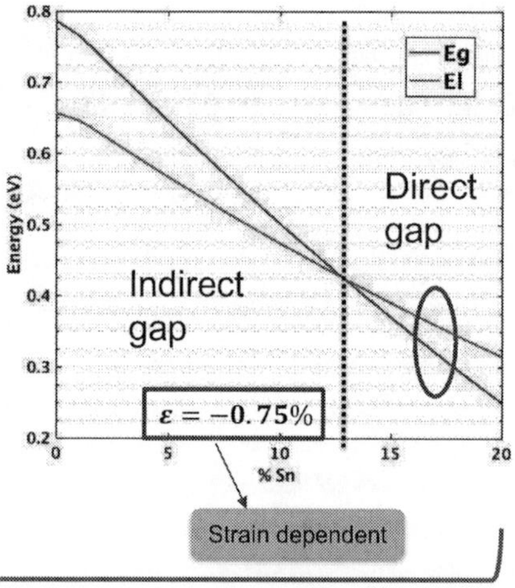

Efficient Ohmic contact mandatory → **Native oxide free surface**

EXPERIMENTAL PROTOCOL

- Stack

XRD Analysis
GeSn 15%

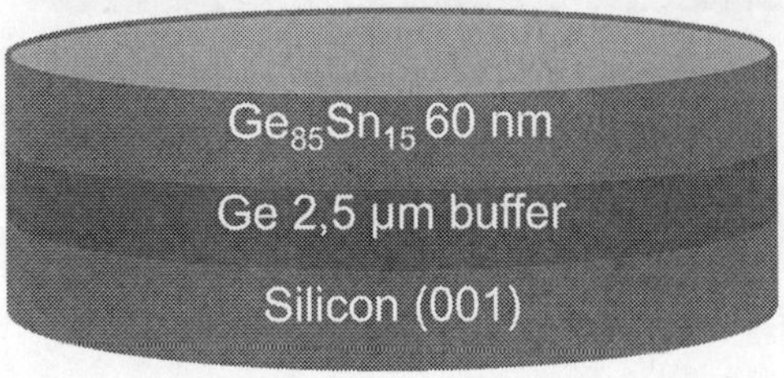

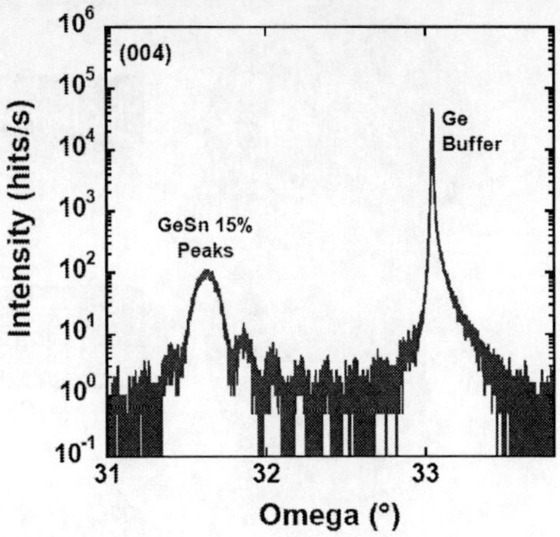

- Characterization

1) Morphology: atomic force microscopy
   Roughness of the surface, pitting…

2) Chemical composition: Plural angular XPS
   $\Theta \approx 75°$ thickness analyzed ≈ 2 nm

OUTLINE

GeSn characterization surface	**Roughness and native oxidation**
WET Surface Preparation	Diluted HF: Morphological impact and oxide removal efficiency
DRY Surface Preparation	Plasma He, Argon : Morphological impact and oxide removal efficiency
WET and DRY Surface Preparation	Alternative WET and DRY combination for surface preparation

OUTLINE

cea tech | leti

- **GeSn characterization surface** — **Roughness and native oxidation**
- **WET Surface Preparation** — Diluted HF: Morphological impact and oxide removal efficiency
- **DRY Surface Preparation** — Plasma He, Argon : Morphological impact and oxide removal efficiency
- **WET and DRY Surface Preparation** — Alternative WET and DRY combination for surface preparation

GESN 15% SURFACE CHARACTERIZATION

cea tech | leti

AFM

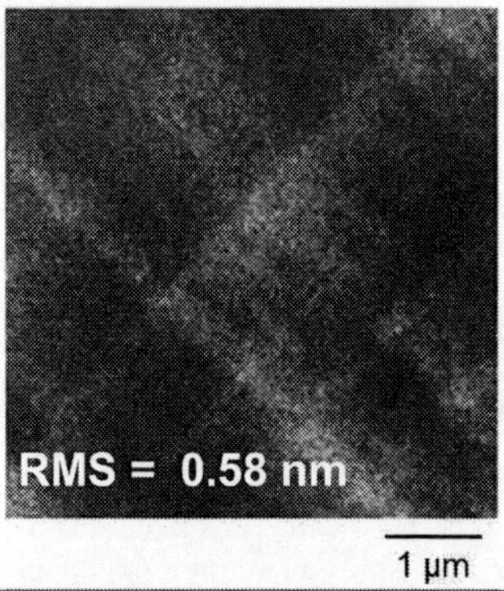

- Low surface roughness

- Sn oxidation almost immediate (<15 min air exposure)
- SnO_x oxidizes to SnO_2 with air exposure

- Ge oxidation slower than that of Sn
- No evolution of the oxide in samples older than 14 days

Grazing XPS Spectra

— $Ge_{0.85}Sn_{0.15}$ As grown
— $Ge_{0.85}Sn_{0.15}$ 8H30 air exposure
— $Ge_{0.85}Sn_{0.15}$ 14 days air exposure

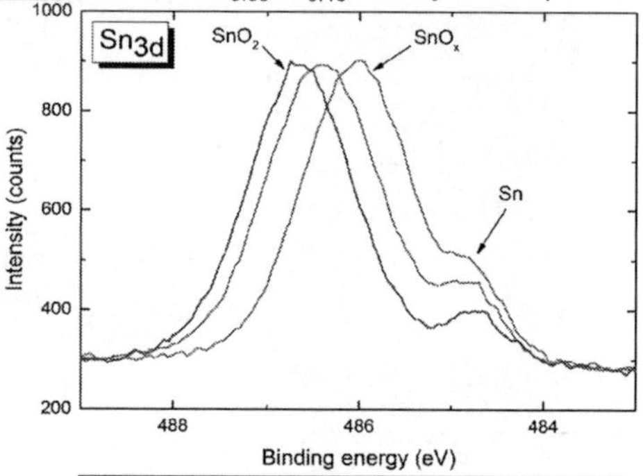

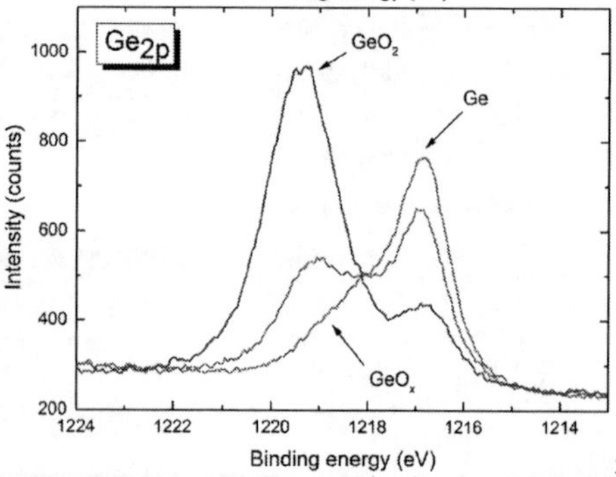

MOTIVATION: SURFACE PREPARATION PRIOR TO METALLIZATION

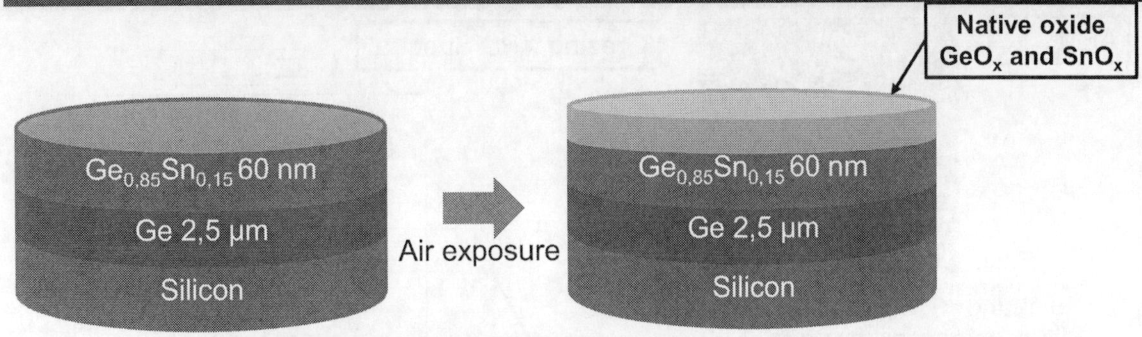

- Oxide grows almost immediately
- Oxide deleterious for Ni and NiPt ohmic contacts
- This oxide should be removed without degrading the surface
- GeSn is a new material : **surface preparation should be investigated**

WET treatment

- HF-Last + N_2 gun drying
- HF + H_2O rinse

DRY treatment

- He
- Argon

OUTLINE

GeSn characterization surface — **Roughness and native oxidation**

DRY Surface Preparation — Plasma He, Argon : Morphological impact and oxide removal efficiency

WET and DRY Surface Preparation — Alternative WET and DRY combination for surface preparation

IMPACT OF DILUTED HF ON GESN 15%

AFM

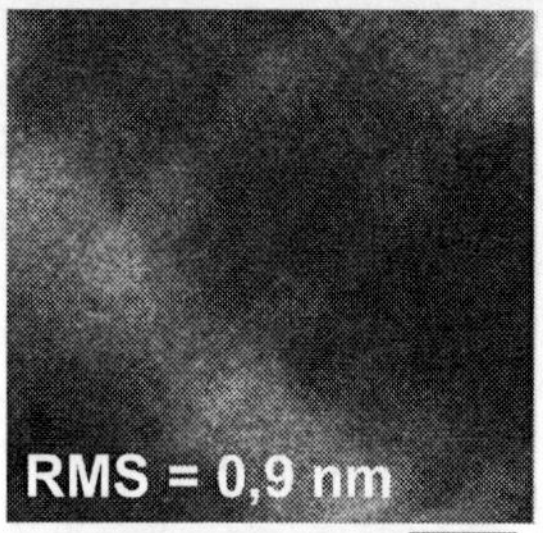

1 µm

- ➢ **No pitting**
- ➢ **Slight roughness increase**

Grazing XPS quantification

	Aged GeSn	HF
Sn/(Sn + Ge)	0.2	0.1
GeOx/Ge	0.5	0.3
SnOx/Sn	0.9	0

- ➢ **[Sn] surface depletion**
- ➢ **Ge top oxide layer**
- ➢ **Oxide free Sn**

Grazing XPS Spectra

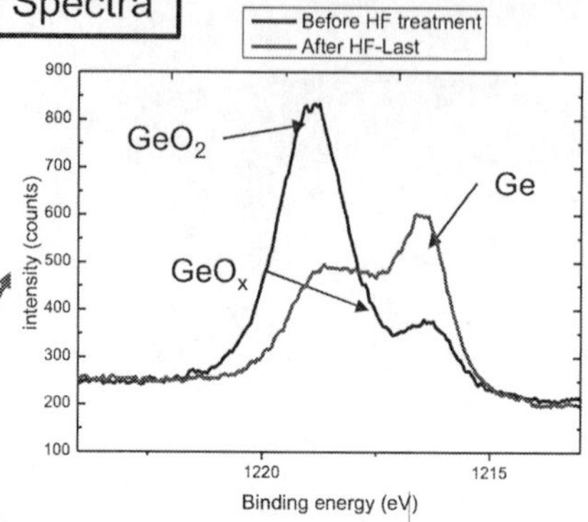

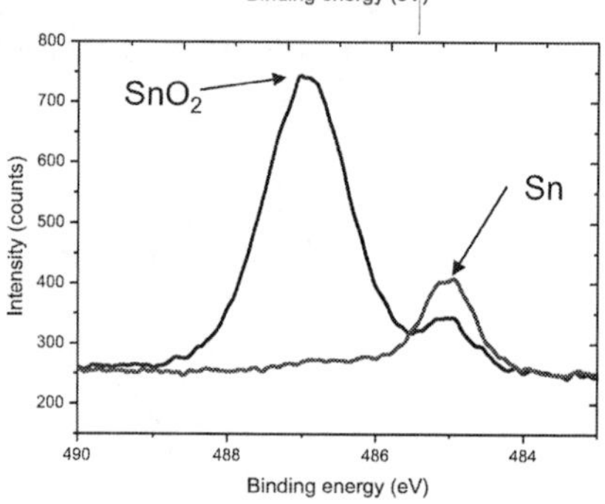

IMPACT OF DILUTED HF + H_2O RINSE ON GESN 15%

AFM

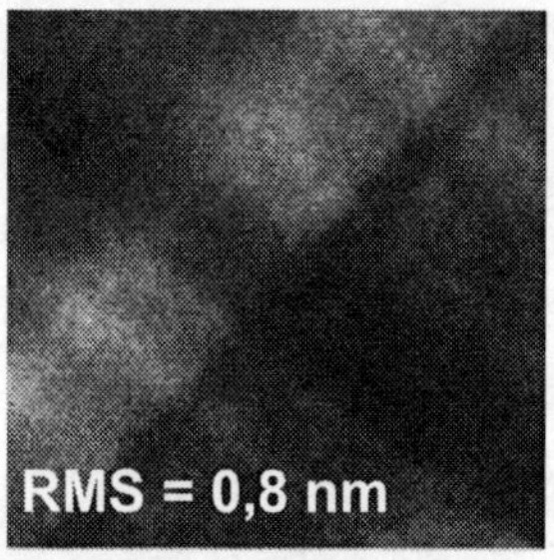

1 µm

- ➢ **No pitting**

Grazing XPS quantification

	Aged GeSn	HF	HF + H_2O
Sn/(Sn + Ge)	0.2	0.1	0.1
GeOx/Ge	0.5	0.3	0.4
SnOx/Sn	0.9	0	0.1

- ➢ **Surface is Sn depleted (as after a dip in HF)**
- ➢ **H_2O rinse post HF step oxidizes slightly Ge and Sn**

Grazing XPS Spectra

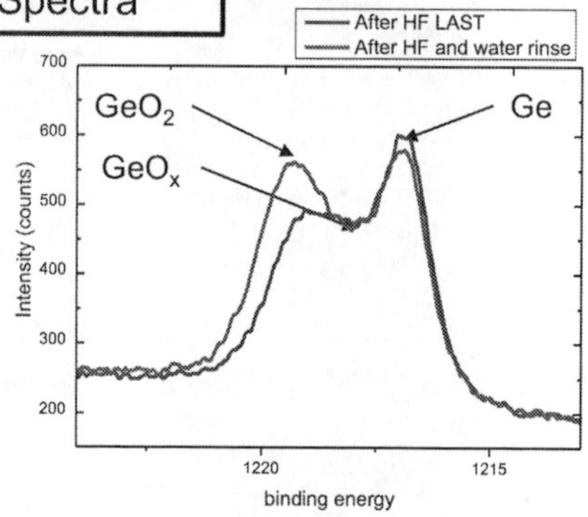

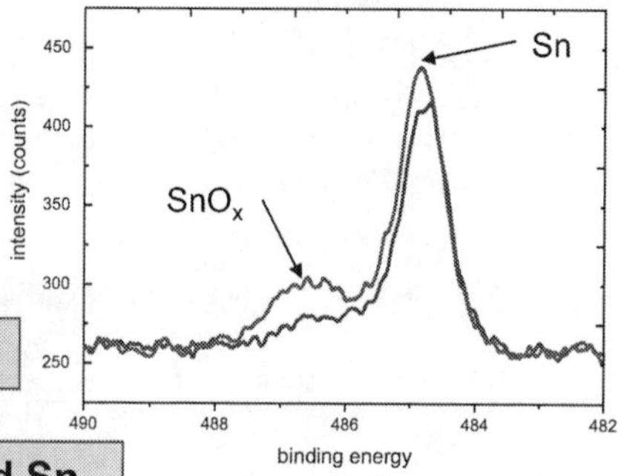

THE AIR BREAK ISSUE

WET surface preparation:

Ex-situ WET cleaning chamber → **Metallic deposition TOOL**

Air break unavoidable

DRY surface preparation:

In-Situ DRY plasma chamber → **Metallic deposition TOOL**

No Air Break

➢ **DRY surface preparation useful for air sensitive material**

OUTLINE

- GeSn characterization surface — **Roughness and native oxidation**
- **WET Surface Preparation** — Diluted HF: Morphological impact and oxide removal efficiency
- DRY Surface Preparation — Plasma He, Argon : Morphological impact and oxide removal efficiency
- WET and DRY Surface Preparation — Alternative WET and DRY combination for surface preparation

DRY IN-SITU PLASMA TREATMENT ON GESN 15%

Grazing XPS quantification

Air break of 15 minutes between the dry treatment and XPS analysis

	Native oxyde	He	Argon soft
Sn/(Sn + Ge)	0.2	0.2	0.1
GeOx/Ge	0.5	0.5	0.4
SnOx/Sn	0.9	0.6	0.6

- Air break has an impact ⟶ native oxide regrowth
- He and Argon plasma seem to reduce Sn oxide

AFM

Native oxide

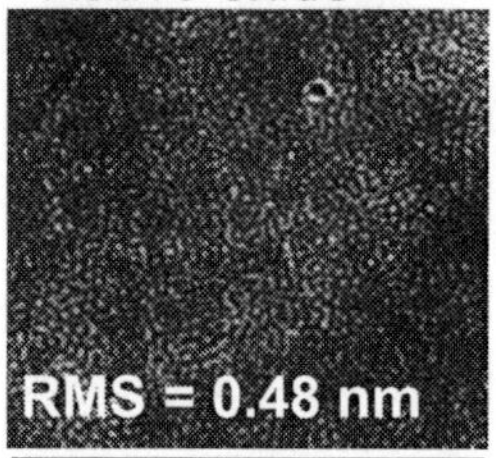

1 µm

He

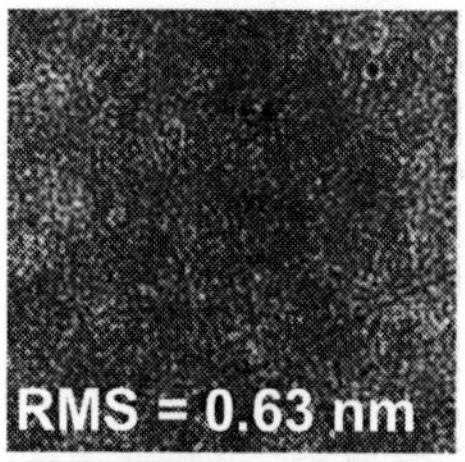

1 µm

Argon soft

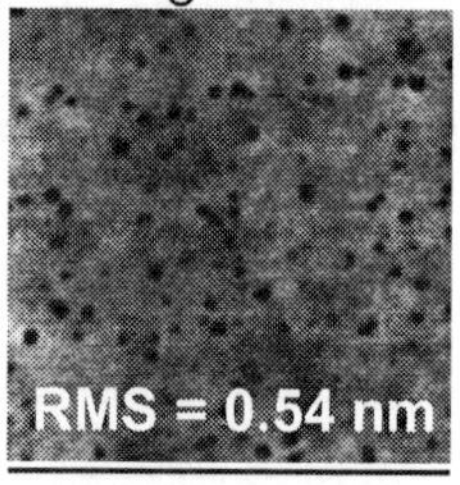

1 µm

- **He: no pitting**
- **Argon soft generates pitting**

TUNING THE ARGON PROCESS ON GESN 15%

Grazing XPS quantification

	Oxyde natif	Argon soft	Argon super soft	Argon extra soft
Sn/(Sn + Ge)	0.2	0.1	0.2	0.2
GeOx/Ge	0.5	0.4	0.4	0.4
SnOx/Sn	0.9	0.6	0.6	0.6

- **Plasma still reduces oxide**

AFM

Plasma power reduction ⟶

Native oxide

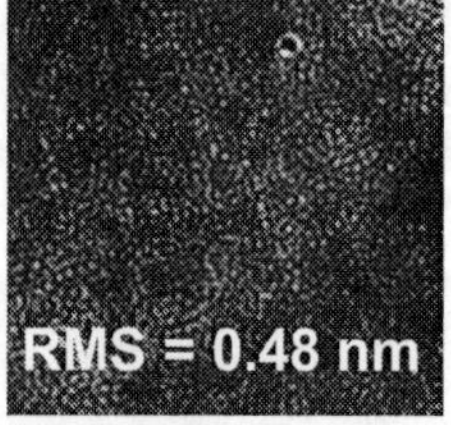

1 µm

Argon soft

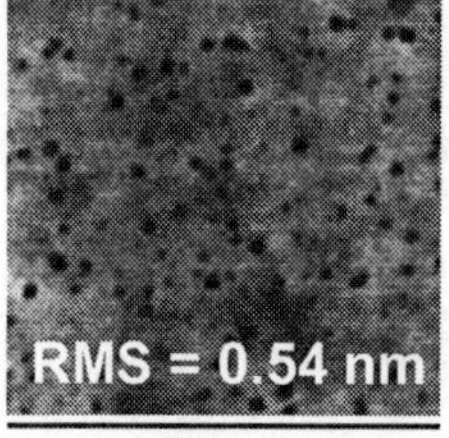

1 µm

Argon super soft

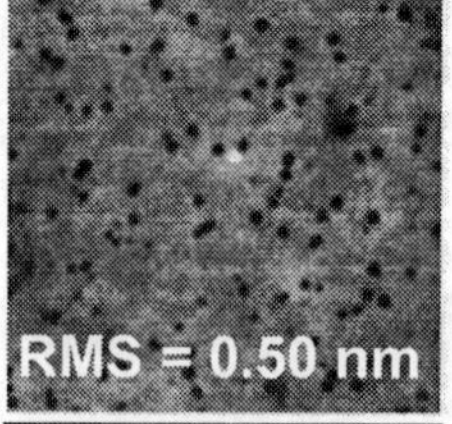

1 µm

Argon extra soft

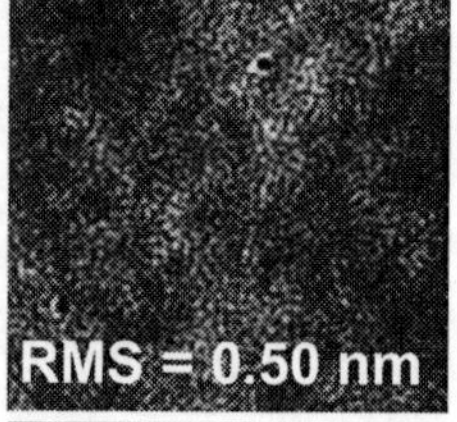

1 µm

- **Fine process tuning: pits are avoided**

OUTLINE

GeSn characterization surface — **Roughness and native oxidation**

WET Surface Preparation — Diluted HF: Morphological impact and oxide removal efficiency

DRY Surface Preparation — Plasma He, Argon : Morphological impact and oxide removal efficiency

WET and DRY Surface Preparation — **Alternative WET and DRY combination for surface preparation**

SPCC 2018 | Raynal Pierre-Edouard| | 15

WET DRY SURFACE PREPARATION COMBINATION

HF Surface preparation → Results in a thin layer of GeO_x without SnO_x

DRY Surface preparation → Reduces Ge and Sn oxides

Combine WET and DRY surface preparation to increase the oxide removal → **WET and DRY treatments**

- HF - He
- HF - Argon

SPCC 2018 | Raynal Pierre-Edouard| | 16

WET DRY SURFACE PREPARATION ON GESN 15%

Grazing XPS quantification

	Native oxide	HF-rinse	HF-He	HF-Argon extra soft
Sn/Ge	0.2	0.1	0.2	0.1
GeOx/Ge	0.5	0.4	0.4	0.4
SnOx/Sn	0.9	0.1	0.6	0.6

- **Surface oxide still present ➡ air break re-oxidation**
- **XPS Analysis needs to be without air break (in progress)**

AFM

Native oxide

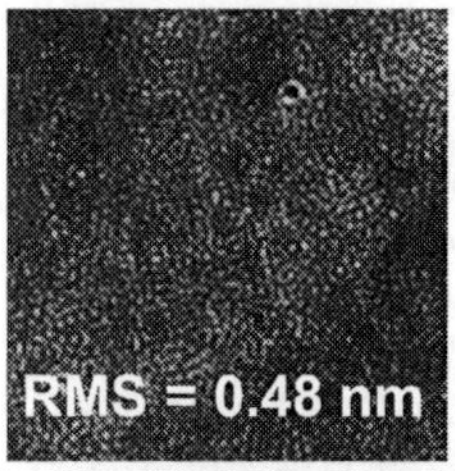

1 µm

HF

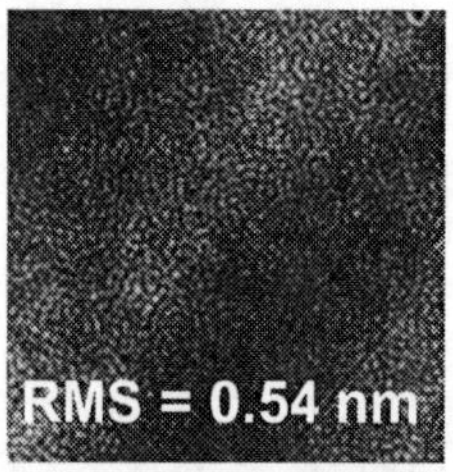

1 µm

HF-He

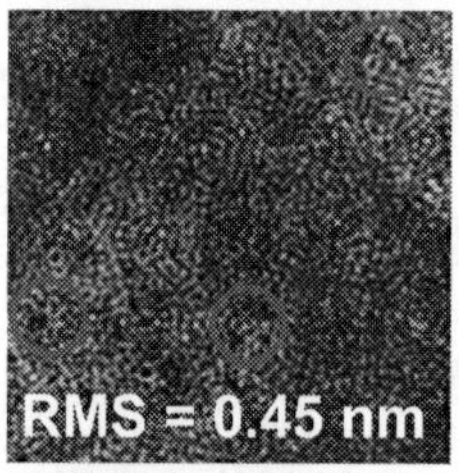

1 µm

HF-Argon extra soft

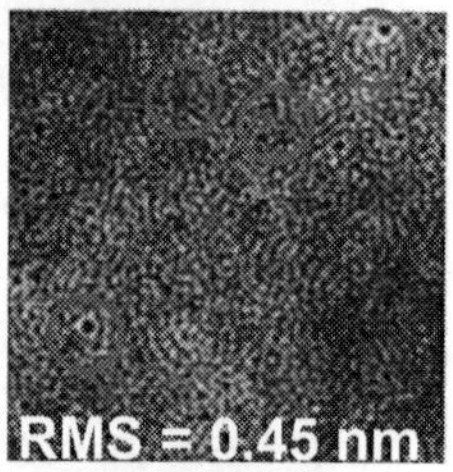

1 µm

- **WET and DRY combinations increase pitting**
- **Shorter exposures to plasma might cure that**

CONCLUSION

1) **GeSn 15%** is very sensitive to air exposure:
   - Formation of a native oxide almost immediately

2) **Diluted HF:**
   - removes the Sn oxide and leaves a thin layer of Ge oxide
   - Does not change the surface roughness

3) **He and Ar plasmas:**
   - reduce the amount of oxide
   - with a fine tuning : no pitting and no roughening

4) **Diluted HF + plasma sequence:**
   - reduces the amount of surface oxide
   - Limits pitting

In-situ characterization (in progress)

ACKNOWLEDGMENT

Thanks to:

LTM: PHD supervisor — L. Vallier

WET: PHD supervisor — V. Loup, P. Besson

Sample — J. Aubin, J.M. Hartmann

DRY — Ph. Rodriguez, A. Quintero

This work was partially supported by:
- the LabEx Minos ANR-10-LABX-55-01
- the French National Research Agency (ANR) under the “Investissements d'avenir” programs: ANR 10-AIRT-0005 (IRT NANOELEC)
- ANR 10-EQPX-0030 (EQUIPEX FDSOI 11) and by Nano2017 project

THANKS YOU FOR YOUR ATTENTION

SPCC 2018 | Raynal Pierre-Edouard| | 19

TEL imec

TOKYO ELECTRON

TEL Technology Center, America, LLC - imec

Selective isotropic etching of Group IV semiconductors to enable gate all around device architectures

SPCC, April 10, 2018

S. Kal[1], C. Pereira[1], Y. Oniki[2], F. Holsteyns[2], J. Smith[1], A. Mosden[1], K. Kumar[1], P. Biolsi[1], T Hurd[1].

[1]TEL Technology Center, America, LLC, USA

[2] Imec, Belgium

Subhadeep.Kal@us.tel.com

TEL Technology Center, America, LLC

Chemical Oxide Removal (COR) Reaction Mechanism: Case for SiO_2 etch

- HF and NH_3 adsorb on the SiO_2 surface, reacting to form $(NH_4)_2SiF_6$ (Ammonium Fluorosilicate - AFS)

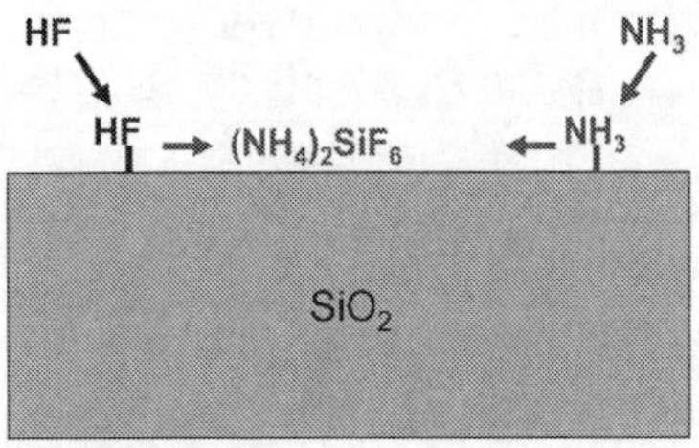

①SiO2 + 4HF $\xrightarrow{1.04eV}$ SiF4 + 2H2O

②SiO2 + 4HF + 4NH3 $\xrightarrow{0.49eV}$ SiF4 + 2H2O + 4NH3

③SiF4 + 2HF + 2NH3 $\xrightarrow{0eV}$ (NH4)2SiF6

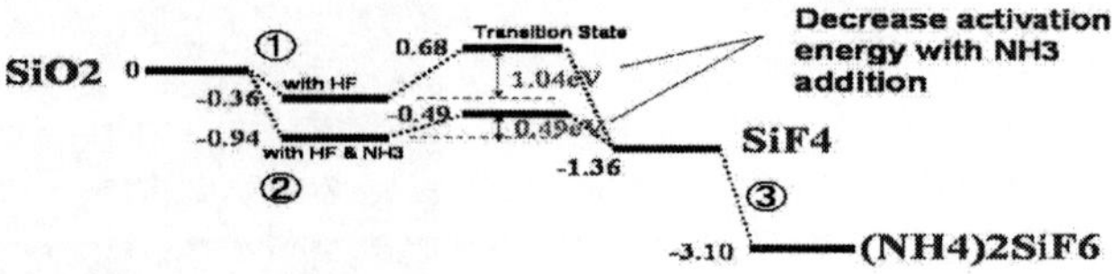

NH_3 catalyzes a desired reaction pathway

Slide courtesy: Tokyo Technology Solutions

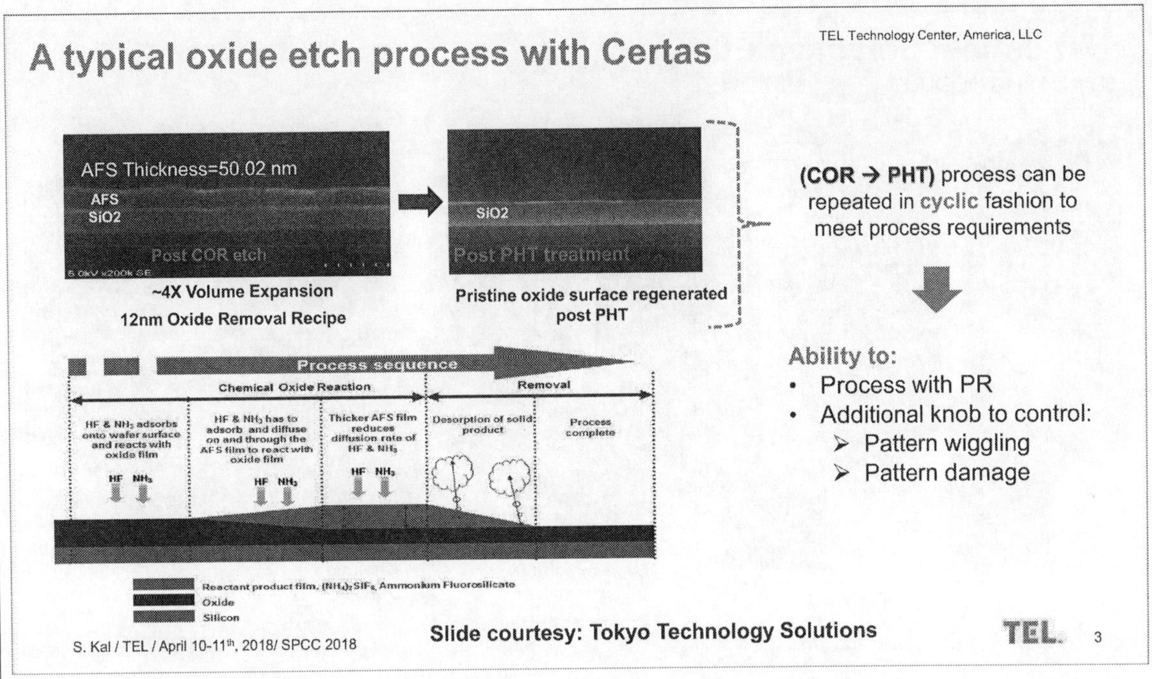

A typical oxide etch process with Certas
TEL Technology Center, America, LLC
AFS Thickness=50.02 nm
AFS
SiO2
Post COR etch
SiO2
Post PHT treatment
~4X Volume Expansion
12nm Oxide Removal Recipe
Pristine oxide surface regenerated post PHT
(COR → PHT) process can be repeated in cyclic fashion to meet process requirements
Ability to:
• Process with PR
• Additional knob to control:
➢ Pattern wiggling
➢ Pattern damage
Process sequence
Chemical Oxide Reaction
Removal
HF & NH3 adsorbs onto wafer surface and reacts with oxide film
HF & NH3 has to adsorb and diffuse on and through the AFS film to react with oxide film
Thicker AFS film reduces diffusion rate of HF & NH3
Desorption of solid product
Process complete
HF NH3
Reactant product film, (NH4)2SiF6 Ammonium Fluorosilicate
Oxide
Silicon
S. Kal / TEL / April 10-11th, 2018/ SPCC 2018
Slide courtesy: Tokyo Technology Solutions
TEL
3

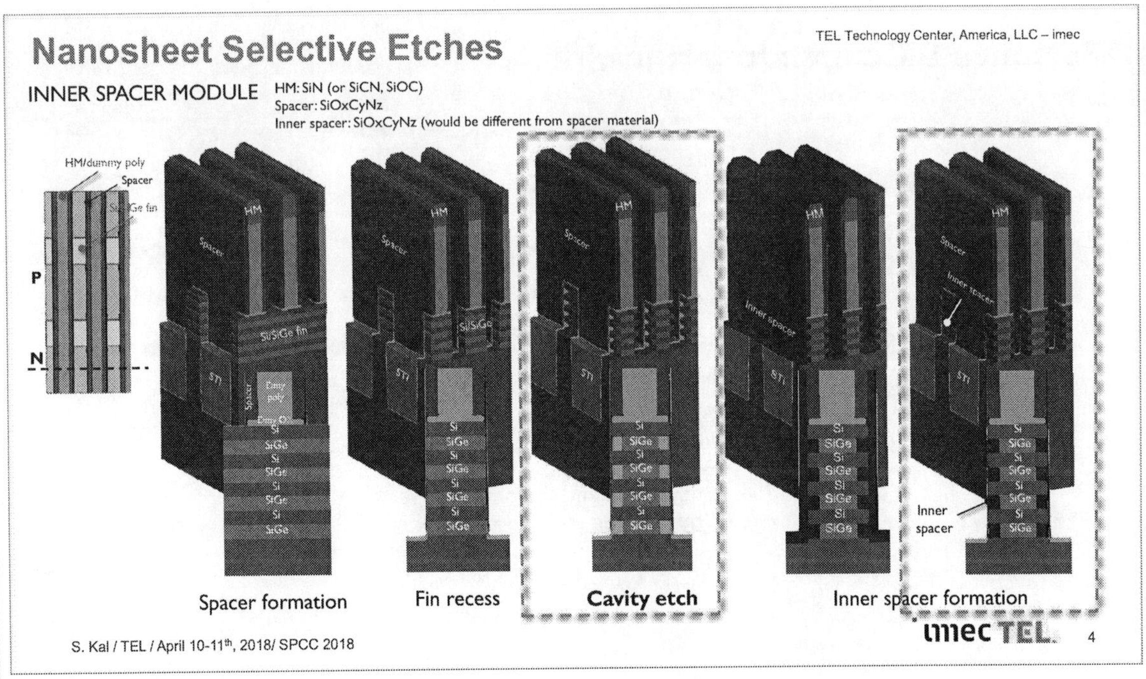

Nanosheet Selective Etches
TEL Technology Center, America, LLC – imec
INNER SPACER MODULE
HM: SiN (or SiCN, SiOC)
Spacer: SiOxCyNz
Inner spacer: SiOxCyNz (would be different from spacer material)
HM/dummy poly
Spacer
P
N
HM
Spacer
STI
Si
SiGe
Inner spacer
Spacer formation
Fin recess
Cavity etch
Inner spacer formation
S. Kal / TEL / April 10-11th, 2018/ SPCC 2018
imec TEL
4

Nanosheet Selective Etches

TEL Technology Center, America, LLC – imec

SD/ILD0/RMG MODULES

N-EPI: SiGe:B, P-EPI: Si:P
CESL: SiN
ILD0: SiO2
Dummy poly (dummy gate): a-Si

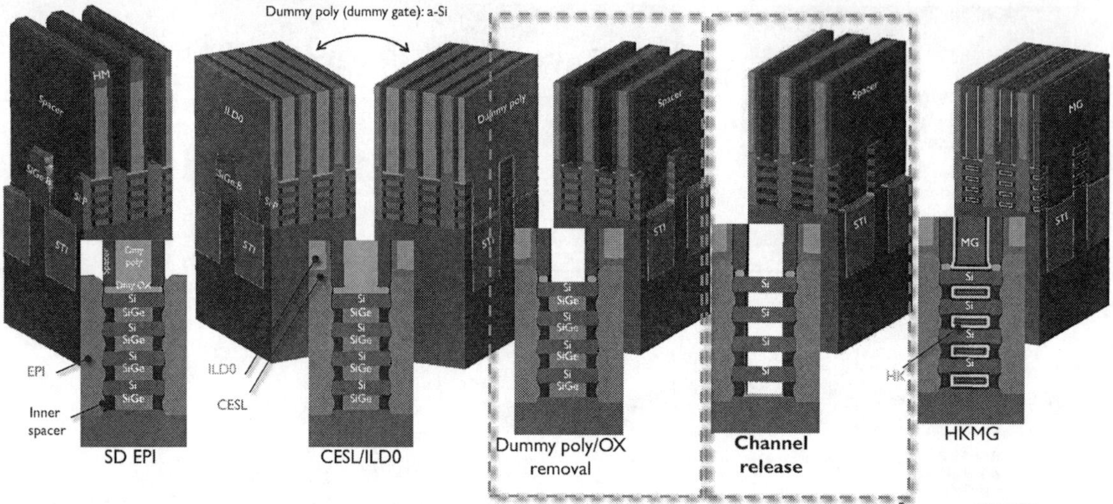

S. Kal / TEL / April 10-11th, 2018/ SPCC 2018

imec TEL. 5

Selective SiGe etch for Nanowire

TEL Technology Center, America, LLC – imec

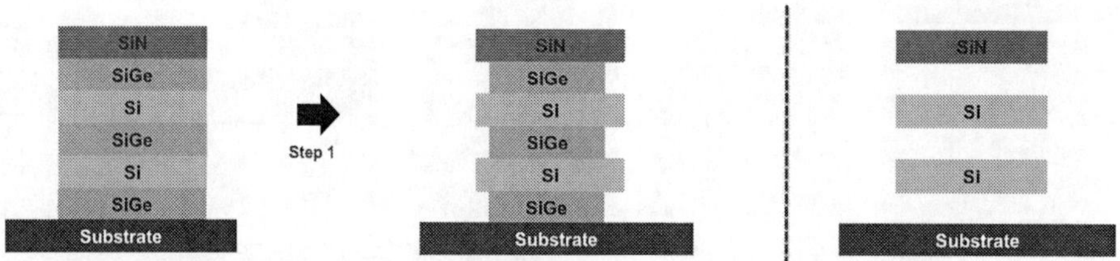

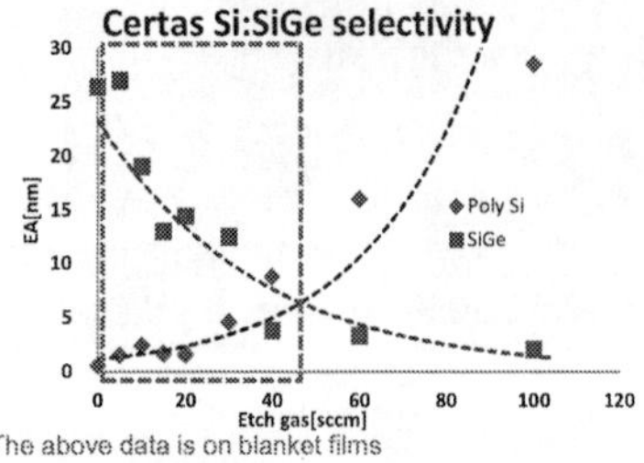

The above data is on blanket films

Partial release :

- SiGe etch =5-6 nm (each side, total = 10-12nm)
- Si loss <1nm
- **Etch target and uniformity >5 A**
- **Square SiGe etch front**

Full release:

- SiGe etch ~25 nm (each side, total ~25nm)
- Si loss <1nm

S. Kal / TEL / April 10-11th, 2018/ SPCC 2018

imec TEL. 6

COR SiGe:Si etch: etch time optimization

TEL Technology Center, America, LLC – imec

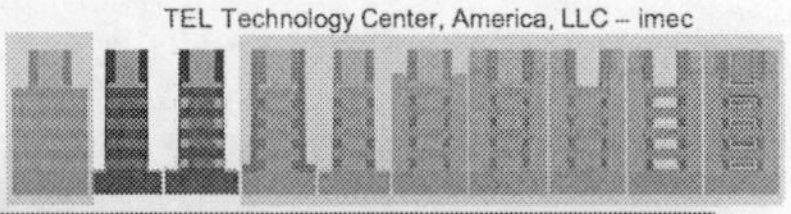

Incoming	POST gas phase etch (recipe A: aggressive etch)		
No process	T1	T2	T3 (T1<<T3)

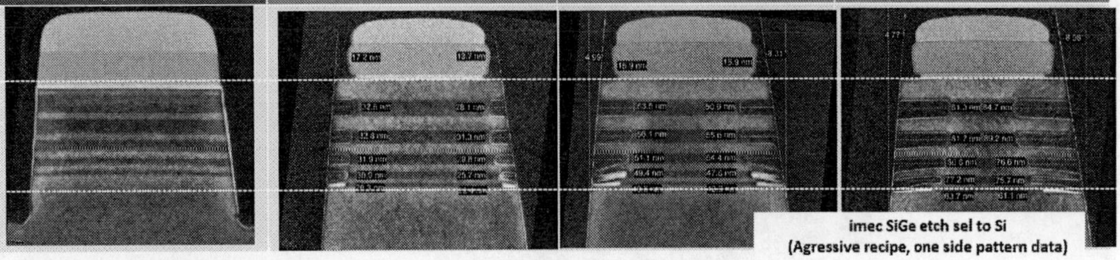

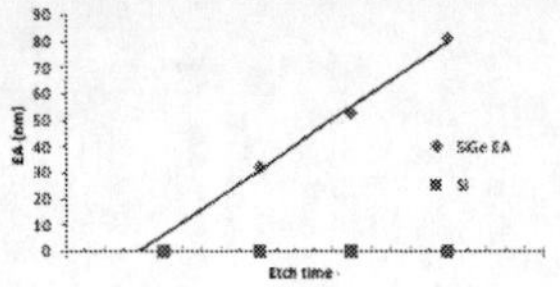

Summary for Si/SiGe stack:

- Selective SiGe: Si etch on imec wafer looks good (SiGe:Si >50:1)
- SiGe etch front looks VERY flat/square
- SiGe EA proportional etch time, without additional Si loss
- Partial SiGe etch uniformity ~3nm for Left /right side & top/bottom layers (incoming tapper may contribute)

S. Kal / TEL / April 10-11th, 2018/ SPCC 2018

imec TEL 7

COR SiGe:Si etch: pressure optimization

TEL Technology Center, America, LLC – imec

Incoming	POST TEL gas phase etch (Recipe B; medium etch)			
No process	P1	P2	P3	P4 (P1<<P4)

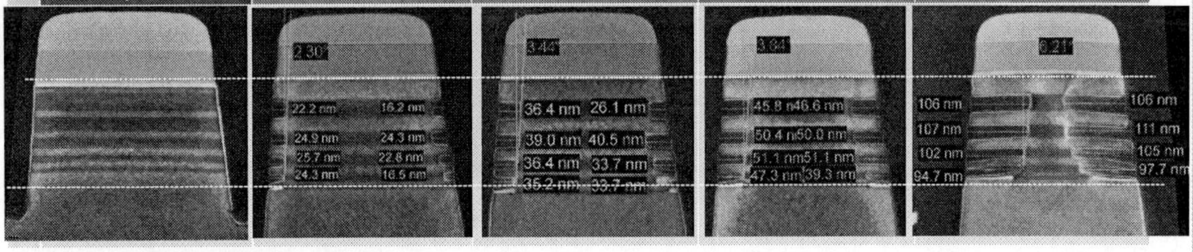

Summary for Si/SiGe stack:

- Selective SiGe: Si etch on imec wafer looks good (SiGe:Si >50:1)
- SiGe etch front looks VERY flat/square
- SiGe EA proportional etch time, without additional Si loss
- Pressure (i.e etch gas partial pressure) is contributing to slower etch rate due to byproduct formation depending on CD → causing left-right and top-bottom non uniformity

S. Kal / TEL / April 10-11th, 2018/ SPCC 2018

imec TEL 8

COR SiGe:Si etch: "cavity" and "channel release"

TEL Technology Center, America, LLC – imec

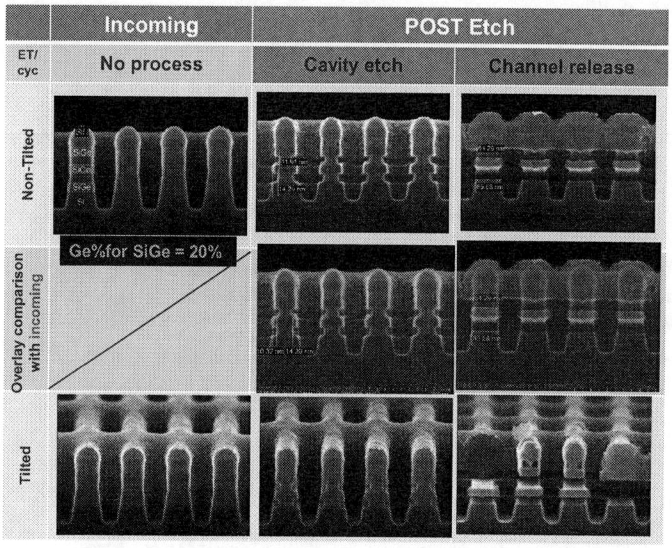

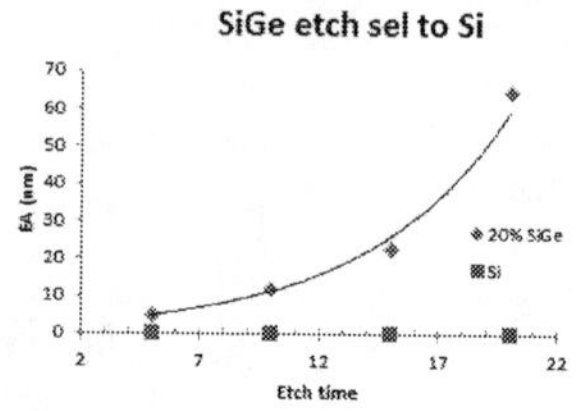

Summary:

- SiGe: Si etch selectivity > 50:1
- No SiN HM loss
- ER = 70nm/min
- SiGe etch front is square

Data based on alternate test structures

S. Kal / TEL / April 10-11th, 2018/ SPCC 2018

imec TEL 9

COR SiGe:Si etch: annealing effect

TEL Technology Center, America, LLC – imec

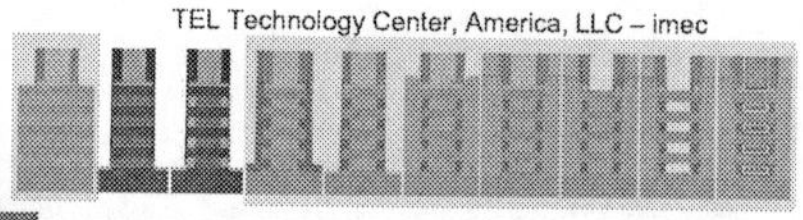

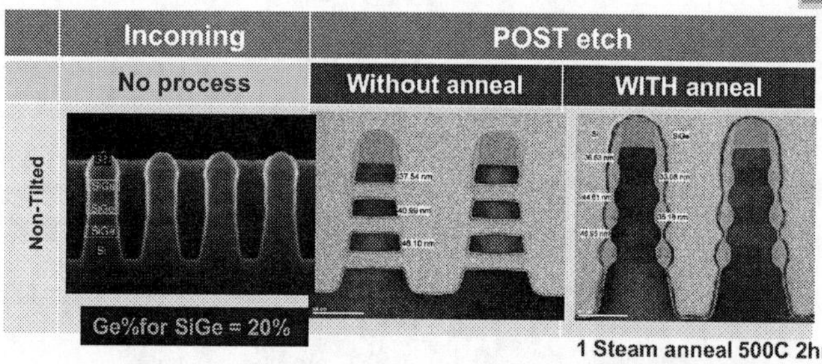

1 Steam anneal 500C 2hrs
2 RTP 850C 1min
3 RTP 850C 5s

Summary :

- Anneal affects the SiGe ER significantly
- Anneal also reduces the SiGe:Si selectivity at the SiGe-Si interface →
    - resulting in Si loss
    - meniscus etch front

Data based on alternate test structures

S. Kal / TEL / April 10-11th, 2018/ SPCC 2018

imec TEL 10

Selective Si etch for Nanowire application

TEL Technology Center, America, LLC – imec

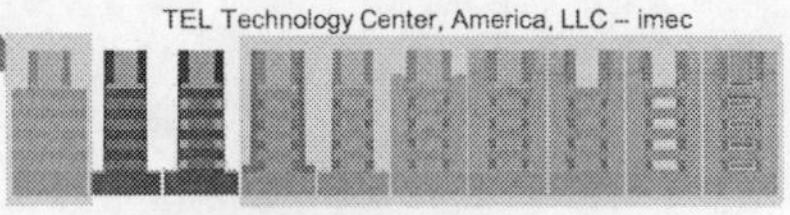

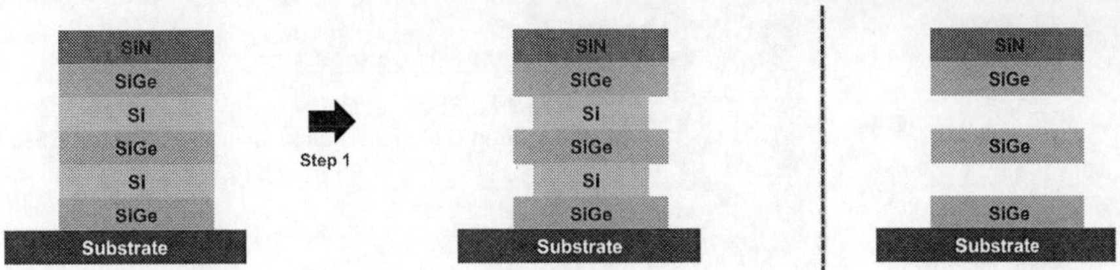

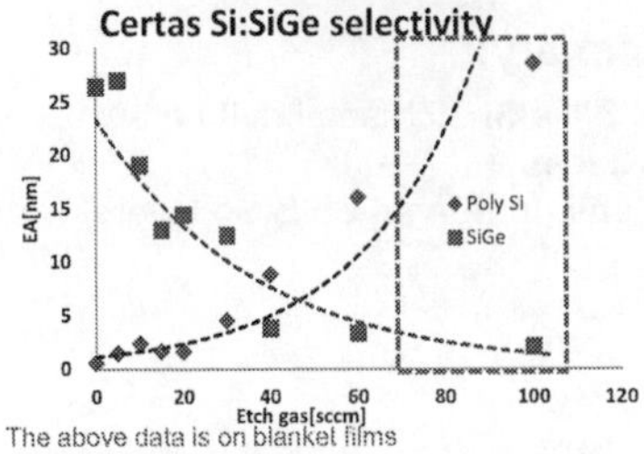

The above data is on blanket films

Partial release :

- Si etch =5-6 nm (each side, total = 10-12nm)
- SiGe loss <1nm
- **Etch target and uniformity >5 A**
- **Square SiGe etch front**

Full release:

- Si etch ~25 nm (each side, total ~25nm)
- SiGe loss <1nm

S. Kal / TEL / April 10-11th, 2018/ SPCC 2018

imec TEL. 11

COR Si:SiGe etch: etch time optimization

TEL Technology Center, America, LLC – imec

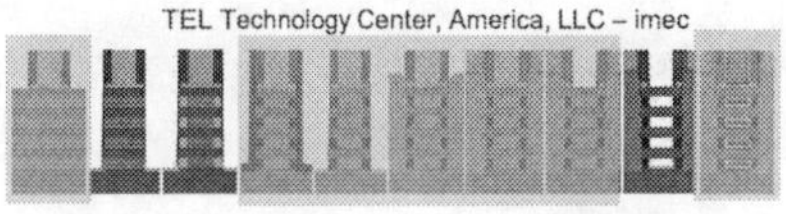

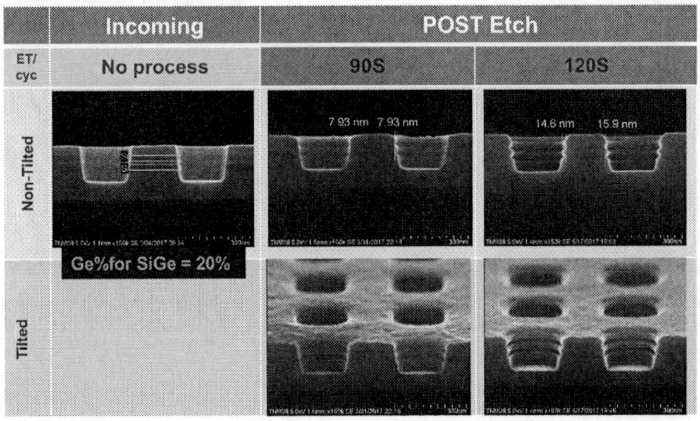

Summary:

- **Si: SiGe etch selectivity > 10:1**
- **No SiN HM loss**
- **ER = 7nm/min**
- **Si etch front is requires further improvement**
- **Post etch surface is smooth**

S. Kal / TEL / April 10-11th, 2018/ SPCC 2018

imec TEL. 12

COR SiN spacer etch

TEL Technology Center, America, LLC – imec

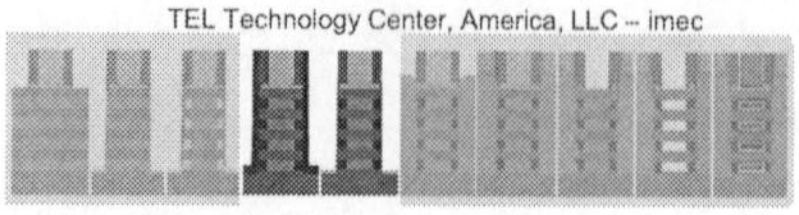

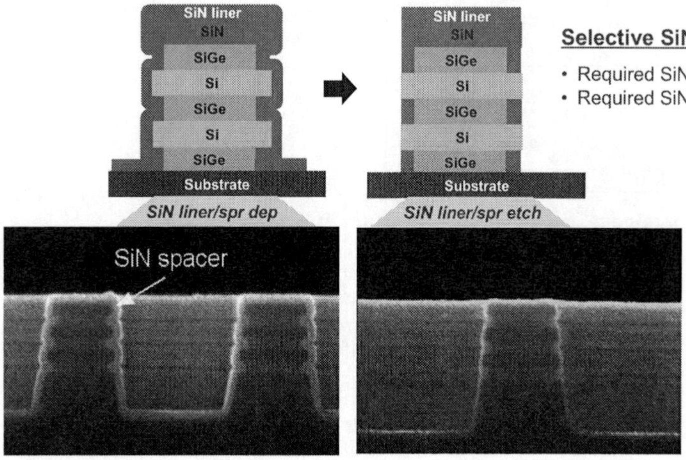

Selective SiN spacer etch:

- Required SiN:Si etch selectivity > 25:1 (no Si loss)
- Required SiN:SiGe etch selectivity > 25:1 (no SiGe loss)

Summary:

- **SiN: SiGe/Si etch selectivity > 50:1**
- **No Si loss**
- **SiN still preserved on SiGe layers**

COR Selective dummy poly (a-Si) pull

TEL Technology Center, America, LLC – imec

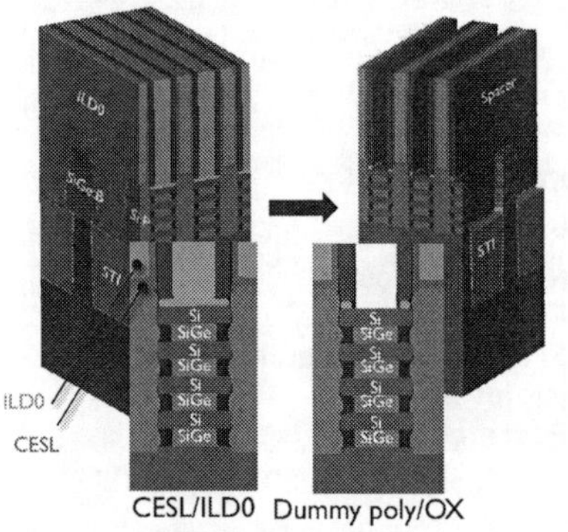

Dummy poly removal :

- Extremely selective Si etch ~100-200nm
- No SiN loss or SiO2 loss

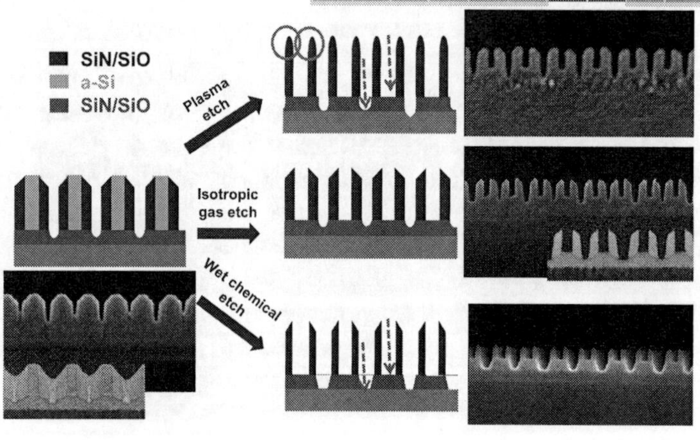

Device structure

Test structure

Nanosheet Selective Etches: *Updated*

TEL Technology Center, America, LLC – imec

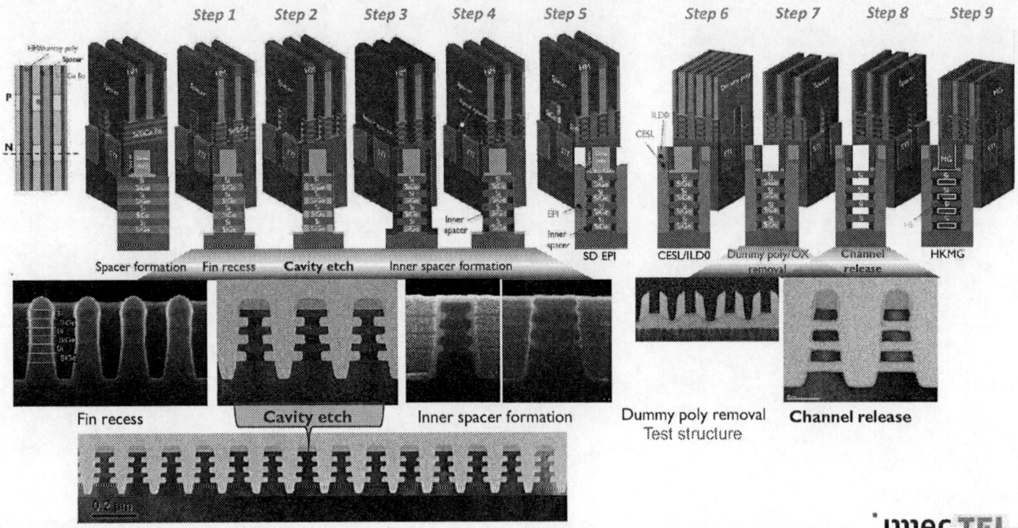

S. Kal / TEL / April 10-11th, 2018/ SPCC 2018

imec TEL 15

Conclusion

TEL Technology Center, America, LLC – imec

- ***Dry plasma free*** etches are advantageous & crucial for Nanowire/CFET integrations applications, due to:
  - High etch selectivity, inherent from the etch mechanism
  - No plasma damage
  - Aspect ratio dependency
  - Cyclic process (potential self limiting capability)

S. Kal / TEL / April 10-11th, 2018/ SPCC 2018

imec TEL 16

TOKYO ELECTRON

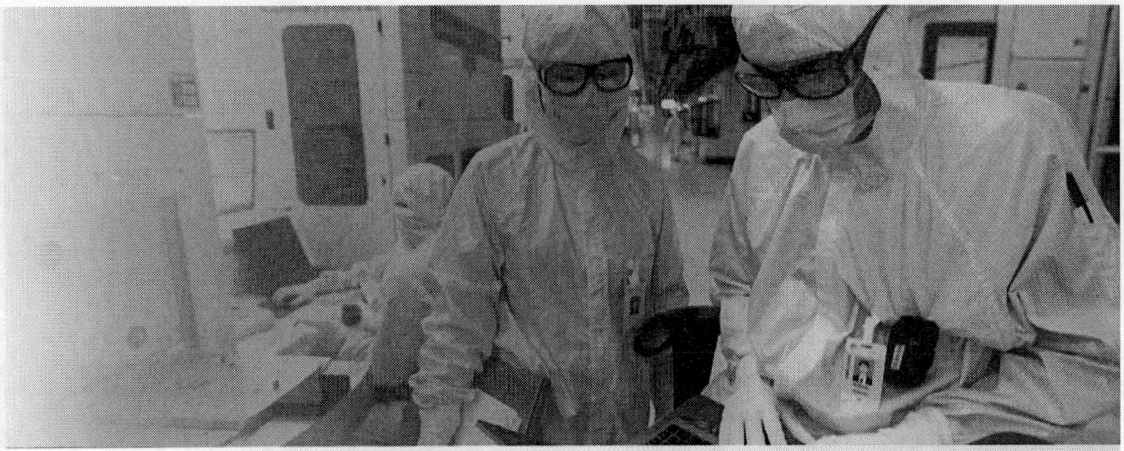

Interconnect Tutorial: A Complex, Important Integration Evolution to sub-14nm Technology Nodes

Kevin Boyd
Deputy Director, BEOL Integration , Advanced Technology Development

GLOBALFOUNDRIES®

SPCC 2018

Agenda

- **Defining Interconnects**
- **Interconnect Fundamentals**
  - Yield, Reliability, Structural Integrity, Performance
  - Metallization evolution
    - Al, Ti, W, Ta, Cu, Co, Ru
  - Dielectric evolution
    - TEOS, FTEOS, Low K, ULK, airgaps
  - Integration evolution
    - Single-Damascene, TFVL, VFTL, TFHM, SADP
    - EUV
- **Interconnect Scaling**
  - Power, Performance, Area, Cost
  - Tradeoffs in the 7nm and 5nm Nodes: Design, Dielectrics, Metals

2

Agenda

- **Defining Interconnects**
- **Interconnect Fundamentals**
  - Yield, Reliability, Structural Integrity, Performance
  - Metallization evolution
    - Al, Ti, W, Ta, Cu, Co, Ru
  - Dielectric evolution
    - TEOS, FTEOS, Low K, ULK, airgaps
  - Integration evolution
    - Subtractive RIE, Single-Damascene, TFVL, VFTL, TFHM, LELE, SADP
    - EUV
- **Interconnect Scaling**
  - Power, Performance, Area, Cost
  - Tradeoffs in the 7nm and 5nm Nodes: Design, Dielectrics, Metals

3

Defining Interconnects

- **Definitions**

BEOL = Back End of Line
Wiring (Global Interconnect)

MOL = Middle of Line
Local Interconnect

FEOL = Front End Of Line
Transistors/Devices

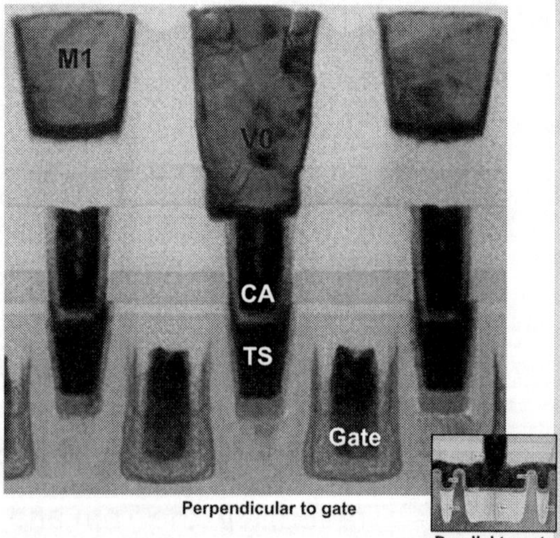

Perpendicular to gate

Parallel to gate

4

Agenda

- **Defining Interconnects**
- **Interconnect Fundamentals**
  - Yield, Reliability, Structural Integrity, Performance
  - Metallization evolution
    - Al, Ti, W, Ta, Cu, Co, Ru
  - Dielectric evolution
    - TEOS, FTEOS, Low K, ULK, airgaps
  - Integration evolution
    - Subtractive RIE, Single-Damascene, TFVL, VFTL, TFHM, LELE, SADP
    - EUV
- **Interconnect Scaling**
  - Power, Performance, Area, Cost
  - Tradeoffs in the 7nm and 5nm Nodes: Design, Dielectrics, Metals

5

Interconnect Fundamentals

- **Interconnect Used in Integrated Circuits**
  - **MOL**
    - **Local Interconnect**
      - Short-length, impacts device performance, connects devices to wiring
  - **BEOL**
    - **Short-range Digital Signals (e.g. between devices within logic blocks)**
      - Minimum-pitch, high frequency, low C desired, low EM risk
    - **Clock Tree**
      - Some minimum-pitch, high frequency, low RC desired, mild EM risk
    - **Long-distance Data Transfer (e.g. memory to CPU)**
      - High-speed, low attenuation/distortion, low RC desired, moderate EM risk
    - **Power Grid**
      - Wide lines, low R desired, high EM risk

Interconnects must meet a variety of (often opposing) circuit wiring needs

6

SPCC 2018

Interconnect Fundamentals

- **Interconnect Impact on Product Performance**

"People talk about reaching the end of Moore's Law, but really, it's irrelevant. Transistors are not a rate-limiting factor in today's computers. We could improve transistors by a factor of 1,000 and it would have no impact on the modern computer. **The rate-limiting parts are how you store and move information.**"

S. Williams (HP), 2013

"Processor chips since around 2000 are power - not area - limited. **All of the power is spent moving data around. It is important to optimize the entire interconnect system – the wire, the circuit, and the NoC together – not just each of the three in isolation."**

B. Dally (NVIDIA), 2012 (also quoting C. Moore (AMD), 2011)

"The interconnect challenges looking into the future are even more daunting than the compute challenges."

S. Borkar (Intel), 2012

Interconnect is now a rate-limiting factor to overall product performance

Compiled by J. Candelaria, SRC, 2014

7

SPCC 2018

Interconnect Fundamentals

- **Interconnect Impact on Product Performance**
  - **RC Performance**

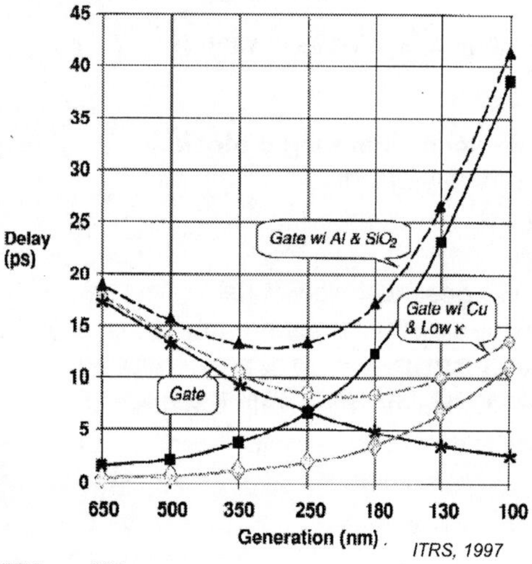

ITRS, 1997

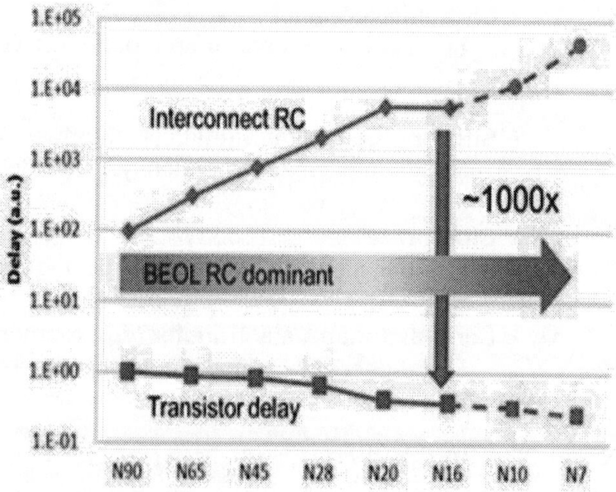

Interconnect RC delay is a dominant factor to overall product performance

8

Defining Interconnects

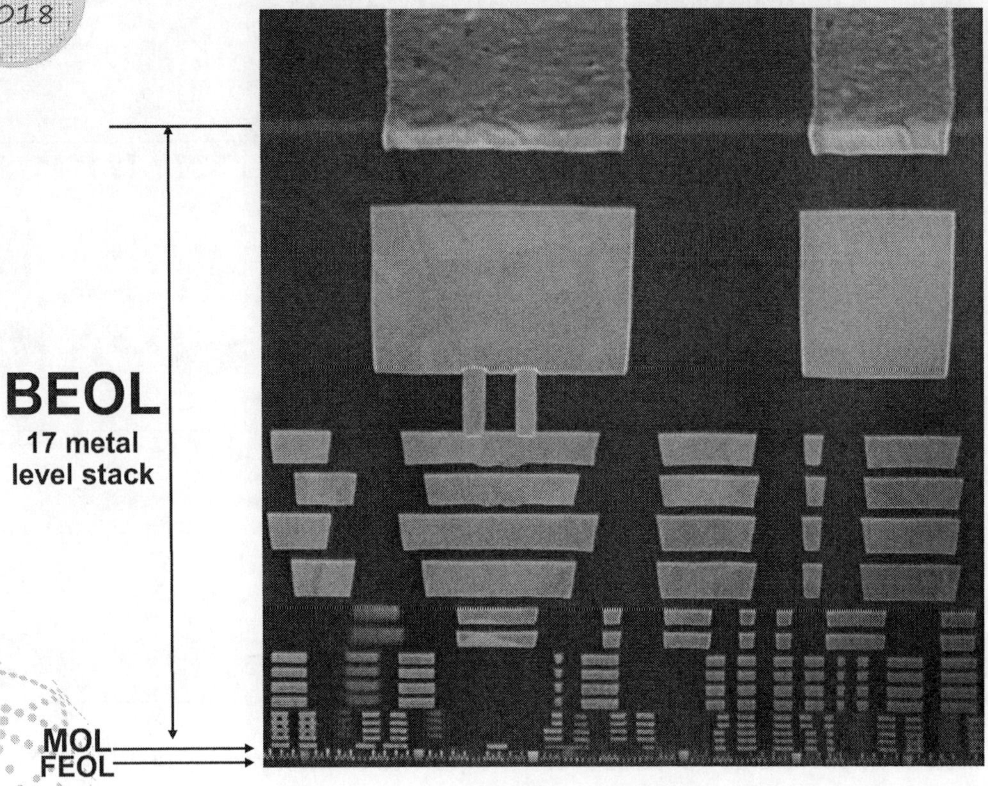

9

Interconnect Fundamentals

- **Interconnect Impact on Product Performance**
  - **Yield**
  - **Reliability**
  - **Structural Integrity (Chip Package Integration = CPI)**
  - **Performance (RC)**

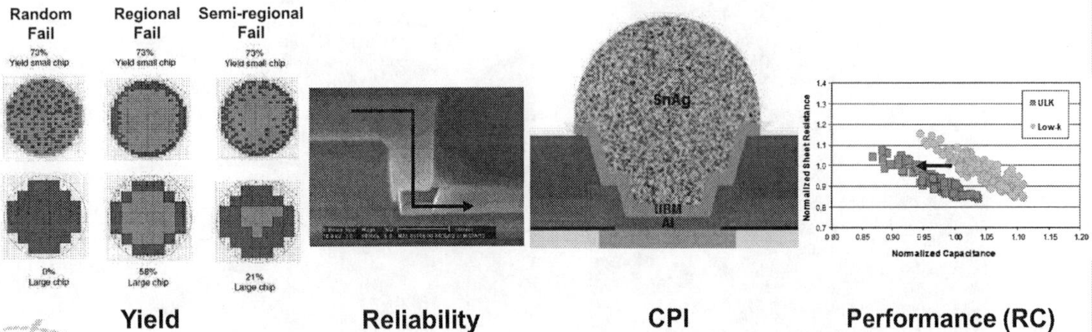

Interconnect directly impacts a broad range of product performance metrics

10

SPCC 2018

Interconnect Fundamentals

- **Interconnect Impact on Product Performance**
  - **Yield**
    - AutoRegressive Integrated Moving Average = **ARIMA**
    - Time To Market = **TTM**

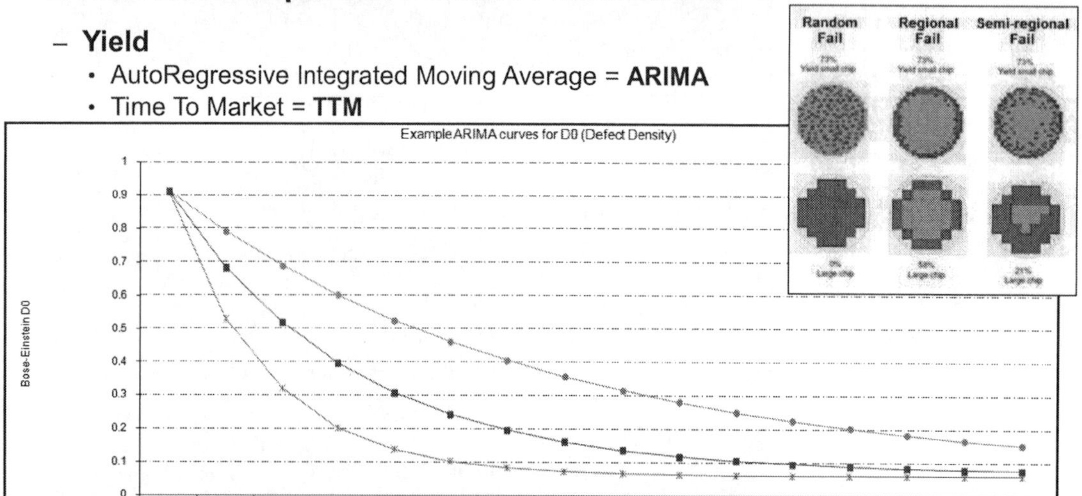

	2012Q1	2012Q2	2012Q3	2012Q4	2013Q1	2013Q2	2013Q3	2013Q4	2014Q1	2014Q2	2014Q3	2014Q4	2015Q1	2015Q2	2015Q3	2015Q4
0.82 ARIMA	0.914	0.531	0.320	0.203	0.139	0.104	0.084	0.073	0.067	0.064	0.062	0.061	0.061	0.060	0.060	0.060
0.95 ARIMA	0.914	0.794	0.690	0.602	0.526	0.461	0.405	0.357	0.316	0.281	0.251	0.225	0.203	0.184	0.168	0.154
0.9 ARIMA	0.914	0.685	0.519	0.397	0.308	0.244	0.197	0.162	0.137	0.119	0.106	0.096	0.089	0.084	0.080	0.077

Interconnect is a key contributor to logic and memory yield, ARIMA (TTM)

11

SPCC 2018

Interconnect Fundamentals

- **Interconnect Impact on Product Performance**
  - **Reliability Failure Modes**
    - **Intrinsic** (wear-out)
    - **Extrinsic** (defects, process integrity issues)

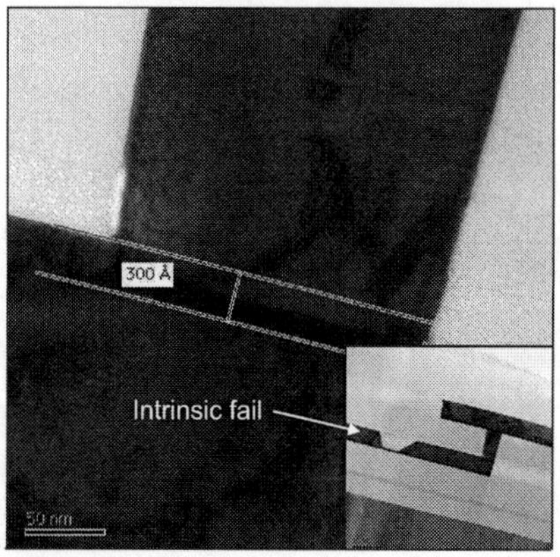

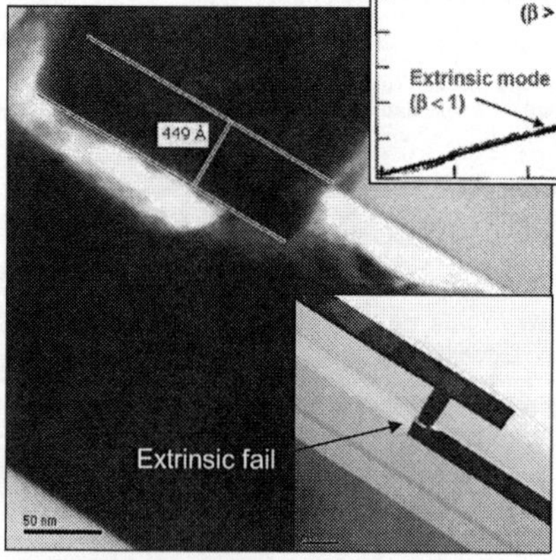

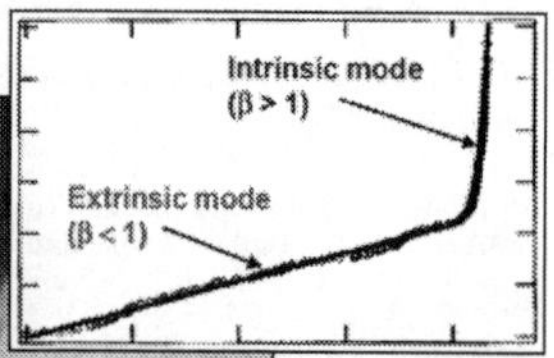

Degraeve et al., TED, 1998

Interconnect must withstand intrinsic and extrinsic fail modes, including EM

12

Interconnect Fundamentals

- **Interconnect Impact on Product Performance**
  - **Intrinsic Reliability (Electromigration = EM)**
    - Elevated temperature and applied voltage direct Cu ions toward vacancies
    - Activation energy (eV) describes the likelihood for diffusion
    - Diffusion depends on temperature, current density, diffusion path and mechanical stress conditions
  - There are generally two modes of EM failure:

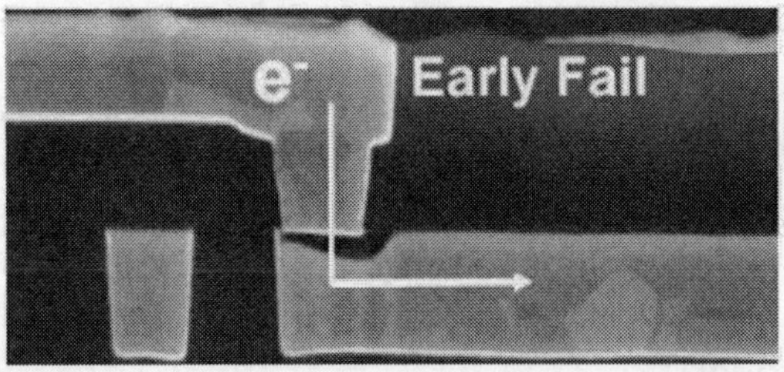

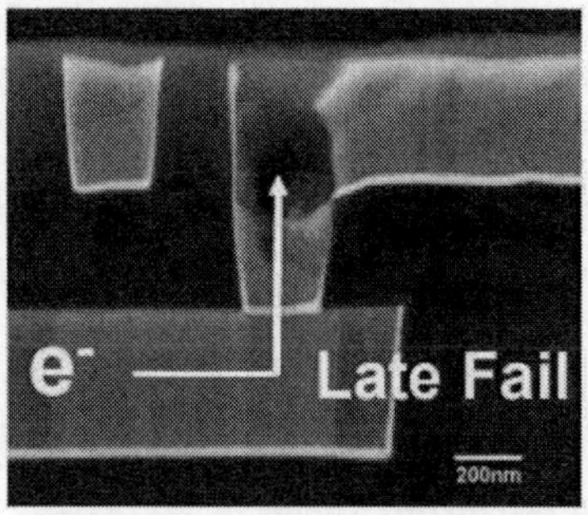

Interconnect must be optimized to alleviate surface and interface-driven EM

13

Interconnect Fundamentals

- **Interconnect Impact on Product Performance**
  - **Intrinsic Reliability (Stress Migration = SM)**
    - Similar fail mode as EM (i.e. void formation), but driving force is physical stress
    - Void typically forms under via (highest mechanical stress gradient)

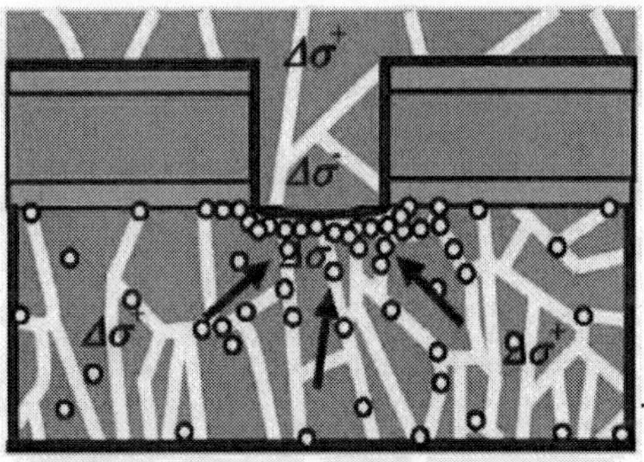

Diffusion of vacancies under via

E. Ogawa, IRPS (2002)

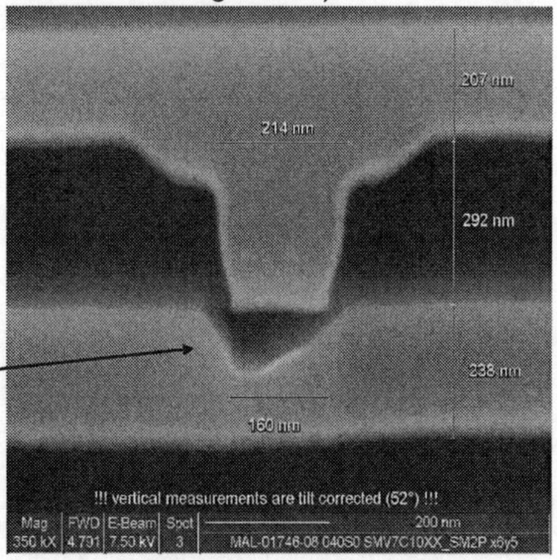

Interconnect must be optimized for SM, especially for vias over wide lines

14

Interconnect Fundamentals

- **Interconnect Impact on Product Performance**
  - **Intrinsic Reliability (Time Dependent Dielectric Breakdown = TDDB)**
    - The loss of insulation between neighboring interconnects → Leakage current and short failures
  - Failure rate depends upon electric field
  - For a given potential difference, **minimum spacing** determines maximum electric field strength
  - Hot charge carriers cause **defects in the dielectric**, which accumulate over time
  - Electric field can cause Cu^{+} **migration** along interfaces or through faulty barriers

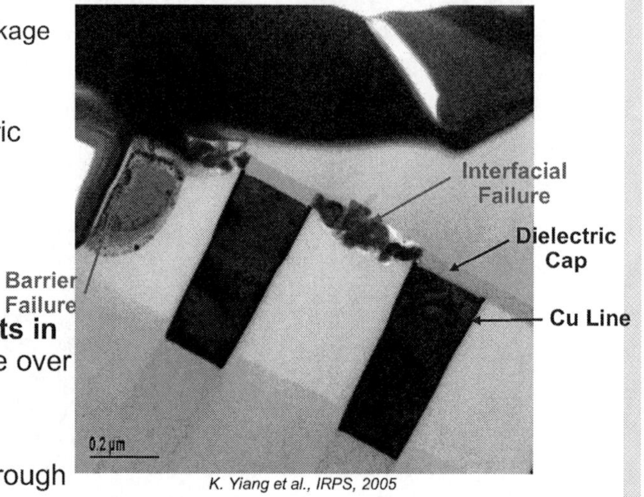

K. Yiang et al., IRPS, 2005

Dielectric and metal interfaces must be optimized to withstand intrinsic TDDB

Interconnect Fundamentals

- **Interconnect Impact on Product Performance**
  - **Structural Integrity (Chip Package Integration = CPI)**

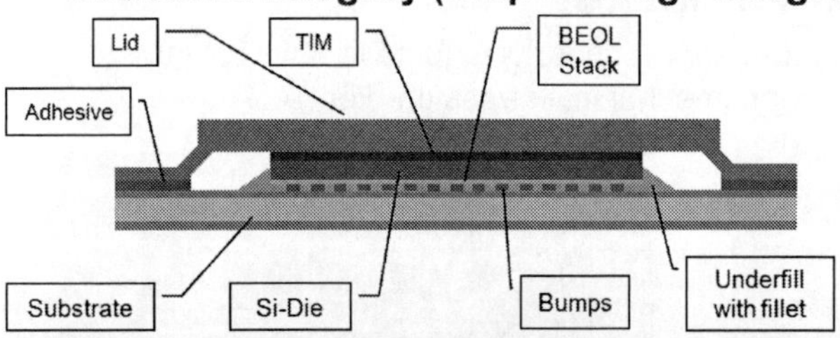

- **Different thermal expansions between Si-Die and Package and Lid (CTE mismatch)**
- **Mechanical stress on Interconnect stack, Bumps, Underfill**
- **Cracks, Delamination**

BEOL Stack must withstand CTE mismatch, mechanical stress in package 16

Interconnect Fundamentals

- **Interconnect Impact on Product Performance**
  - **RC Performance (Cu resistivity as a function of linewidth)**

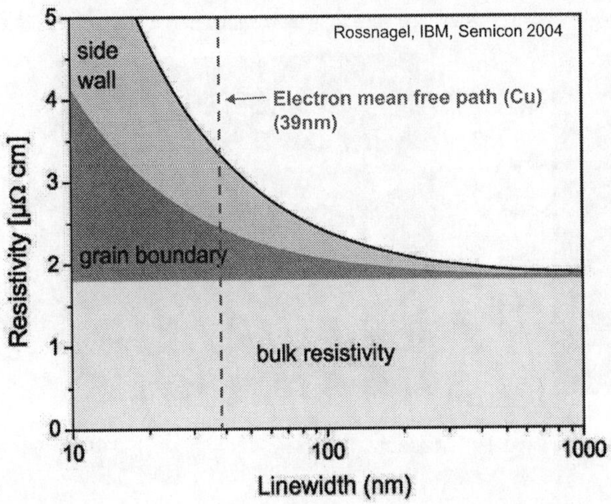

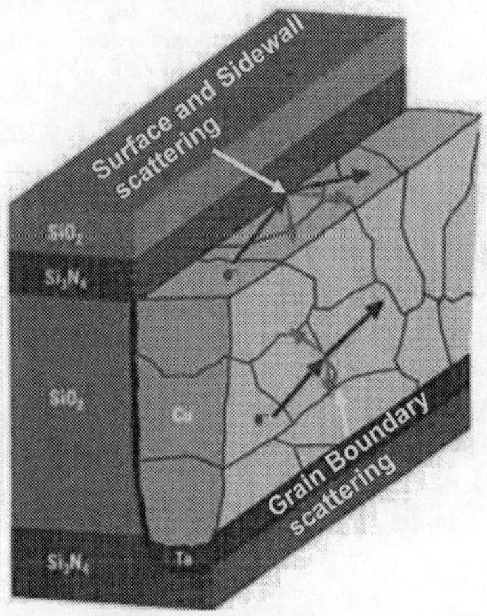

G. Schindler et al., AMC., 2002.

Sidewall/grain boundary scattering, barrier thickness affect Cu line resistivity

17

Interconnect Fundamentals

- **Interconnect Impact on Product Performance**
  - **RC Performance (narrow Cu linewidths)**
    - **Sidewall Scattering**
      - As Cu linewidth approaches mean free path (~39nm), sidewall scattering effects become more pronounced
    - **Grain Boundary Scattering**
      - More important in narrow lines with small grains
      - Less important for wide lines, bamboo structure
    - **Barrier Thickness Scaling**
      - Barrier (high resistivity) thickness must scale sectional area to avoid R increase

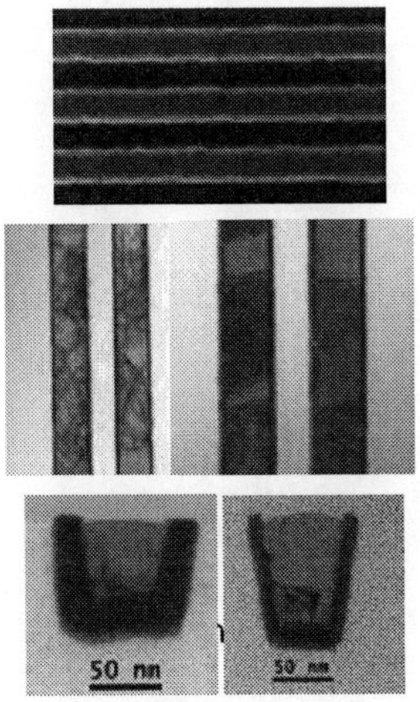

Sidewall/grain boundary scattering, barrier thickness affect Cu line resistivity

18

Interconnect Fundamentals

- **Metallization Evolution**
  - **BEOL: Al → Cu**

A Motorola μprocessor with 5 Al layers (1996)

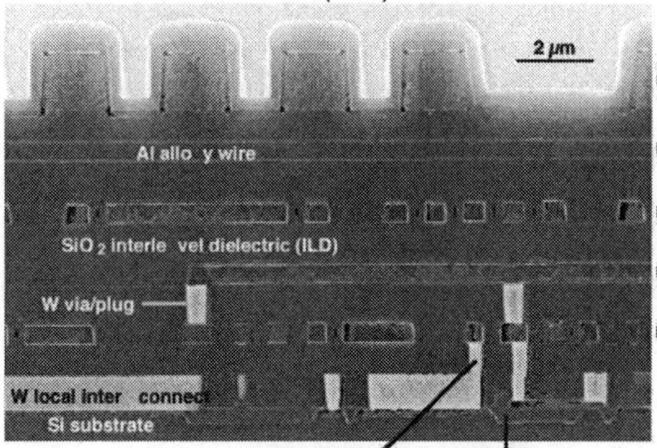

IBMs 7 layer Copper Metallization Technology (1998)

http://web.stanford.edu/class/ee311/NOTES/Interconnect_Al.pdf, K. Saraswat

Cu replaces Al/W to mitigate interconnect RC delay in ≤180nm nodes

Interconnect Fundamentals

- **Metallization Evolution**
  - **BEOL: Cu → Enhanced Cu (Mn)**

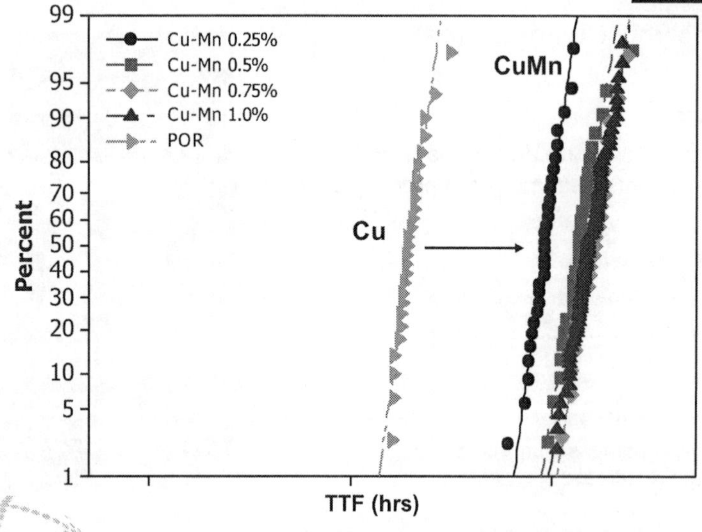

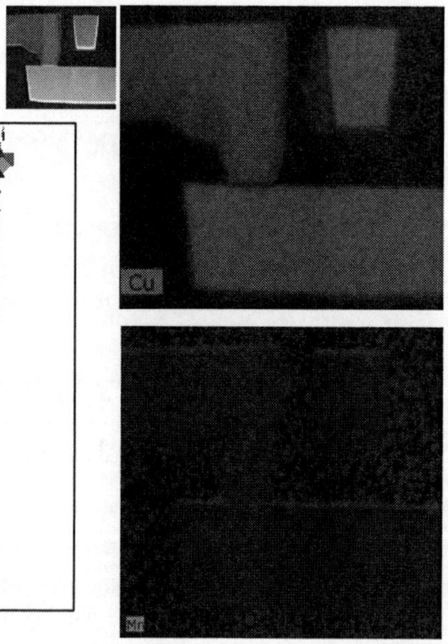

CuMn seed improves EM performance, but with increase in line resistance

Interconnect Fundamentals

- **Metallization Evolution**
  - **BEOL: Cu → Enhanced Cu (Co)**
    - CVD Co liner greatly enhances seed wetting and nucleation, improving sidewall coverage and void-free fill
    - Selective Co cap enhances resistance to electromigration

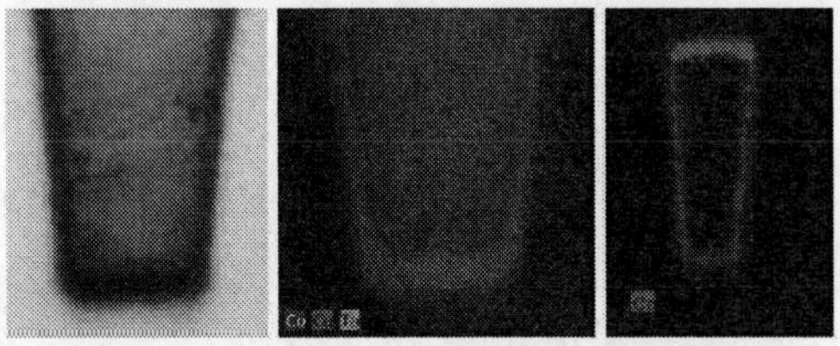

Without Co Liner **With Co Liner** **Without Co Liner** **With Co Liner**

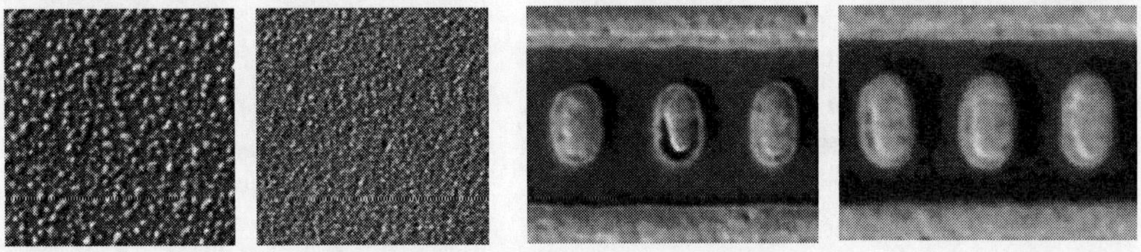

Co liner improves Cu seed wetting and nucleation, Co cap improves EM

21

Interconnect Fundamentals

- **Dielectric Evolution**
  - **TEOS → FTEOS → Low K → ULK**

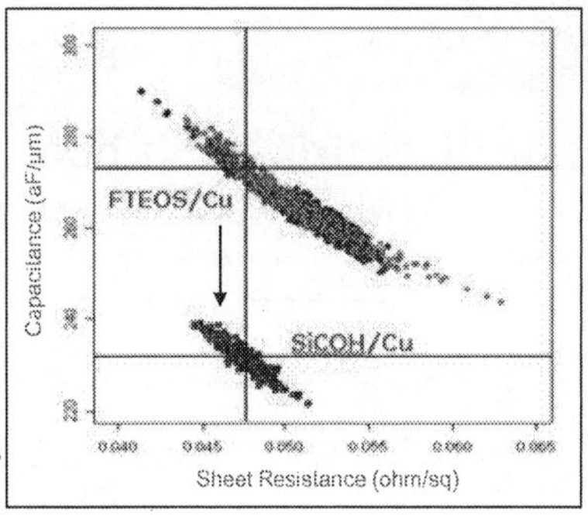

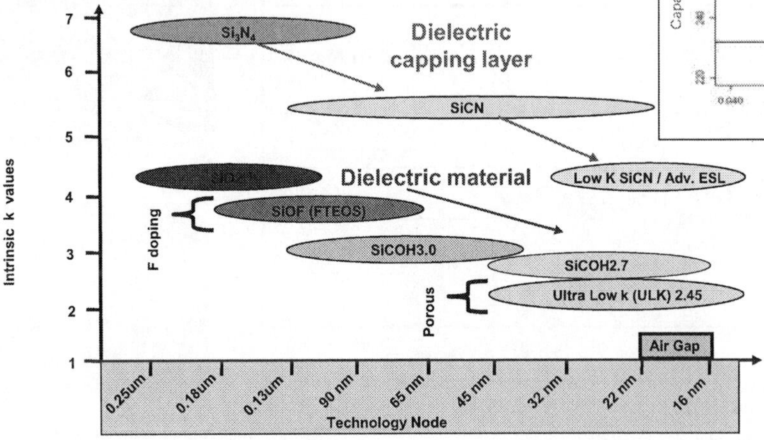

Low dielectric constant (κ) materials (e.g. low density, porous) reduce C_{total}

22

Interconnect Fundamentals

- **Dielectric Evolution**
  - TEOS → FTEOS → Low K → ULK

- **To reduce κ from ~4.2 to ~2.7**
  - F doping (~3.9 - ~3.6)
  - C doping (~3.2 - ~2.7)
  - Reduction in film density
- **To reduce κ below 2.7**
  - New dielectric materials
  - Introduce/increase porosity for SiO_2-based system
- **Lowering κ value degrades thermomechanical materials properties**
  - Modulus, hardness, stress, and thermal conductivity

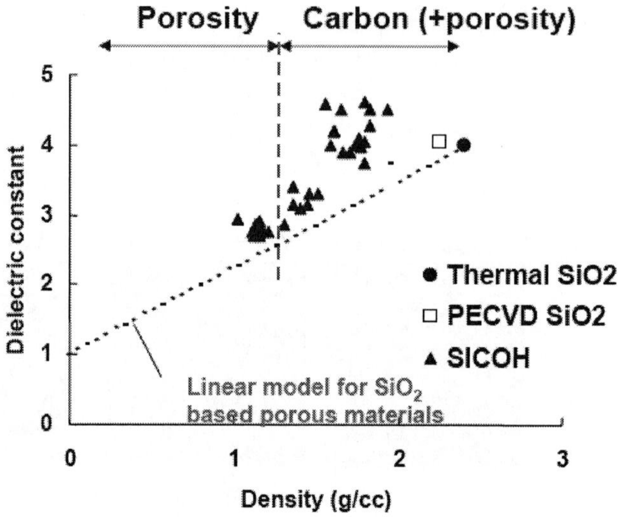

Lowering dielectric constant (κ value) degrades thermomechanical strength

23

Interconnect Fundamentals

- **Dielectric Evolution**
  - Low K → Ultra Low K (ULK) → Airgaps

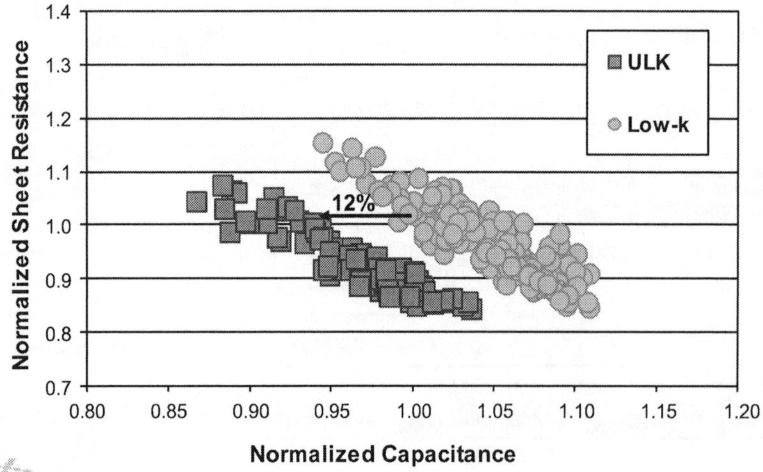

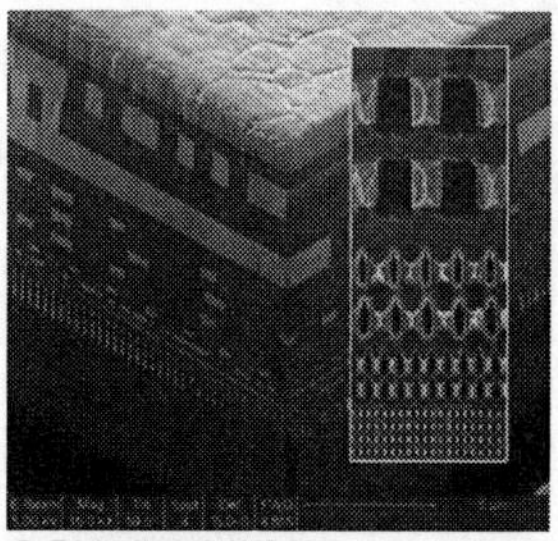

D. Edelstein et al, AMC 2005
Nitta et al., IITC 2008

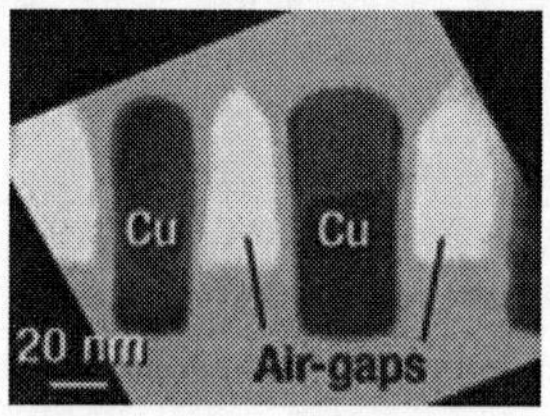

http://www.electroiq.com/articles/sst/print/volume-53/issue-6/features/interconnects_low-/air-gaps_for_interconnects.html

Porous ULK reduces C over Low κ, airgaps reduce further (but CPI risk)

24

Interconnect Fundamentals

- **Process Integration Evolution**
  - **Single-Damascene (Cu) → Dual-Damascene TFVL/VFTL**
    - Trench First Via Last = **TFVL**
    - Via First Trench Last = **VFTL**

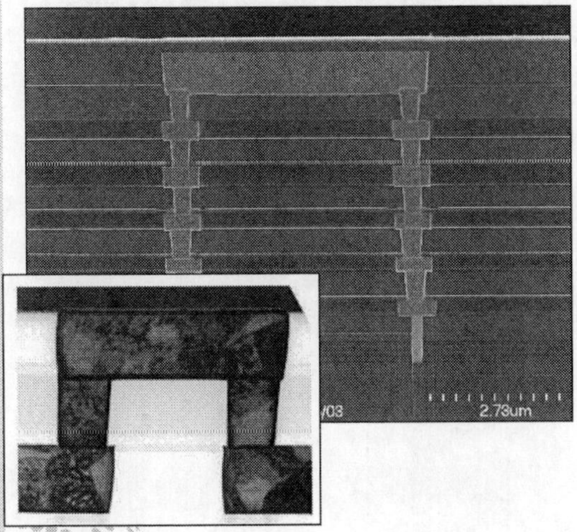

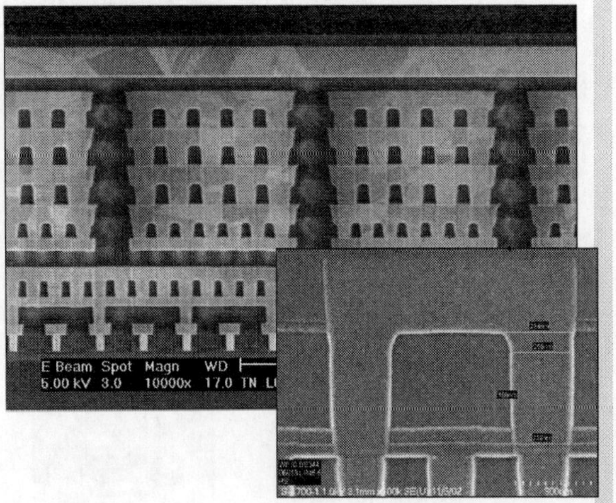

Dual-damascene integrations more cost-effective than single-damascene

Interconnect Fundamentals

- **Process Integration Evolution**
  - **VFTL → TFMH→ SADP (next page)**

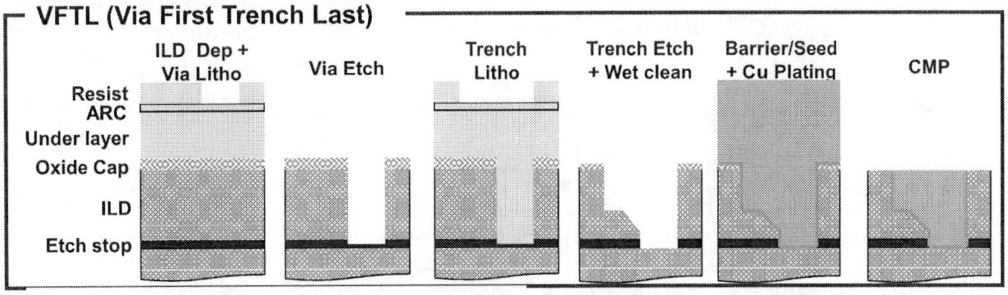

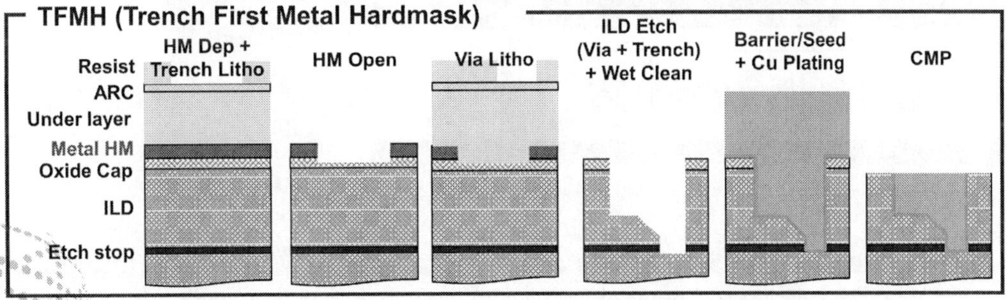

Advanced technology nodes require more complex integration *Matthias Lehr*

SPCC 2018

Interconnect Fundamentals

- **SADP Integration**

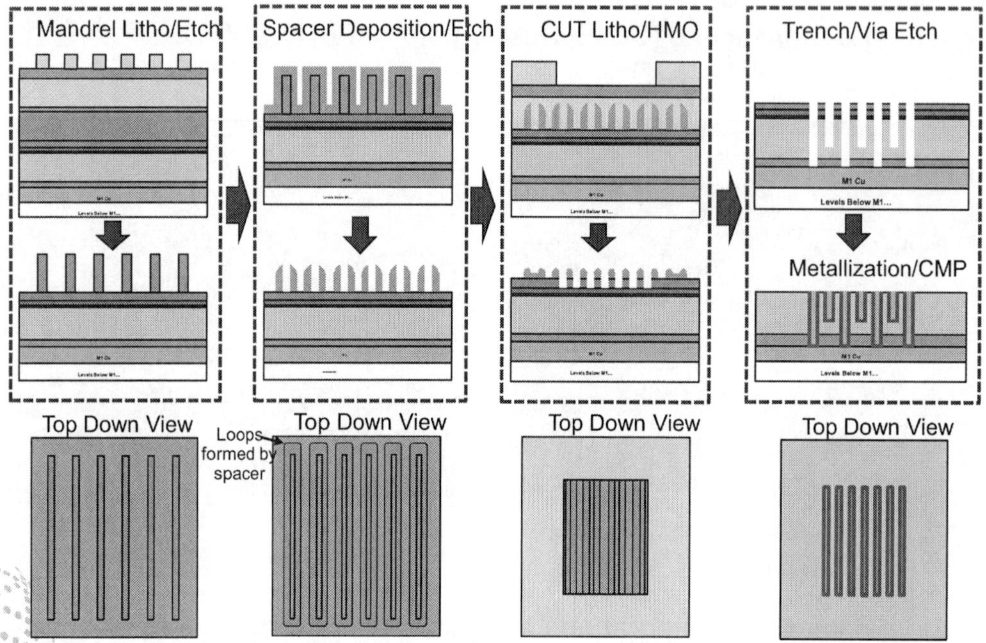

Advanced technology nodes require more complex integration *Matthias Lehr*

SPCC 2018

Interconnect Fundamentals

- **Process Integration Evolution**
  - Extreme UltraViolet Lithography = **EUV**

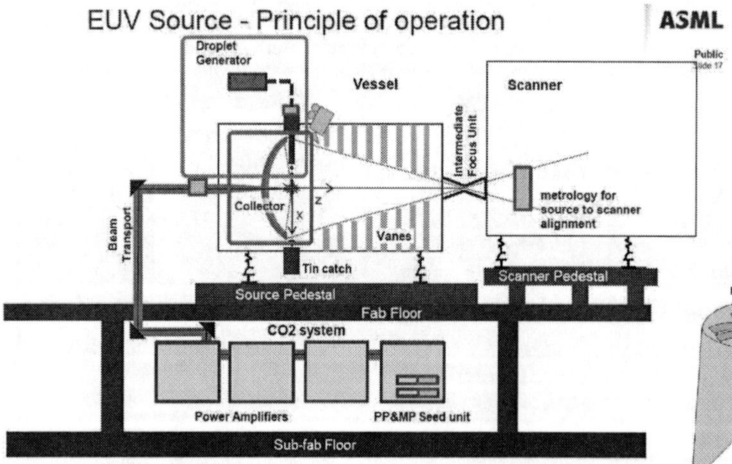

Resolution	16nm
Full wafer CDU	≤ 1.3nm
DCO	≤ 1.5nm
MMO	≤ 2.5nm
Focus control	≤ 70nm
Productivity	≥ 125 WPH

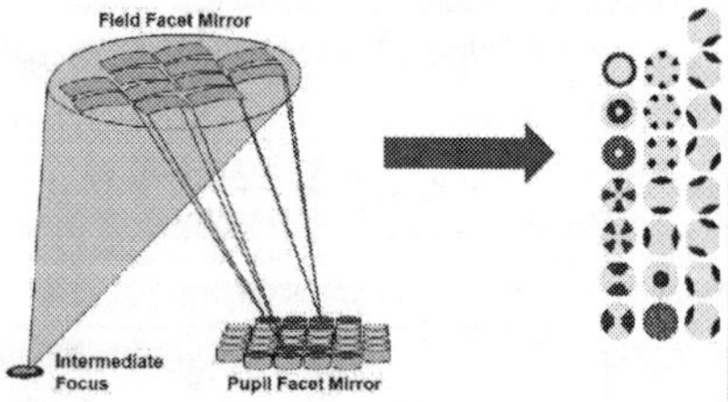

Reference: ASML tech symposium and http://www.nature.com/articles/srep09235

EUV needed at 7nm/5nm nodes to reduce complexity of optical integrations

EUV challenges: droplet generator, collector lifetime, tool uptime, pellicle/mask, resist

28

Agenda

- **Defining Interconnects**
- **Interconnect Fundamentals**
  - Yield, Reliability, Structural Integrity, Performance
  - Metallization evolution
    - Al, Ti, W, Ta, Cu, Co, Ru
  - Dielectric evolution
    - TEOS, FTEOS, Low K, ULK, airgaps
  - Integration evolution
    - Subtractive RIE, Single-Damascene, TFVL, VFTL, TFHM, LELE, SADP
    - EUV
- **Interconnect Scaling**
  - Power, Performance, Area, Cost
  - Tradeoffs in the 7nm and 5nm Nodes: Design, Dielectrics, Metals

29

Interconnect Scaling

- **Power, Performance, Area, Cost**

14nm, Area =1 **10nm, Area =0.56** **7nm, Area =0.33** **5nm, Area =0.17**

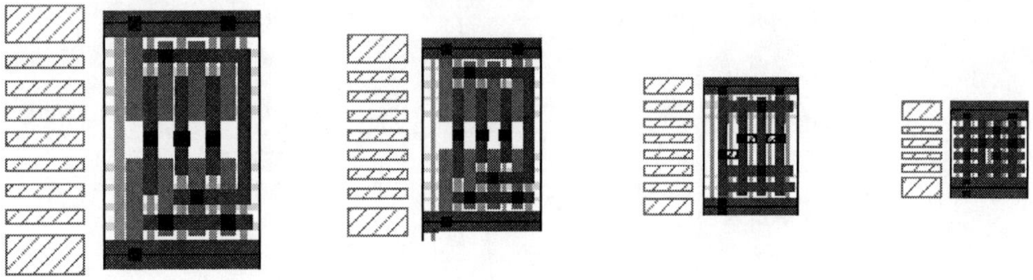

- **0.7X pitch scaling equates to ~50% area scaling**
- **Minimum-pitch lines and wide power rails exist within same metal layer**
- **Sub-14nm node patterning requires additional design restrictions (e.g. bidirectional to unidirectional)**

Interconnect scaling into the sub-14nm nodes requires architectural change

30

SPCC 2018

Interconnect Scaling

- **Tradeoffs in the 7nm and 5nm Nodes: Design**

Unrestricted **Partially-restricted** **More-restricted**

- **Partial layout restrictions can greatly reduce metal complexity while maximizing line end extensions**
- **Completely unidirectional routing allows fewest design-sensitive systematics, enables LELE or SADP**

Layout restrictions are increasingly important in lower metal layers due to tradeoffs in area, track utilization, and pin count

31

SPCC 2018

Interconnect Scaling

- **Tradeoffs in the 7nm and 5nm Nodes: Design**

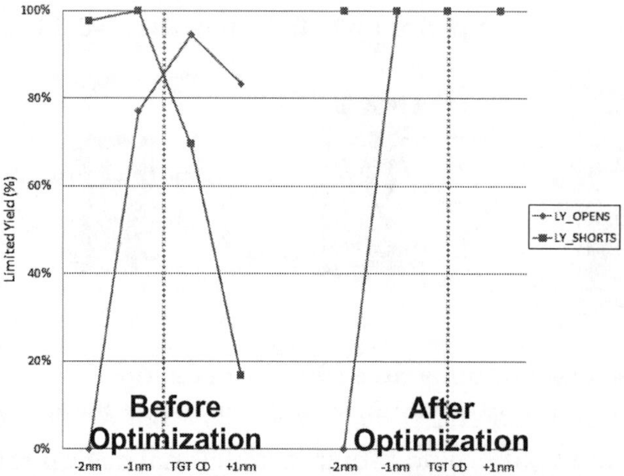

- **Even with significant design and process optimization, process windows can span only several nanometers**

Product-level logic yields more restricted by systematic design weakpoints

Interconnect Scaling

- **Tradeoffs in the 7nm and 5nm Nodes: BEOL (Dielectrics)**

Reduced Sensitivity To Process Damage

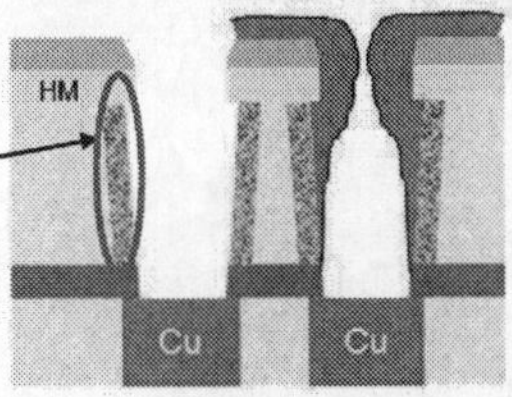

Minimized Hard Mask Undercut

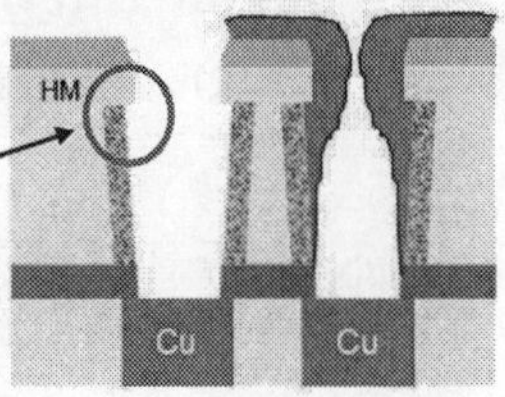

Parameter	ILD A	ILD B	ILD C	ILD D	ILD E
k for bulk	2.7	2.55	2.4	2.4	2.4
E (GPa)	8	6.64	7.34	7.0	8.0
Process Induced Damage Ratio (lower is better)	0.33	0.57	0.65	0.35	0.31
Relative Patterning Process Window					

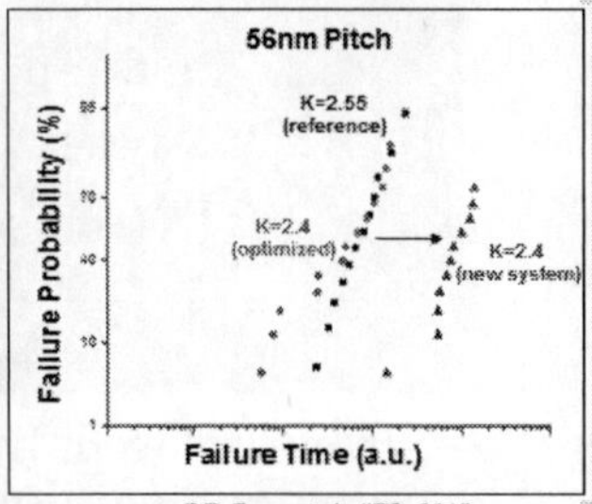

E.T. Ryan et al., IITC, 2015

Improvements in dielectric mechanical strength and resistance to process damage allow for more robust process integration, improved reliability (TDDB)

33

Interconnect Scaling

- **Tradeoffs in the 7nm and 5nm Nodes: BEOL (Dielectrics)**

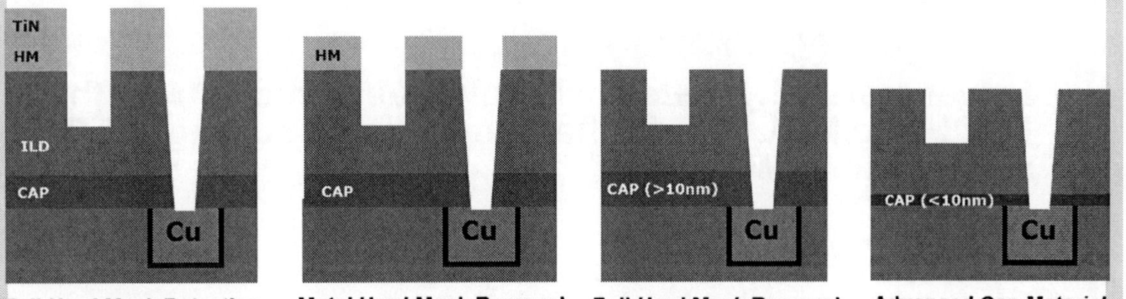

Full Hard Mask Retention AR ~ 7.5 | **Metal Hard Mask Removal** AR ~ 7 | **Full Hard Mask Removal** AR ~ 6 | **Advanced Cap Material** AR < 6

New dielectric hard mask and capping (and/or etch stop layers) materials can reduce integrated aspect ratios, greatly enhance metallization process window

34

Interconnect Scaling

- **Tradeoffs in the 7nm and 5nm Nodes: BEOL (Metals)**

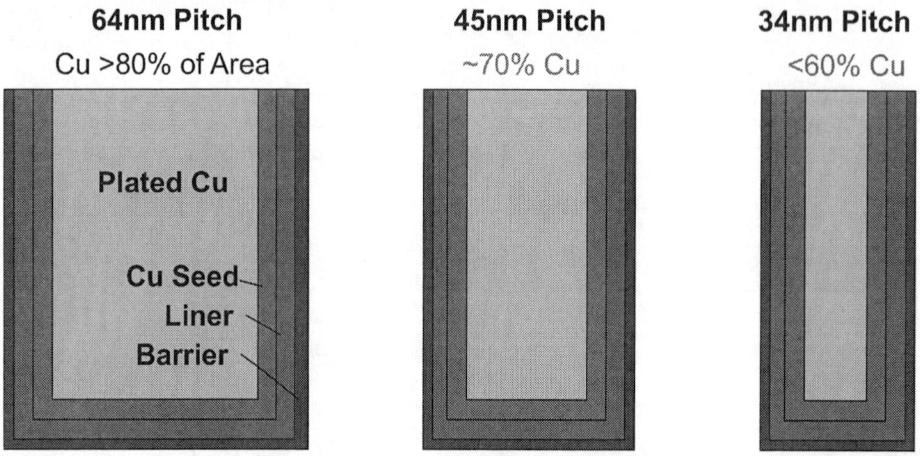

  - **Structural (e.g. increased aspect ratio) and Materials (e.g. CVD seed enhancement/liners/alloys/metals) changes are needed to enable pitches below 45nm; more radical changes may be necessary for pitches <34nm**

Structural or materials changes are needed to continue Cu interconnect scaling

35

Key Messages

- **Interconnect (MOL/BEOL) is both increasingly complex and increasingly important in sub-14nm Technology Nodes**
- **Lower resistivity metals will continue to move from the BEOL into MOL with further dimensional scaling (BEOL will consume MOL)**
- **Even with EUV, CD and overlay control will continue to become increasingly challenging**
- **New materials, methods, and integrations will be needed to continue performance improvement and differentiation at product/system-level**

36

SPCC 2018 Acknowledgements

- Robert Fox
- Rod Augur
- Seungman Choi
- E. Todd Ryan
- Keith Tabakman
- Bill Taylor
- André Labonté
- Patrick Justison
- Matthias Lehr
- Oliver Aubel
- Luke England

37

SPCC 2018 References

- "The Struggle to Keep Scaling BEOL, and What We Can Do Next", R. Augur, IEDM, 2016.
- "Moving Boundaries: Material Innovations for Future BEOL Interconnects", E. Todd Ryan, N. LiCausi, L. Liebman, B. Briggs, X. Zhang, X. Lin, J. Kelly, S. Nguyen, MRS, February 2017.
- "Process Window Challenges in Advanced Manufacturing: New Materials and Integration Solutions", R. Fox, R. Augur, C. Child, M. Zaleski, AMC, September 2015.
- "A Survey Addressing on High Performance On-Chip VLSI Interconnect", C. Mohamed Yousuff, V. Mohamed Yousuf Hasan, and M. R. Khan Galib, International Journal of Electronics and Telecommunications, 2013, Vol. 59, No. 3, pp. 307–312.
- "Electromigration - A Brief Survey and Some Recent Results", James R. Black, IEEE Trans. Elec. Dev. 16 (4): 338–347, April 1969.
- "Strategies to Ensure Electromigration Reliability of Cu/Low-k Interconnects at 10 nm", Anthony S. Oates, ECS J. Solid State Sci. Technol.2015 volume 4, issue 1.
- "Addressing Cu/Low-k Dielectric TDDB-Reliability Challenges for Advanced CMOS Technologies", F. Chen, M. Shinosky, IEEE Transactions on Electron Devices, volume 56, issue 1, Jan. 2009.
- "Progress in the development and understanding of advanced low k and ultralow k dielectrics for very large-scale integrated interconnects—State of the art", A. Grill, S. M. Gates, E. T. Ryan, S. V. Nguyen and D. Priyadarshini, Appl. Phys. Rev. 1, 011306 (2014).
- "Process technology scaling in an increasingly interconnect dominated world", J. S. Clarke, C. George, C. Jezewski, A. Maestre Caro, D. Michalak, J. Torres, VLSI Technology, Digest of Technical Papers, 2014.
- "Optimizing ULK Film Properties to Enable BEOL Integration with TDDB Reliability", E.T. Ryan, IITC 2015.
- "Electrical Reliability Challenges of Advanced Low-k Dielectrics", C. Wu, Y.Li, M.R. Baklanov, K. Croes, ECS J. Solid State Sci. Technol. 2015 volume 4, issue 1.

38

SPCC 2018

References

- "Characterization of 'Ultrathin-Cu'"Ru(Ta)/TaN Liner Stack for Copper Interconnects", C.-C. Yang, S. Cohen, T. Shaw, P.-C. Wang, T. Nogami, D. Edelstein, IEEE ELECTRON DEVICE LETTERS, VOL. 31, NO. 7, JULY 2010.
- "Comprehensive study of the resistivity of copper wires with lateral dimensions of 100 nm and smaller", W. Steinhögl, G. Schindler, G. Steinlesberger, M. Traving, M. Engelhardt, J. Appl. Phys. 97, 023706 (2005).
- "Electron scattering at surfaces and grain boundaries in Cu thin films and wires", J. S. Chawla, F. Gstrein, K. P. O'Brien, J. S. Clarke, and D. Gall, Phys. Rev. B 84, 235423.
- "Effects of microstructure on interconnect and via reliability: Multimodal failure statistics", C. V. Thompson, H. Kahn, J. of Electronic Materials, June 1993, Vol. 22, Issue 6, pp 581–587.
- "Improved Reliability of Copper Interconnects Using Alloying", J.P. Gambino, Proc.17th IEEE IPFA, pp 1-7 (2010).
- "Electromigration-resistance enhancement with CoWP or CuMn for Advanced Cu Interconnects", C. Christiansen, B. Li, Matthew Angyal, T. Kane, V. McGahay, Y. Y. Wang, S. Yao, IRPS 2011.
- "Co Capping Layers for Cu/Low-k Interconnects", C.-C.Yang, P. Flaitz, B. Li, F. Chen, C. Christiansen, D. Edelstein, S.-Y. Lee, P. Ma, AMC 2010.
- "Effects of cap layer and grain structure on electromigration reliability of Cu/low-k interconnects for 45 nm technology node", L. Zhang, J. P. Zhou, J. Im, P. S. Ho, O. Aubel, C. Hennesthal, E. Zschech, IRPS 2010.
- "Plasma Etch Challenges for Porous Low k Materials for 32nm and Beyond", Cathy Labelle, C. Sandow, S. Schmidt, S. Richter, W. Yu, B. Zhang, Q. T. Zhao and S. Mantle, CSTIC, 2011.
- "Photonic Integration for Interconnect", W. Bogaerts, P. Absil, Photonics Integration Forum, Eindhoven, 22 June 2011.

39

Characterization of Post-etch Residue Clean By Chemical Bonding Transformation Mapping

Muthappan Asokan, Oliver Chyan*,

Interfacial Electrochemistry and Materials Research Lab,
University of North Texas

Cheng-Hsien Wu[+], Chih-Cheng Shih, Ting-Chang Chang,

Departments of Physics,
[+]Department of Materials and Optoelectronic Science,
National Sun Yat-Sen University,
Kaohsiung, Taiwan

Present to SPCC-2018, *chyan@unt.edu*

Interfacial Electrochemistry & Material Research Lab UNT

Objective: Utilize MIR-IR to optimize RIE etching and post RIE cleaning with minimum dielectric damages

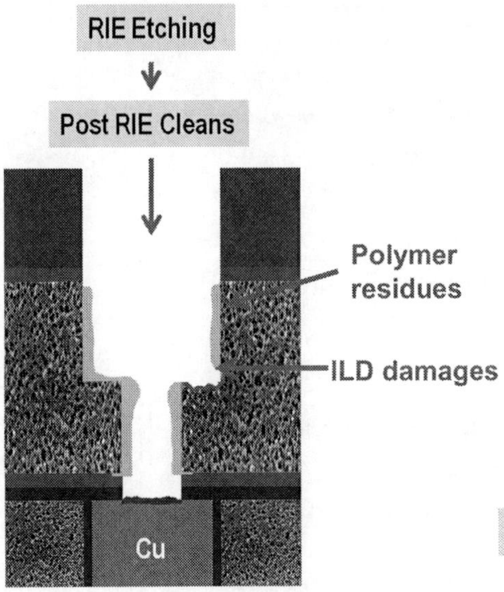

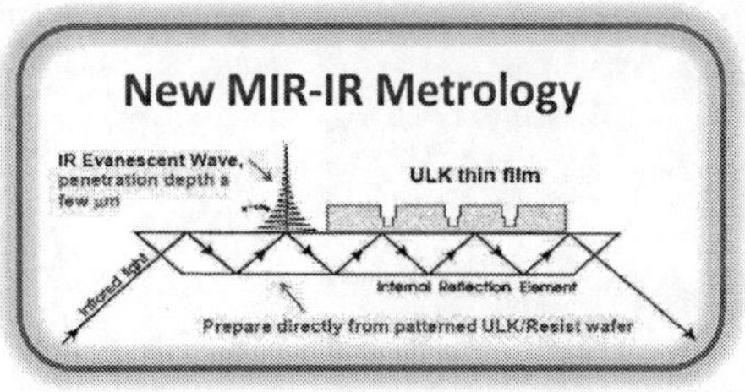

"Method for Wafer Characterization", *U.S. Patent 9366601*, 2016.

Physical constrains: small & deep

22nm SRAM array, IMEC

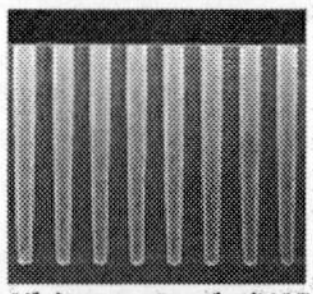
High-aspect-ratio (HAR)

Mechanical constrains: low modulus

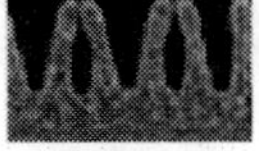
Collapse

Wiggle line

Interfacial Electrochemistry & Material Research Lab UNT

Common Characterization Tools for ILD Nanostructure

Physical Tools

SEM: Decorated DHF Etch

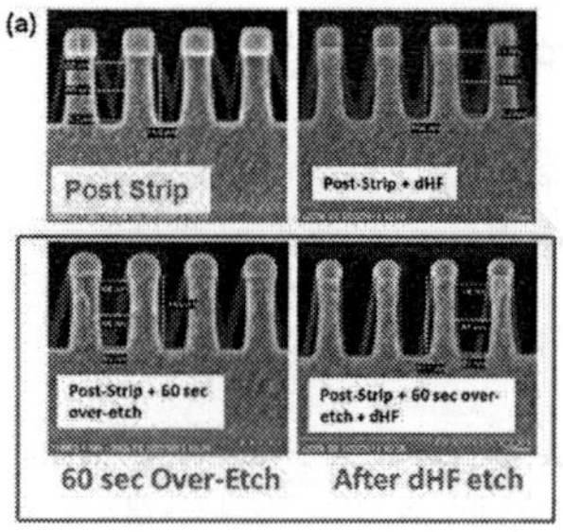

No Chemical Info

XPS : best for blanket wafer

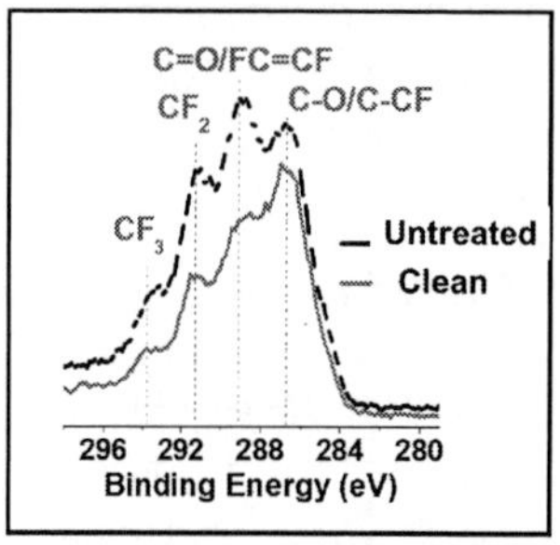

Atomic concentration % to assess F removal, take-off angle analyses to assess part of the side wall.

TEM/EEL

Labor intensive,
Destructive.

Electrical Tools

Hg Probe

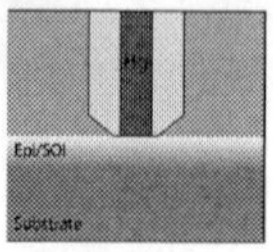

Non-contact,
Contamination issues,
For blanket wafer.

MOS Structure

Costly,
Difficult locating sources of defects.

Interfacial Electrochemistry & Material Research Lab
UNT

Ultra-clean Silicon Wafer Surface Preparation

Front end (FEOL) Cleans

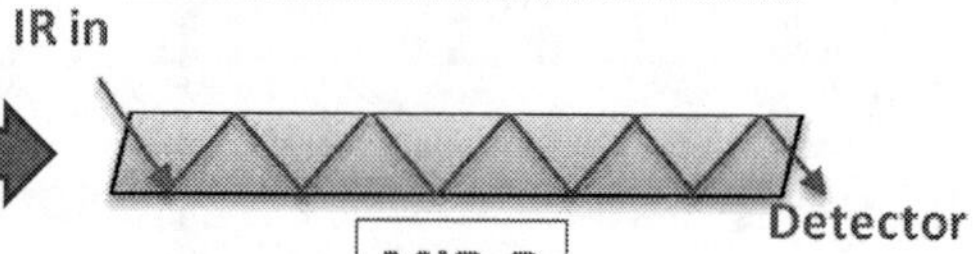

MIR-R

Multiple Internal Reflection Infrared Spectroscopy (MIR-IR)

Patterned Si wafer = IR waveguide

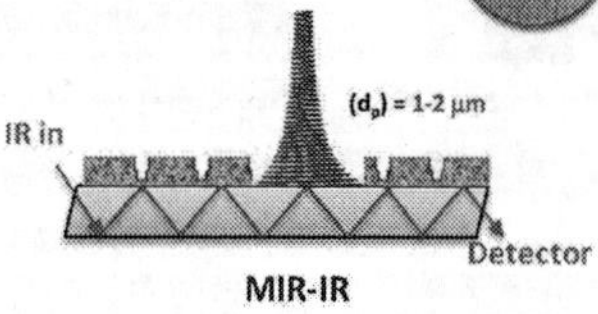

MIR-IR Advantages

- Analyze Patterned Wafer Directly
- Reproducible
- Superior Sensitivity (<10% monolayer)

"Method for Wafer Characterization", *U.S. Patent 9366601*, 2016.

Interfacial Electrochemistry & Material Research Lab
UNT

Detect Trace organics (< ppb) Contamination in HF Solution

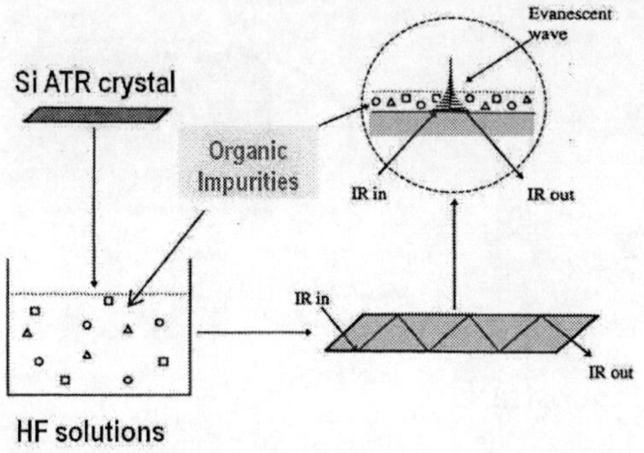

Anal. Chem. **1997**, *69*, 2434–2437

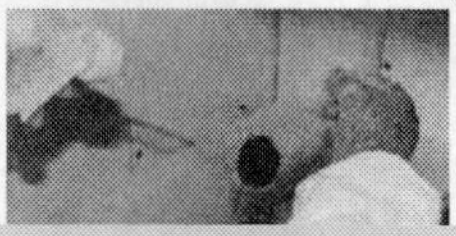

Organics contamination

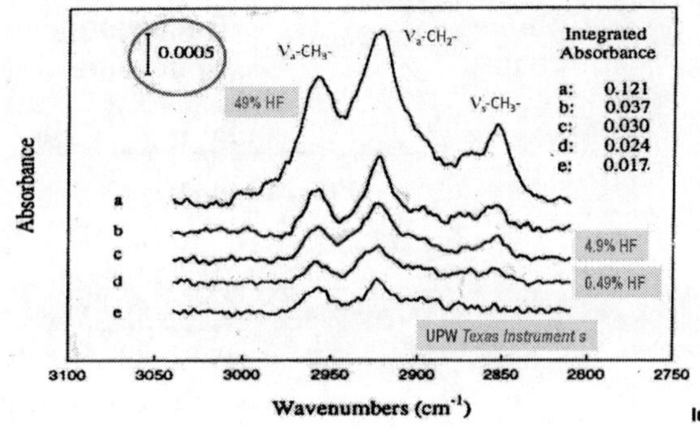

MIR-IR

Front end (FEOL) Cleans

Interfacial Electrochemistry & Material Research Lab UNT

MIR-IR : Apply to BEOL Processing

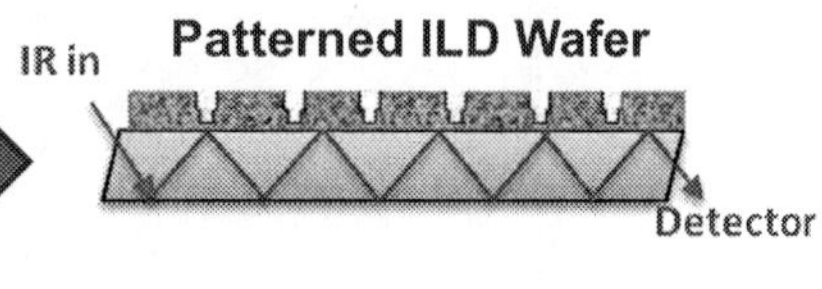

Interfacial Electrochemistry & Material Research Lab UNT

MIR-IR is a Very Powerful Substantive Tool---
Isolate Low-*k* Film Stack only Spectra

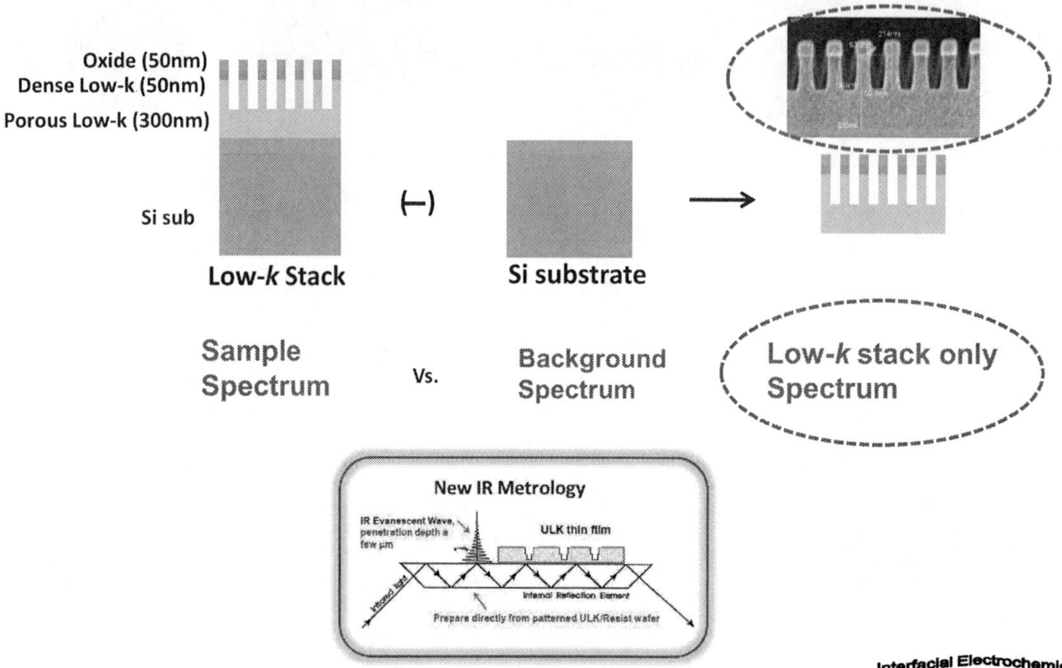

7

Chemical Bonding Transformation Map for Porous Low-k NanoStructure

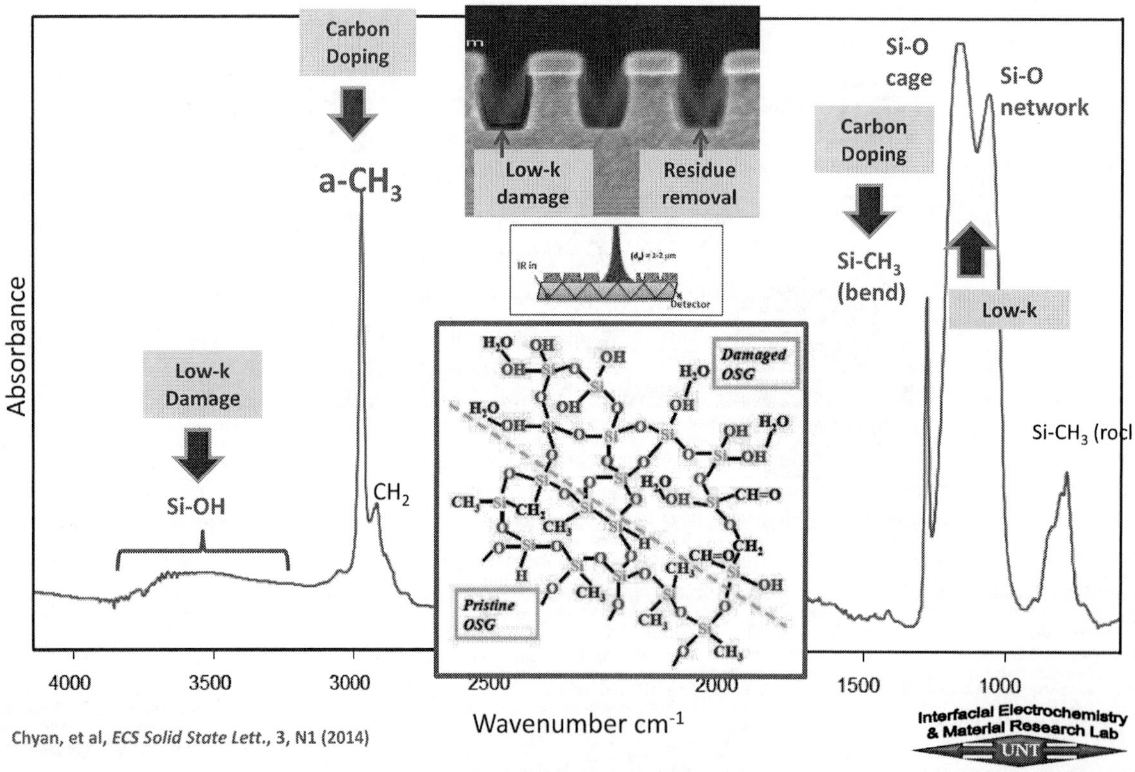

Chyan, et al, *ECS Solid State Lett.*, 3, N1 (2014)

Identify Lowest Low-*k* Damage from four Strip Processes

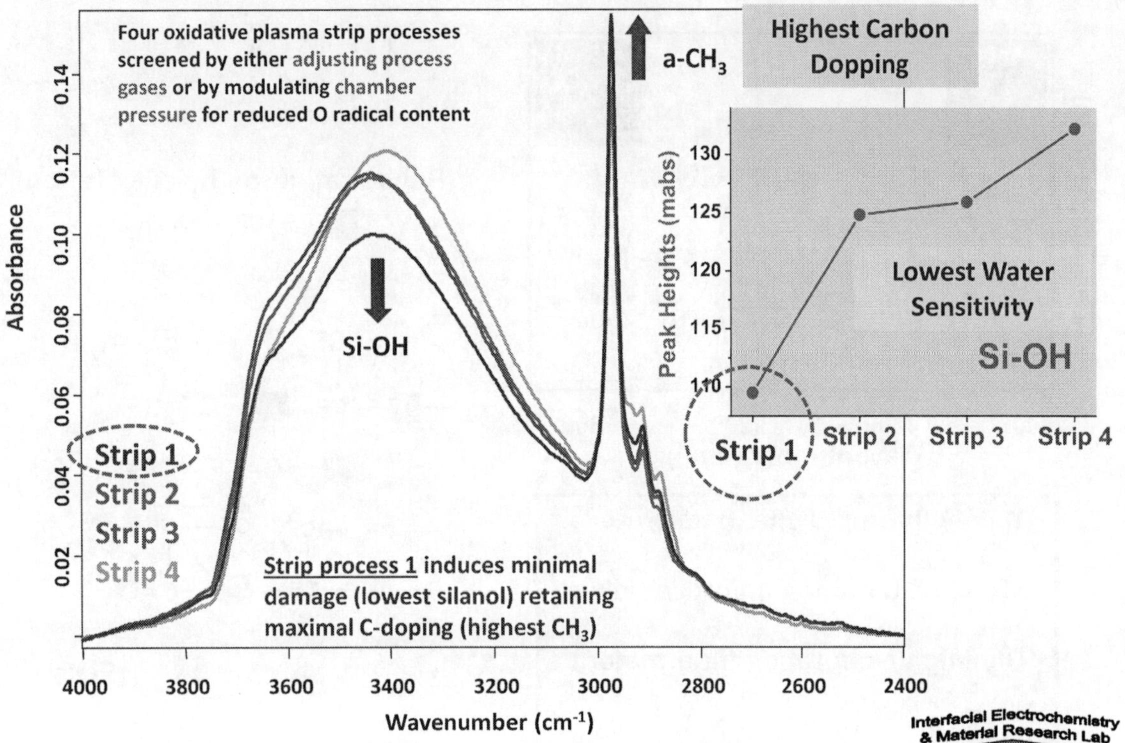

Interfacial Electrochemistry & Material Research Lab
UNT

Identify Chemical Bonding Structure and Explore Cleaning Processes for Post-Etch Residues

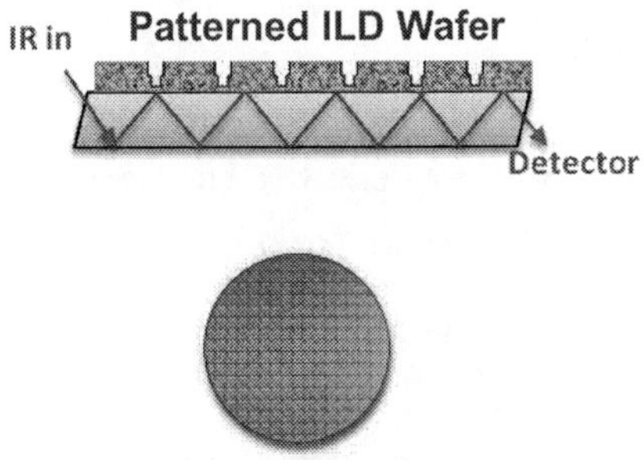

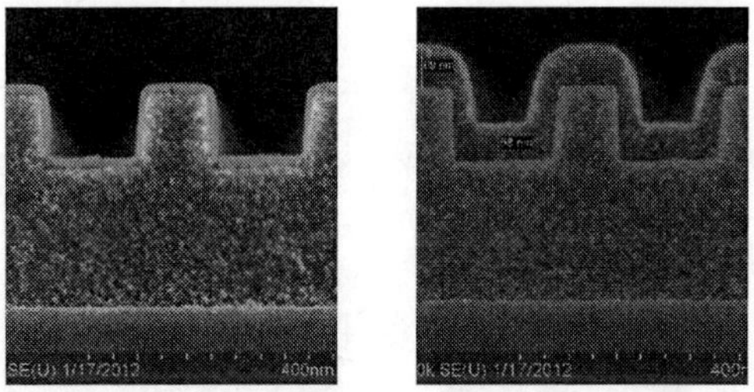

Interfacial Electrochemistry & Material Research Lab
UNT

Chemical Bonding Structure of Etch Residues

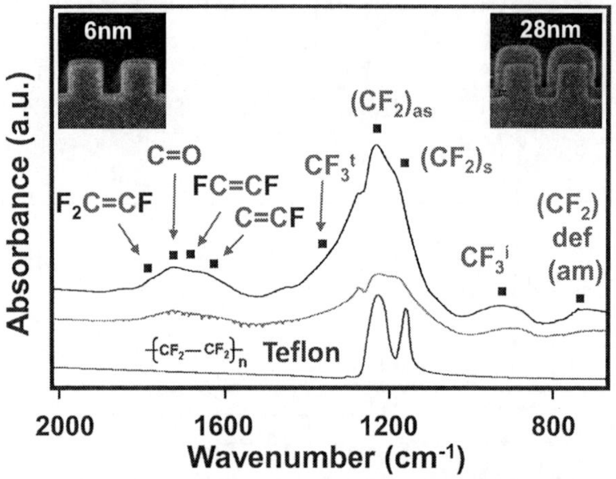

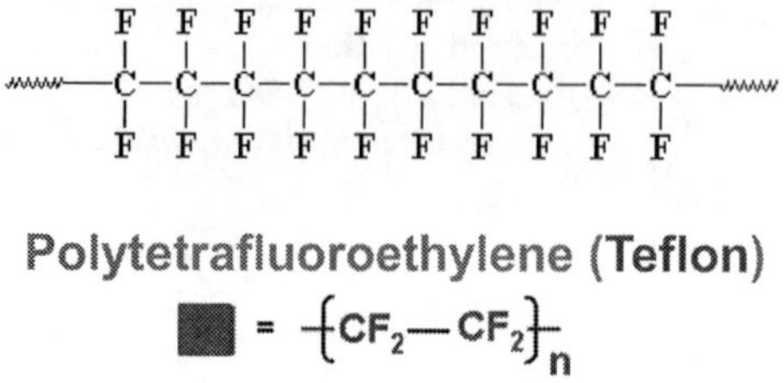

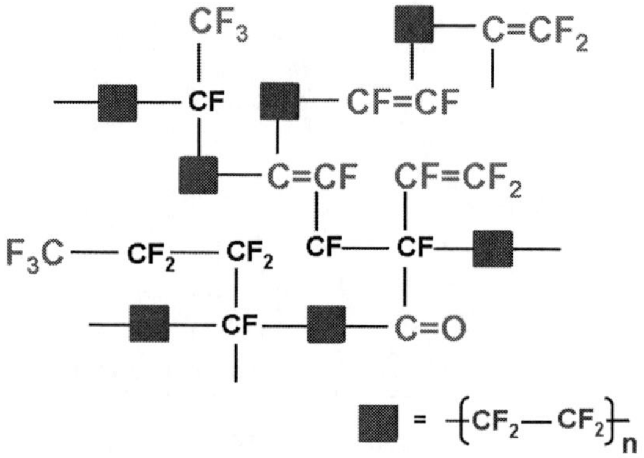

- Mainly fluoropolymer backbone
- Significant branching/cross-linking
- Olefinic unsaturation (fluorinated)
- Carbonyl functional groups

ECS Solid State Lett., 2, N11 (2013)

Monitor Wet Clean Efficiency on Post-etch Residue

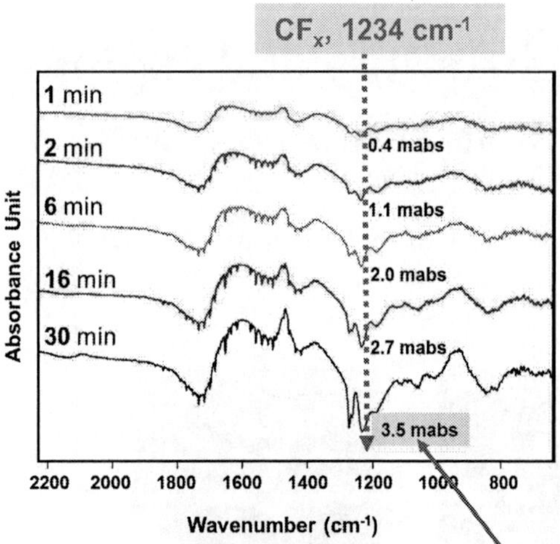

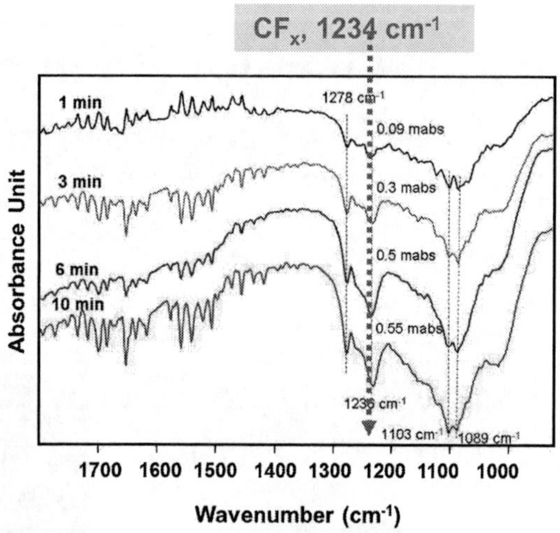

Ca. 30 % removal after 30 min

Interfacial Electrochemistry & Material Research Lab UNT

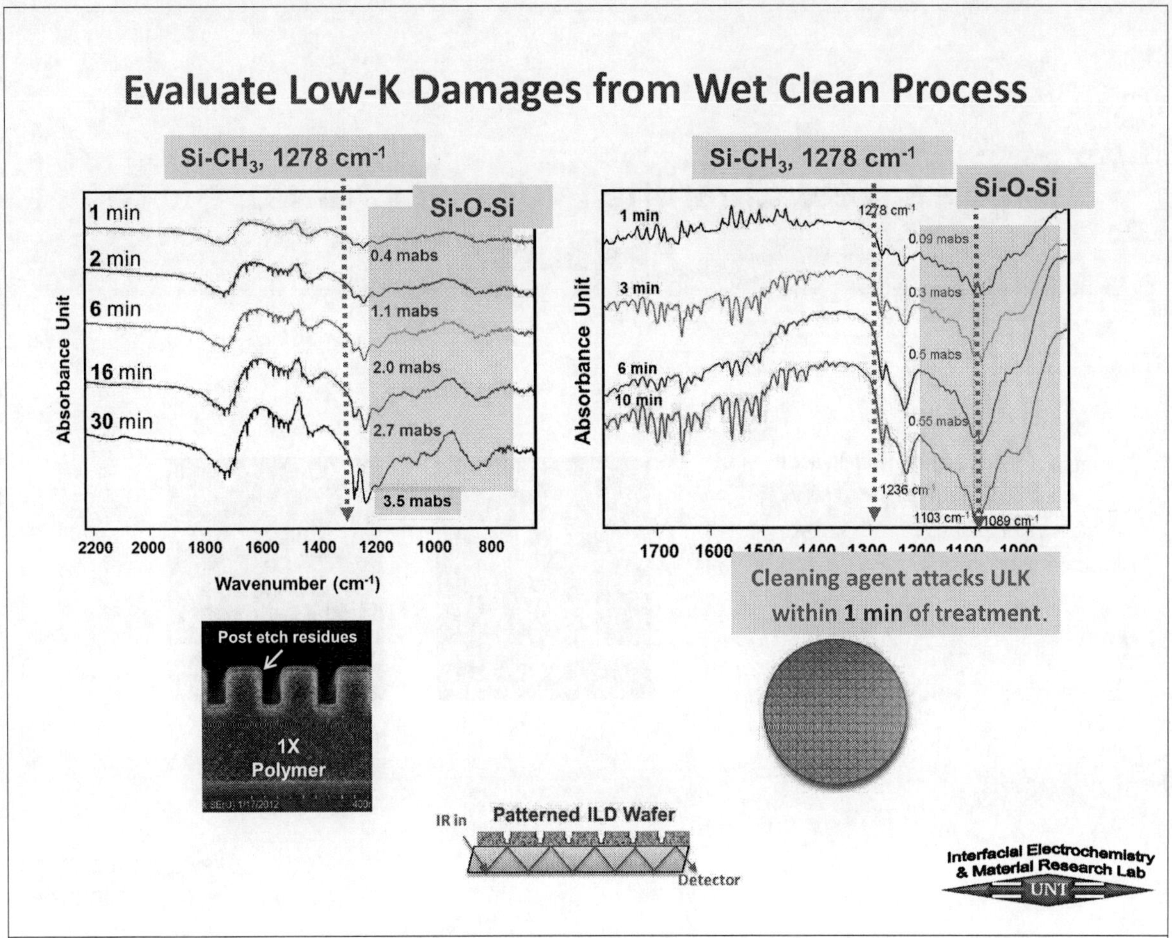
Evaluate Low-K Damages from Wet Clean Process
Si-CH3, 1278 cm-1
Si-O-Si
1 min
2 min
6 min
16 min
30 min
0.4 mabs
1.1 mabs
2.0 mabs
2.7 mabs
3.5 mabs
Absorbance Unit
Wavenumber (cm-1)
Si-CH3, 1278 cm-1
Si-O-Si
1 min
3 min
6 min
10 min
1278 cm-1
0.09 mabs
0.3 mabs
0.5 mabs
0.55 mabs
1236 cm-1
1103 cm-1
1089 cm-1
Cleaning agent attacks ULK within 1 min of treatment.
Post etch residues
1X Polymer
IR in
Patterned ILD Wafer
Detector
Interfacial Electrochemistry & Material Research Lab
UNT

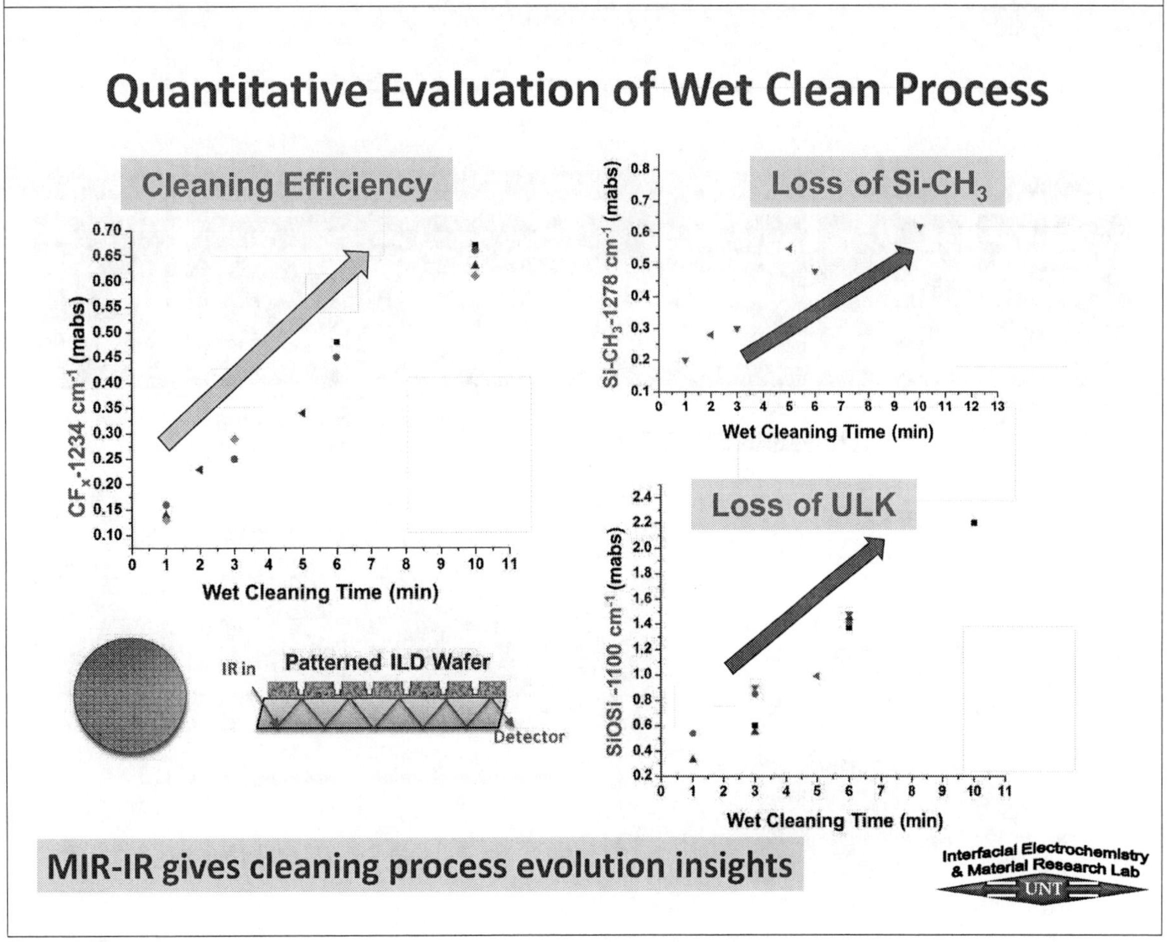
Quantitative Evaluation of Wet Clean Process
Cleaning Efficiency
CFx-1234 cm-1 (mabs)
Wet Cleaning Time (min)
Loss of Si-CH3
Si-CH3-1278 cm-1 (mabs)
Wet Cleaning Time (min)
Loss of ULK
SiOSi -1100 cm-1 (mabs)
Wet Cleaning Time (min)
IR in
Patterned ILD Wafer
Detector
MIR-IR gives cleaning process evolution insights
Interfacial Electrochemistry & Material Research Lab
UNT

Explore New Cleaning Processes for Post-Etch Residues

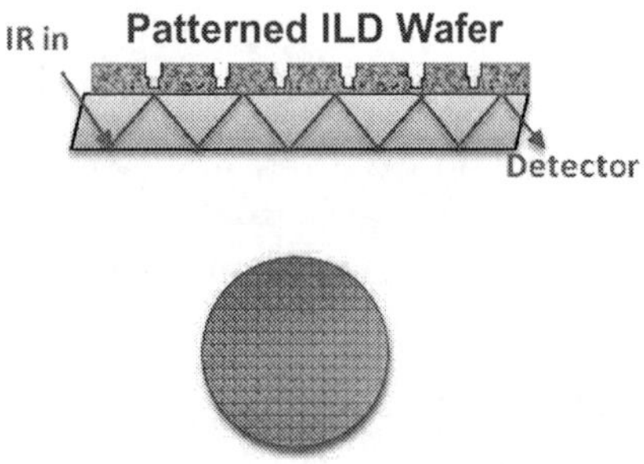

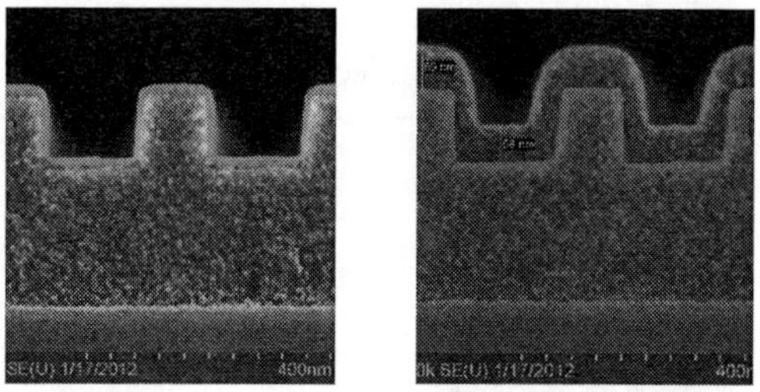

Interfacial Electrochemistry & Material Research Lab UNT

UV-assisted Cleaning on Post-etch Residues

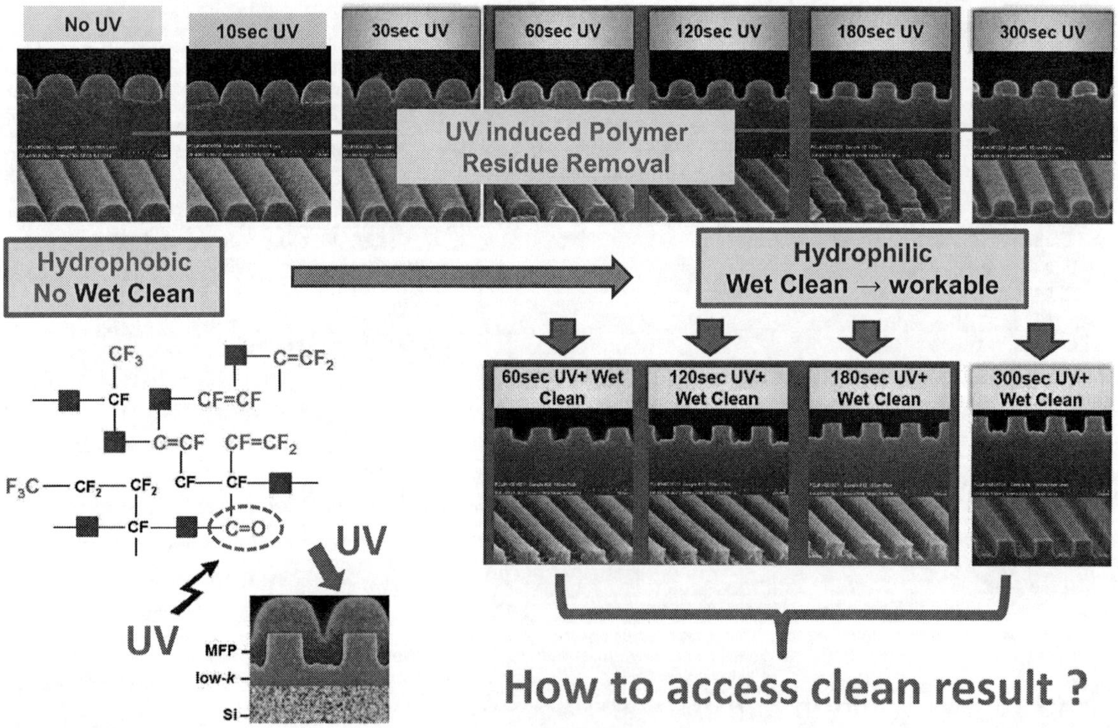

MIR-IR can Assess UV Cleaning Process

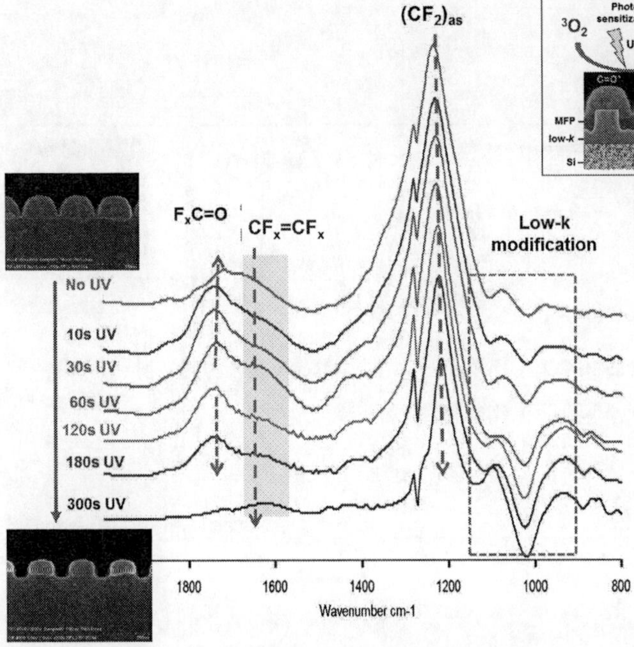

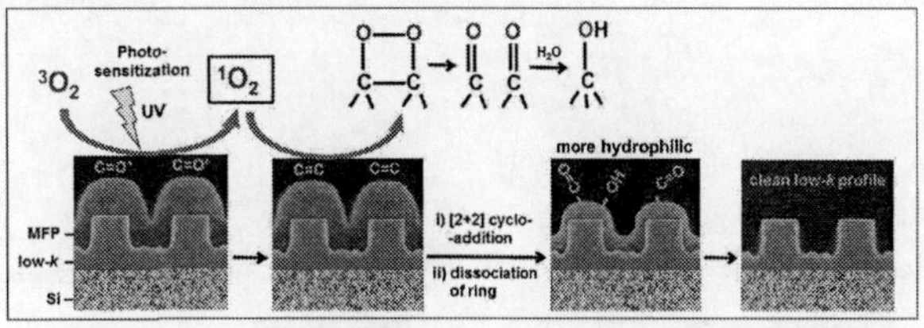

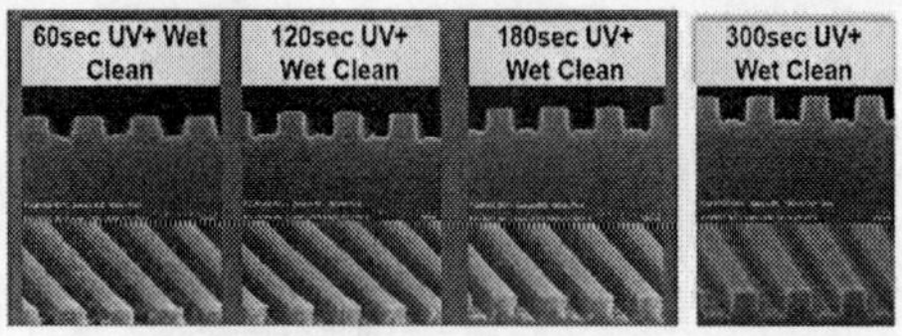

Hydrophilic → Wet Clean workable
- How effective?
- Low-k damages?

Process evolution insights shorten R&D cycles

Interfacial Electrochemistry & Material Research Lab UNT

Super Critical Cleaning for Post-etch Residues

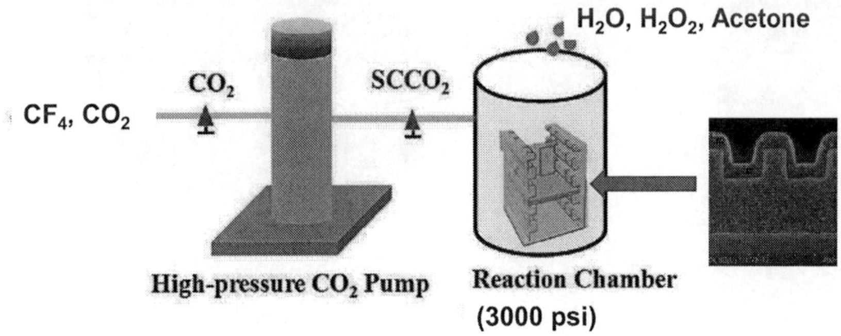

(a) Reference: RSC Adv., 2017, 7, 11585–11590
(b) Reference: Appl. Phys. Lett. 104, 243508 (2014)

國立中山大學 National Sun Yat-sen University

Interfacial Electrochemistry & Material Research Lab UNT

Comparison of three SCC-CO_2 Cleans (Differentials)

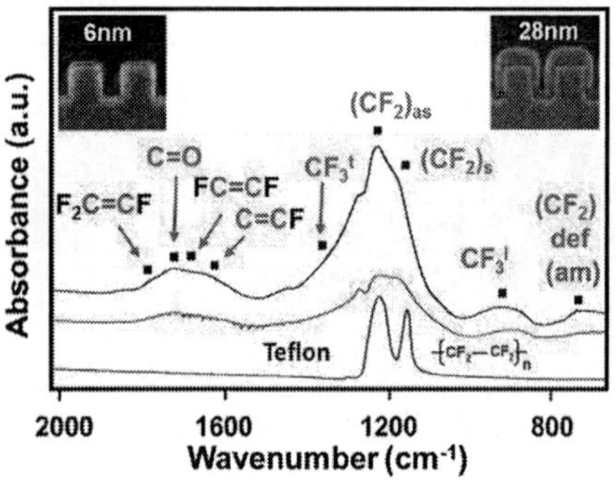

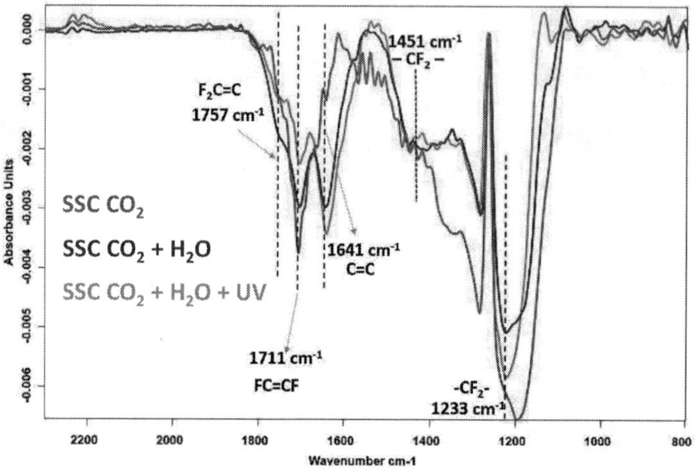

Interfacial Electrochemistry & Material Research Lab UNT

Exploring new SCC-CF_4 Cleans (TEM Data)

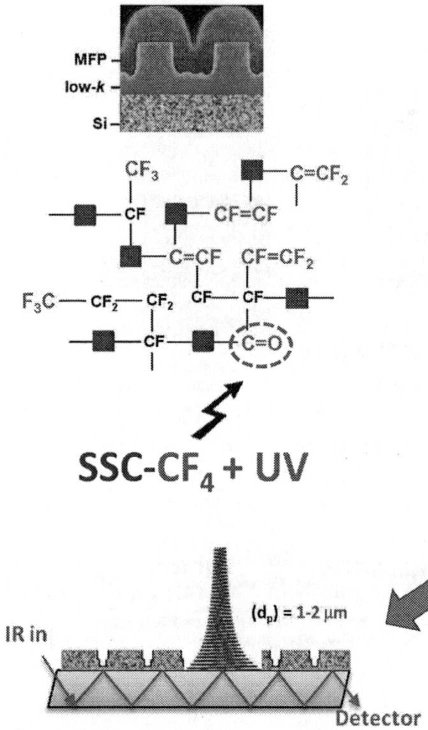

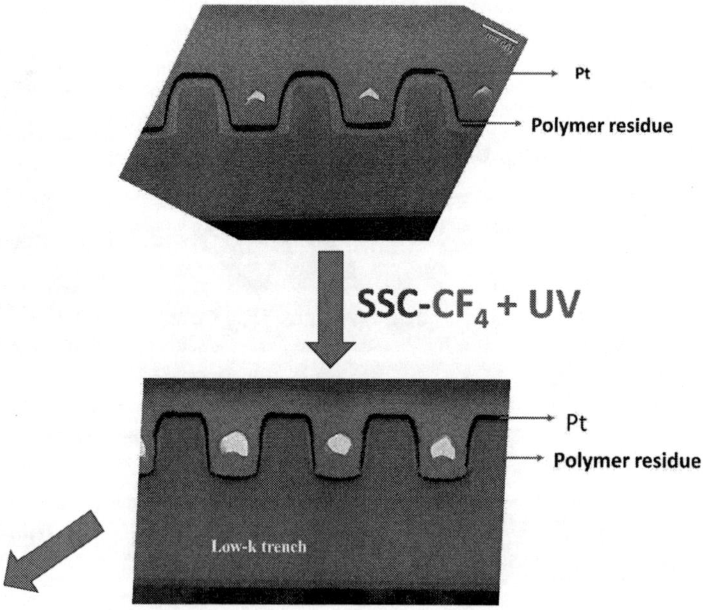

Interfacial Electrochemistry & Material Research Lab UNT

Goal: Achieve better understanding of fundamental materials properties at the critical interfaces of practical applications.

Interfacial Electrochemistry & Material Research Lab
UNT

Group members:

Muthappan Asokan

Josh Caperton

Issac Goutham,

國立中山大學
National Sun Yat-sen University

intel TEL ATMI Lam RESEARCH

Kanwal JitSingh	*Intel*
Alan M Myers	*Intel*
Ian Brown	*TEL (now with SCREEN)*

21

MIR-IR Applications in Advanced IC Fabrication

- FEOL, BEOL etching and cleaning, formulation development
- Monitor Post-etch residue removal & minimize Low-k damages
- Optimize Plasma Etch/Strip/Clean Process integration
- Monitor Low-k damages and optimize restoration process
- UV curing on porous Low-k dielectrics materials
- Evaluate TiN hard mask for Low-k pattern fabrication
- Flowable low-k dielectrics for gap filling in nanostructure
- Atomic Layer Deposition/Etching : provide critical interfacial chemical bonding info for better atomic layer control.

PI: Oliver Chyan*, *Chyan@unt.edu*

Interfacial Electrochemistry & Material Research Lab
UNT

BEOL PRE-METALLIZATION WET CLEAN: POST-ETCH RESIDUE REMOVAL AND METAL COMPATIBILITY

Q.T. LE*, E. KESTERS*, Y. AKANISHI**, A. IWASAKI**, AND F. HOLSTEYNS*
* IMEC, LEUVEN, BELGIUM
** SCREEN SEMICONDUCTOR SOLUTIONS Co., LTD., JAPAN
Email address: QuocToan.Le@imec.be

SPCC 2018, Boston, April 9-11th, 2018

PUBLIC

OUTLINE

- Post-etch residues formation on OSG 2.55: Angle-resolved XPS characterization
  - Introduction and objectives
  - Types of residues formed during/after dielectric patterning
  - Effect of post-etch treatment and wet clean on residue removal
- Effect of wet clean on ECD cobalt
  - Etch quantification
  - Co oxides formation after cleaning
- Impact of dissolved oxygen in dilute HF on cobalt etch
  - Etch quantification
  - Thickness of surface Co oxides

INTRODUCTION TO BEOL POST-ETCH RESIDUE FORMATION AND ITS REMOVAL

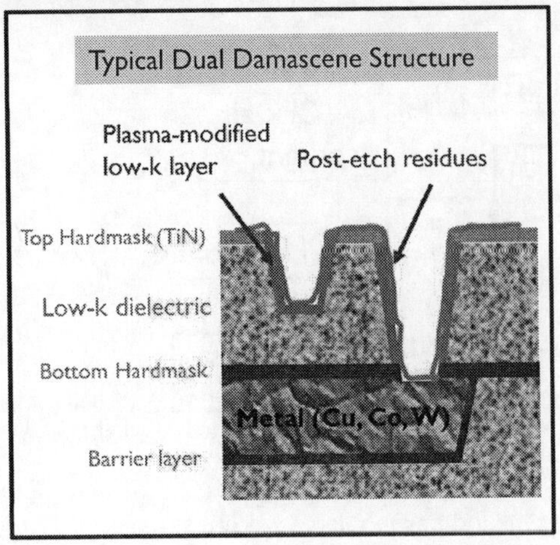

- Post-etch residues need to be removed before metallization, including barrier deposition + metal fill and CMP
- Challenges
  - Remove (or preserve) TiN HM
  - Compatibility requirements:
    - Dielectrics, including advanced porous low-k → preserve properties
    - Cu, Co, W, liner and barrier → minimum etching, no corrosion induced

OBJECTIVES

- Types of etch residues formed after plasma etch
- Removal of residues by wet clean

imec Q. T. Le et al., SPCC 2018, Boston, April 9-11th, 2018 3

FORMATION OF POST-ETCH RESIDUES ON PATTERNED OSG

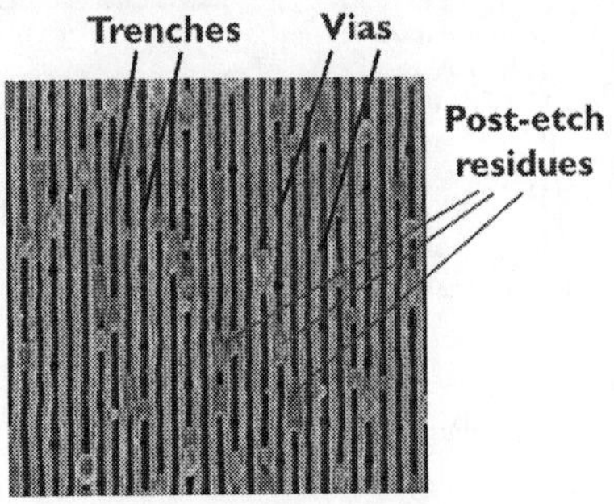

Residues observed on *DD TiN/OSG* structure (22.5 nm half pitch)

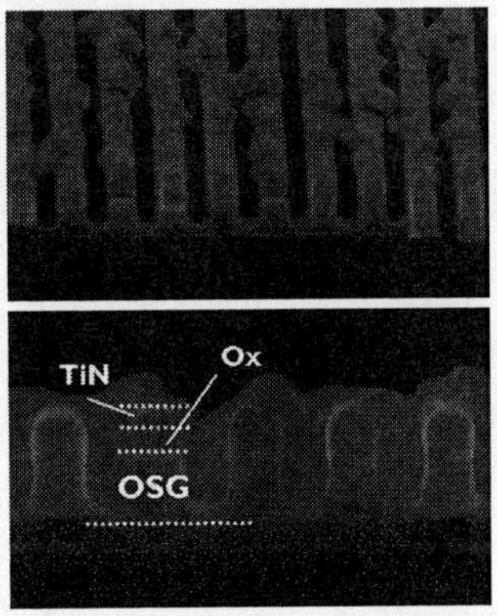

45 nm half pitch

Residues also observed on SD *TiN/OSG* structure

Use this structure as a short-loop test vehicle for characterization

- A dry post-etch treatment is commonly performed, followed by a wet clean step to remove these residues

imec Q. T. Le et al., SPCC 2018, Boston, April 9-11th, 2018 4

XPS CHARACTERIZATION

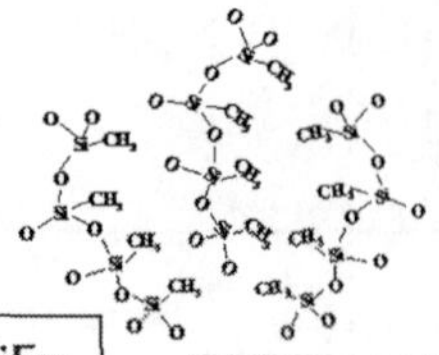

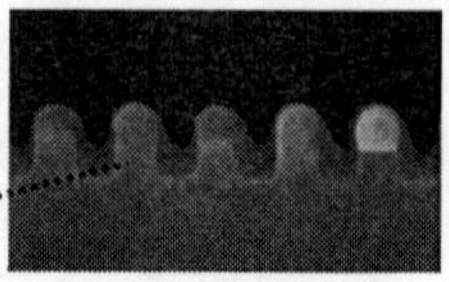

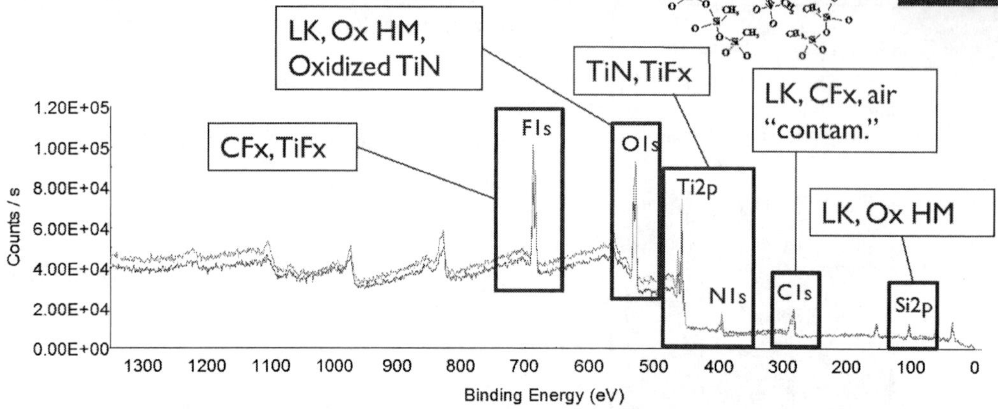

General scans from two measurements: parallel and perpendicular to lines
➔ Intensity is slightly different but same elements detected

XPS CHARACTERIZATION: TYPICAL F1s AND Ti2p

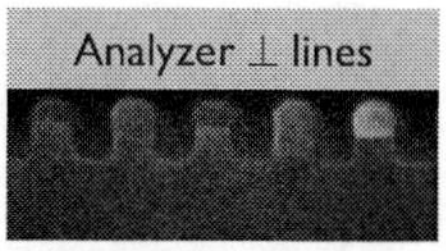

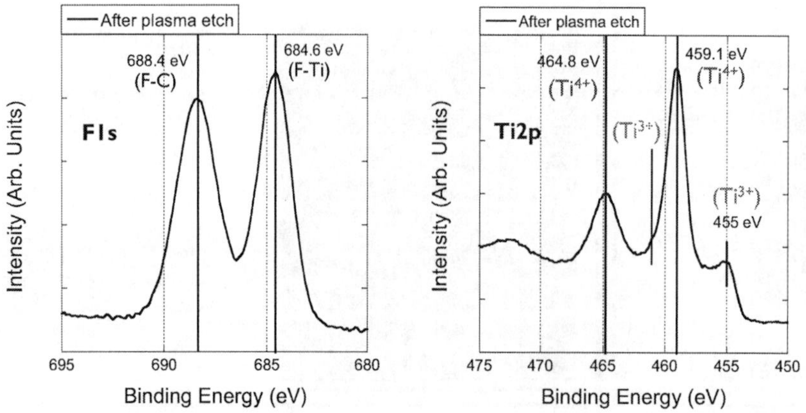

- P. J. Matsuo et al., J. Vac. Sci. Technol. B 17, 1435 (1999).
- Q. T. Le et al., ECS J. Solid-State Sci. Technol. 5, N5 (2016).

- Presence of both CFx ("polymer" component) and TiFx (metal-containing) residues
- Ti detected at surface mainly from TiO_2 and TiFx

XPS CHARACTERIZATION OF BLANKET AND PATTERNED SURFACE

RELATIONSHIP BETWEEN ELECTRON TAKE-OFF ANGLE AND PROBING DEPTH & AREAS

Sampling depth depends on the *take-off angle (TOA, θ)* and *inelastic mean free path (λ)* of photoelectrons

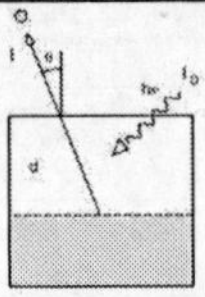

$$I = I_0 \exp\left(\frac{-d}{\lambda \cos\theta}\right)$$

$$\ln\left(\frac{1}{I_0}\right) = \frac{-d}{\lambda \cos\theta}$$

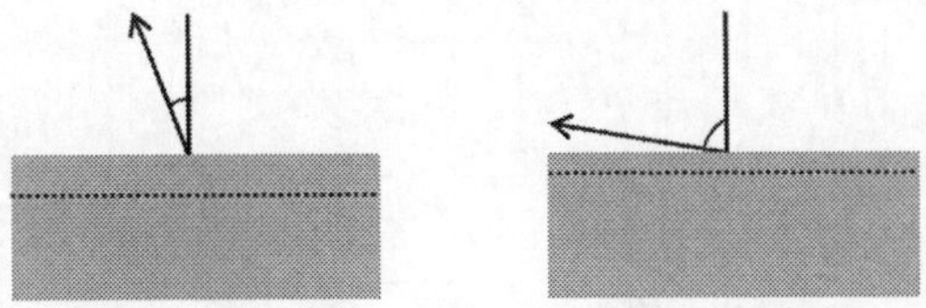

TOA = 22 deg.
Less surface sensitive

TOA = 78 deg.
More surface sensitive

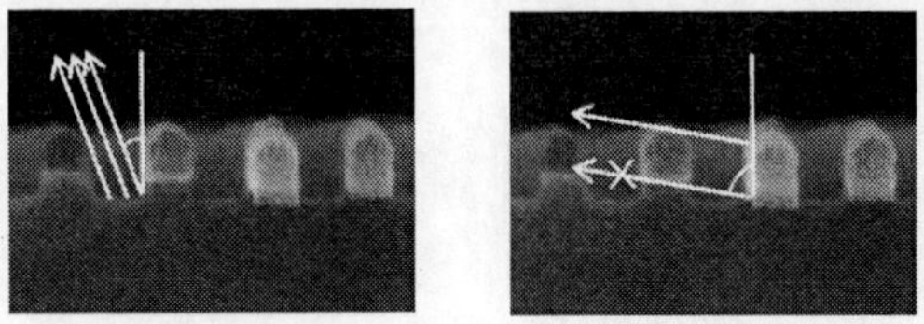

TOA = 22 deg:
Photoelectrons from *top surface, sidewall, and trench bottom*

TOA = 78 deg:
Photoelectrons *mainly* from *top surface*

imec   *Q. T. Le et al., SPCC 2018, Boston, April 9-11th, 2018*   7

XPS CHARACTERIZATION: CORE LEVEL SPECTRA

AFTER ETCH

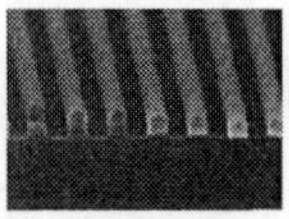

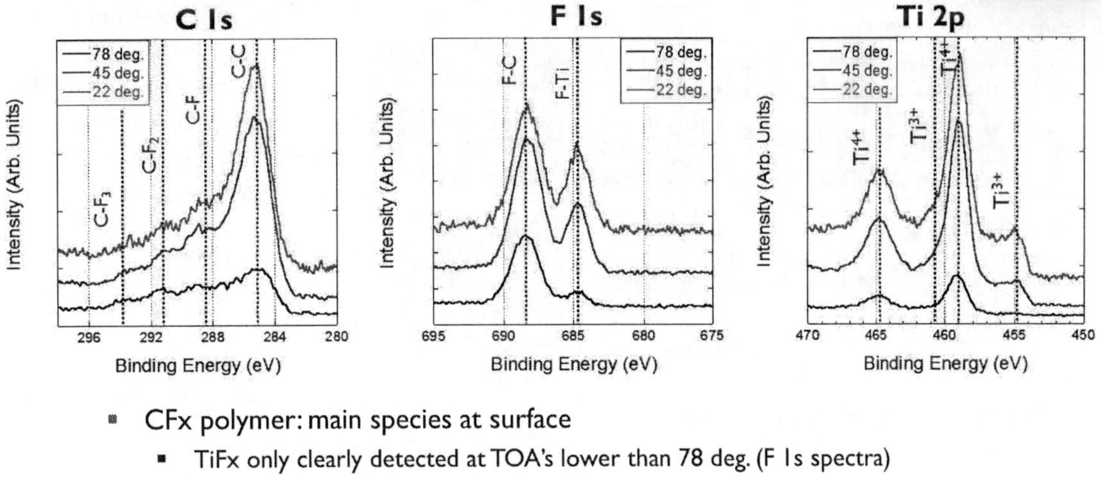

- CFx polymer: main species at surface
  - TiFx only clearly detected at TOA's lower than 78 deg. (F 1s spectra)
- Ti detected at surface mainly from TiO_2

imec   *Q. T. Le et al., SPCC 2018, Boston, April 9-11th, 2018*   8

XPS CHARACTERIZATION: Ti 2p

WITH SUBSEQUENT DRY ETCH TREATMENT OR WET CLEAN

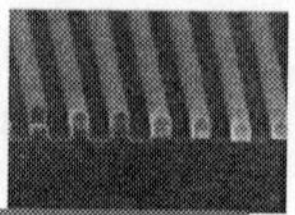

Dry etch treatment

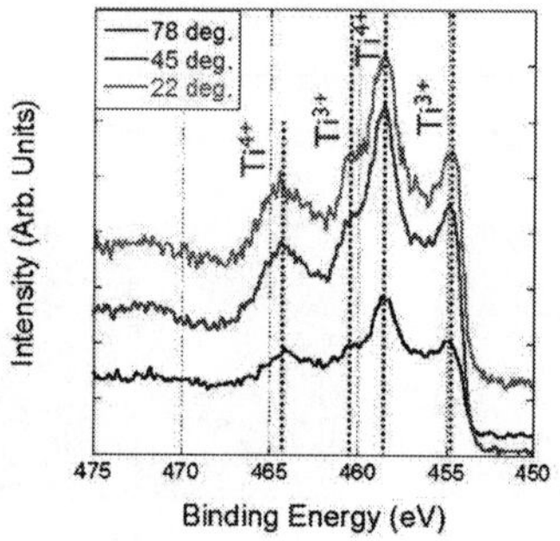

- Removal of CFx residues from the surface
  ➔ TiN (Ti^{3+}) clearly detected for all TOA's

Wet clean

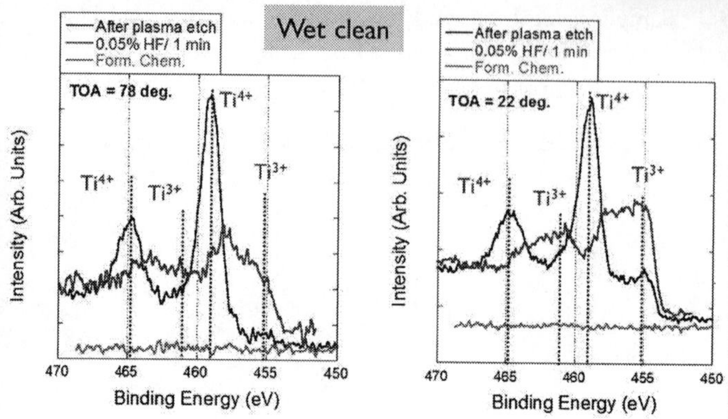

- 0.05% HF solution dissolved TiO_2
- Form. Chem. completely removed TiN

MAIN LEARNING

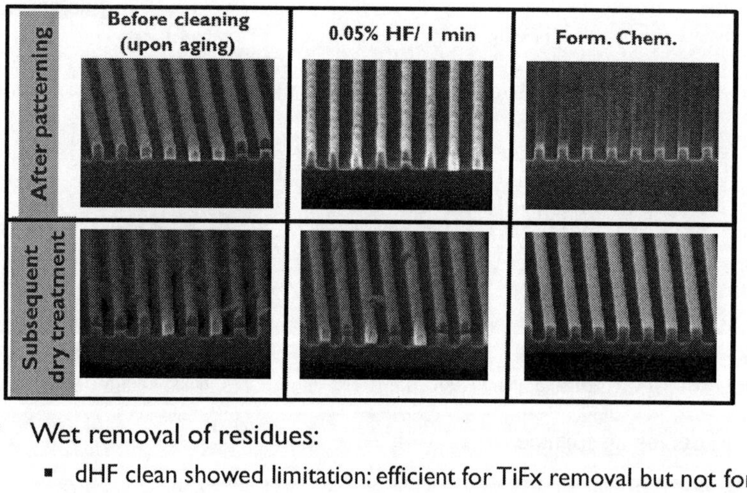

- Dry etch treatment performed after patterning: removed CFx residues and exposed the top hard mask surface
  - Subsequent formation and growth of TiFx species
  - Aging time affects the amount and density of Ti-containing residues (TiFx, TiOxFy, ...)
- Wet removal of residues:
  - dHF clean showed limitation: efficient for TiFx removal but not for CFx removal; high risk of SiO_2 HM and low-k attack for extended cleaning time

EFFECT OF WET CLEAN ON ECD COBALT

EXPERIMENTAL

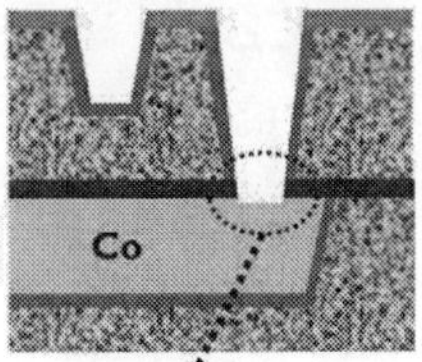

Use blanket Co to study interaction between Co (surface) and wet clean chemicals

Objective: quantification of wet etching for blanket ECD Co and growth of Co oxides

Chem 1	Chem 1 Dilution	Chem 2	pH
0.05% HF	1:1000		~3
SC1	1:4:50		10.8
SC1	1:4:100		11
-	-	Formulated chem.	7-10
0.05% HF	1:1000	Formulated chem.	

- *Beaker set-up*
- 0.05% HF at RT (*saturated dissolved oxygen concentration*)
- APM at RT
- Formulated mixture at 50 C
- ECD Co; nominal thickness ~220-230 nm
- Thickness quantification: 4-pt probe
- CoO thickness and surface characterization: Spectroscopic Ellipsometry, XPS

imec Q. T. Le et al., SPCC 2018, Boston, April 9-11th, 2018 12

BLANKET Co: ETCH QUANTIFICATION

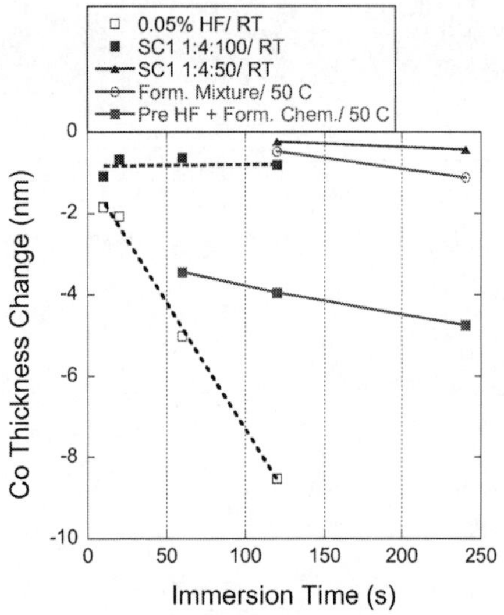

- Co etch rate
  - 0.05% HF > Form. chem. with HF pre-treat > SC1 1:4:100 ~ SC1 1:4:50 ~ Form. chem.
  - In acidic medium

$$Co + H_2O \rightarrow Co(H_2O)_{ads} \rightarrow Co(OH)^+ + H^+$$
$$\downarrow$$
$$Co^{2+} + H_2O$$

- Co etch rate appears to increase significantly if the existing Co oxides layer was removed (by 0.05% HF pre-treatment) prior to Form. chem. clean
  ➔ Co etch amount (etch rate) depends on the surface Co oxides properties

imec   *Q. T. Le et al., SPCC 2018, Boston, April 9-11th, 2018*   13

BLANKET Co: THICKNESS OF SURFACE Co OXIDES

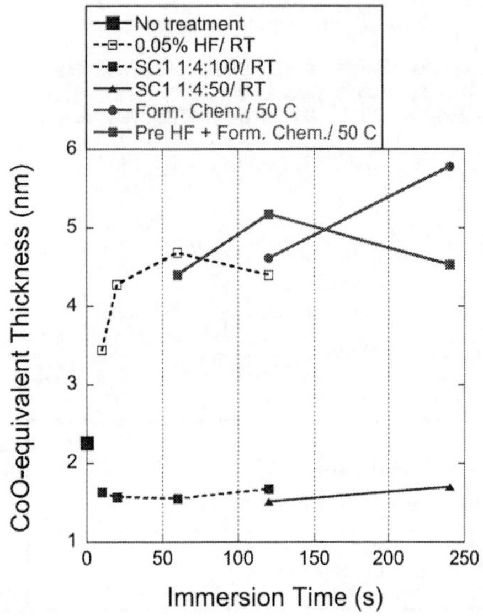

- CoO thickness at surface of ECD Co ~2.3 nm
- Surface CoO-equivalent thickness Form. Chem. > 0.05% HF > SC1
  - In SC1: high pH, formation of $Co(OH)_2$ that passivates the surface ➔ limited surface layer thickness

$$Co + OH^- \rightarrow CoO + H^+ + 2e^-$$
$$Co + 2OH^- \rightarrow Co(OH)_2 + 2e^-$$

M. Zhong et al., JES 161, C138 (2014)

  - Samples treated with Form. Chem.
    - Low etch rate for Co (previous slide)
    - Thickness of Co oxides measured most likely reflexes the presence of a Co-corrosion inhibitor complex layer

imec   *Q. T. Le et al., SPCC 2018, Boston, April 9-11th, 2018*   14

BLANKET ECD Co: XPS RESULTS

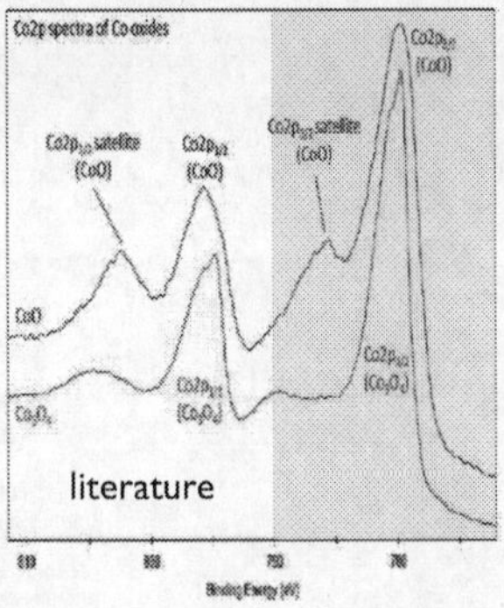

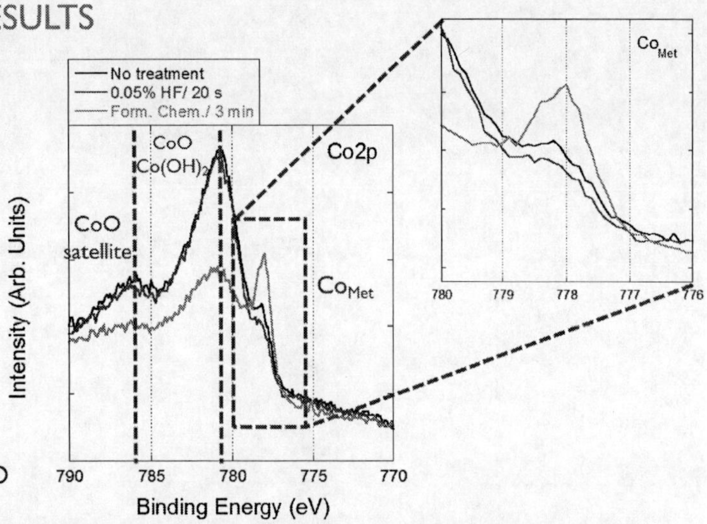

- Higher intensity of metallic Co component detected = thinner CoO layer at surface

Thickness of surface CoO layer: 0.05% HF > No treatment > Form. Chem.

imec Q. T. Le et al., SPCC 2018, Boston, April 9-11th, 2018 15

SUMMARY

- Blanket ECD Co etch: 0.05% HF > SC1 1:4:100 ~ SC1 1:4:50 ~ Form. Chem.
- SC1 at RT has low etch rate for Co
  - Could be used for pre-metallization clean
  - However, *post-etch residue removal* aspect needs to be considered as well
- Formulated mixture represents another option to control Co etch rate, surface properties, and post-etch residue removal

imec Q. T. Le et al., SPCC 2018, Boston, April 9-11th, 2018 16

IMPACT OF DISSOLVED OXYGEN IN DILUTE HF ON COBALT ETCH

EFFECT OF DISSOLVED OXYGEN (DO) IN dHF SOLUTION ON Co ETCH

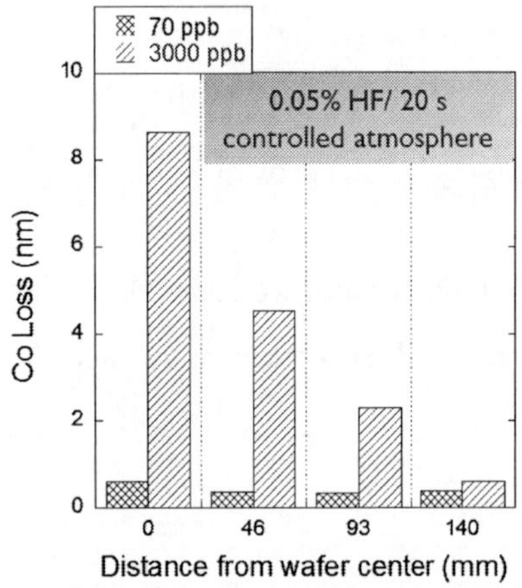

$2Co + O_2 + 4H^+ + \rightarrow 2Co^{2+} + 2H_2O$

- SU-3200 platform, SCREEN Single wafer cleaning tool
- Material: ECD Co (300 nm nominal thickness)
- Cleaning solution: 0.05% HF
- DO: 70-3000 ppb range

- Co loss substantially increased with increasing DO concentration in dHF
- Low DO vs. high DO
  - Co loss <<1 nm
  - Good uniformity between center and edge
- *If HF is used for pre-metallization clean, DO concentration must be controlled to have a good control of the Co loss*

imec

Q. T. Le et al., SPCC 2018, Boston, April 9-11th, 2018

18

EFFECT OF FLUID DYNAMICS AND PARTIAL OXYGEN PRESSURE

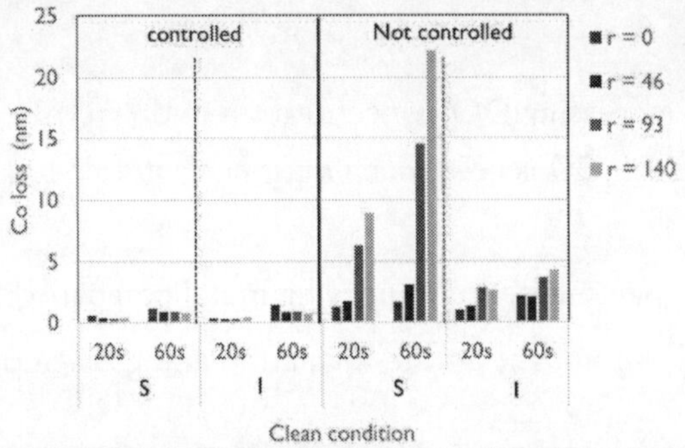

0.05% HF
DO: 70 ppb

S: Reference fluid dynamics

I: Improved fluid dynamics

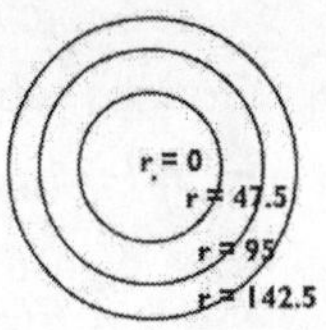

- Controlled atmosphere
  - Co loss <1 nm for 20 s clean
  - Good uniformity between center and edge
- Improved fluid dynamics: drastic reduction of Co loss
  - Fluid dynamics can improve metal compatibility even for non-controlled atmosphere case

imec Q. T. Le et al., SPCC 2018, Boston, April 9-11th, 2018 19

EFFECT OF FLUID DYNAMICS AND PARTIAL OXYGEN PRESSURE
CoO THICKNESS

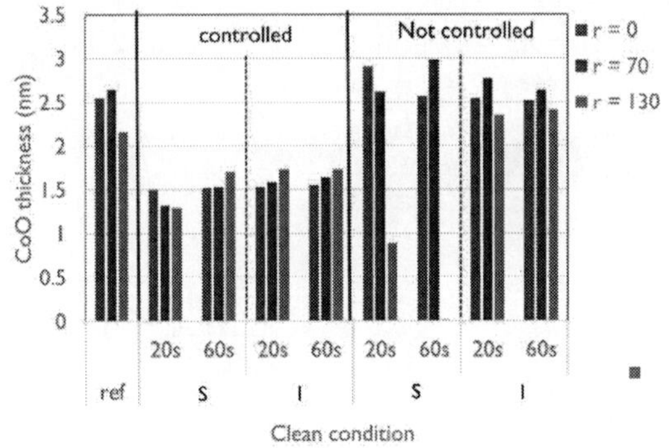

0.05% HF
DO: 70 ppb

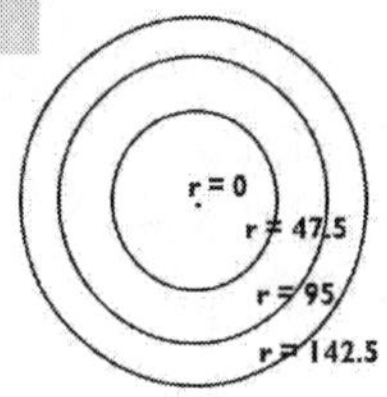

- Oxygen in atmosphere has a significant impact on the formation and growth of the Co oxides

- Controlled vs. non-controlled atmosphere
  - Thinner layer of CoO measured at the surface
  - Similar CoO thickness measured regardless of cleaning time and fluid dynamics

imec Q. T. Le et al., SPCC 2018, Boston, April 9-11th, 2018 20

SUMMARY

- Co loss substantially increased with increasing DO concentration in dHF (*)
  ➔ *If HF is used for pre-metallization clean, DO concentration must be controlled to have a good control of the Co loss*
- Fluid dynamics represents a relevant parameter for improving metal compatibility
- The atmosphere oxygen has a significant impact on the formation and growth of the Co oxides

(*) Similar behavior also observed for Cu:
E. Kesters *et al.*, SPCC presentation, 2017

imec

Q. T. Le et al., SPCC 2018, Boston, April 9-11th, 2018

21

ACKNOWLEDGEMENT

- S. Decoster
- T. Conard
- S. Braun
- A. Klipp
- A. Mizutani

THANK YOU!

imec

22

imec
embracing a better life
imec
PUBLIC

TECHNIC

A Global Leader in

Specialty Chemicals

Surface Finishing Equipment

Engineered Powders

Analytical Controls

Advanced Engineered Solutions

Selective Cleaner of High-K based Dry Etch residues

Dr. J. DAVIOT
Technology & Innovation Director
Wet Surface Cleaning , Stripping and Etching Group

Jerome.daviot@technic.fr

outlines

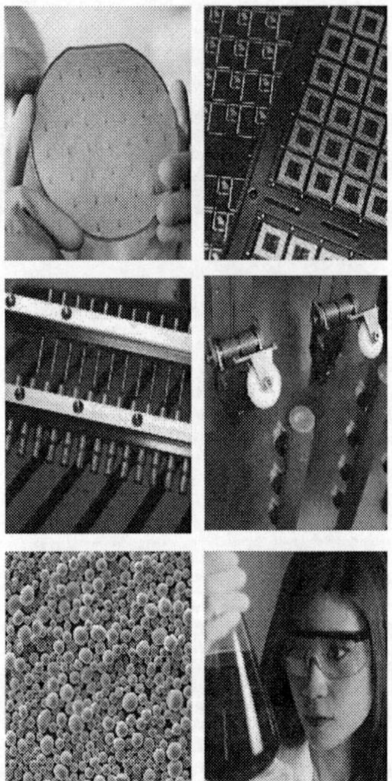

- **Background & Motivation**
- **Formulation Concept & Design**
- **Implementation in High-Volume Manufacturing**

TECHNIC

Domain of applications

- **MIM CAPACITORS**
  - Metal-Insulator-Metal (MIM) capacitors in
    - radio frequency (RF) circuits
    - analog integrated circuits (ICs)
    - microprocessor units (MPUs)
    - dynamic random access memory (DRAM) cells

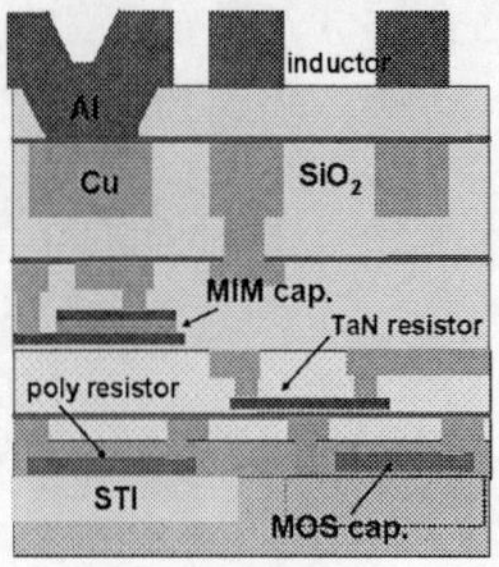

MIM capacitor advantages and applications
High linearity and dynamic range
Reducing the coupling of high frequency noise (Digital/Analog circuit region)
Decoupling capacitor
Oscillators and phase-shift networks
Coupling capacitance and bypass capacitance

Source: sematech, application of MIM capacitor, August 31, 2000

TECHNIC

Cleaning Process Challenges for High-K-based Devices

AlCu PAD on TaN barrier

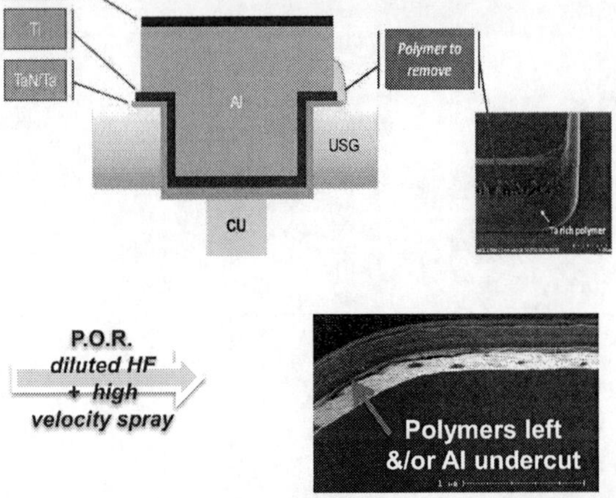

MIM capacitors Ta_2O_5 high-K

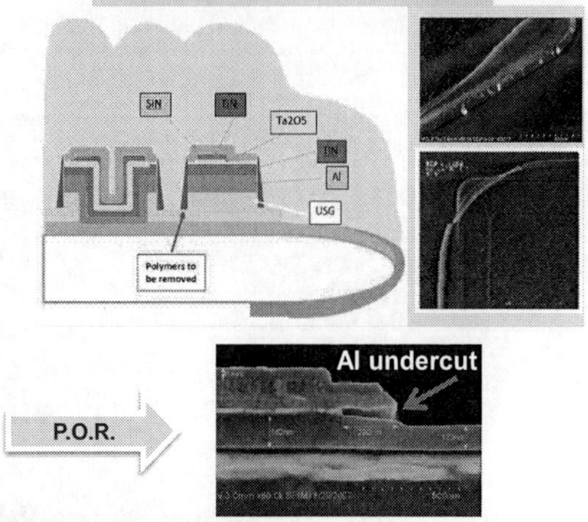

- **Remove resilient Ta, Zr, Hf based PER created during plasma etching**
- **Hinder Al, Cu metal attack**
- **Prevent surface contamination**

TECHNIC

Current State of the Art

5

Solutions	DHF w/wth Nanospray	TMAH based stripper	HF/H2SO4	SAC	Standard Al/Cu PER
Equipment	SWU	Raider	Zeta	MP3001	Spray batch
	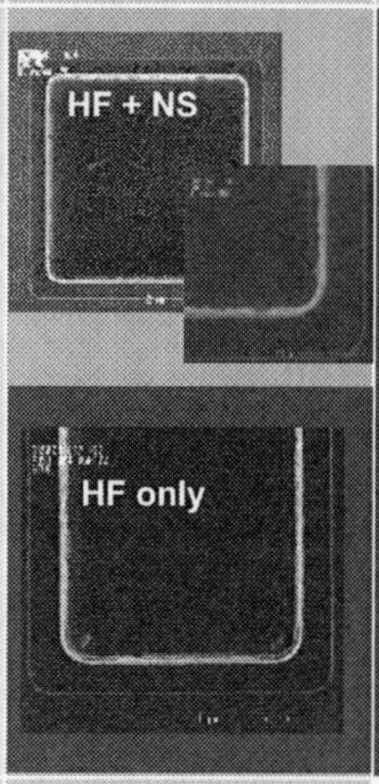		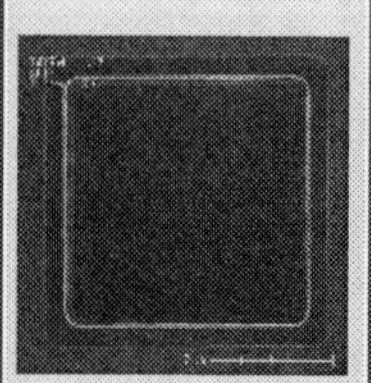	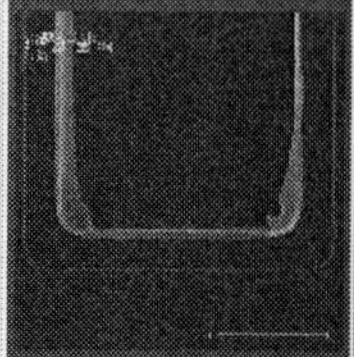	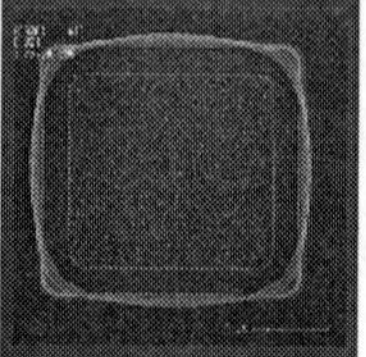
Al undercut nm	10-70	N/A	~ 100	OK	OK
comment	Cleaning mostly by physical action w/ limited efficiency		high Al attack	TaO2 residues untouched	TaO2 residues untouched

TECHNIC

Current State of the Art

6

Solutions	DHF w/wth Nanospray	TMAH based stripper	HF/H2SO4	SAC	Standard Al/Cu PER
Equipment	SWU	Raider	Zeta	MP3001	Spray batch

Al undercut nm	10-70	N/A	~ 100	OK	OK
comment	Cleaning mostly by physical action w/ limited efficiency		high Al attack	TaO2 residues untouched	TaO2 residues untouched

TECHNIC

High-k PER Cleaning Conundrum

Std Ti, Al based PER residues
→ removed

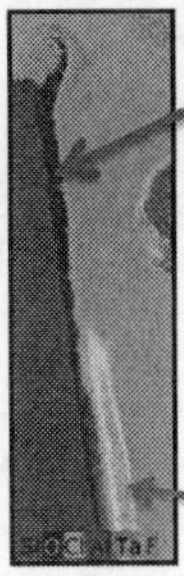

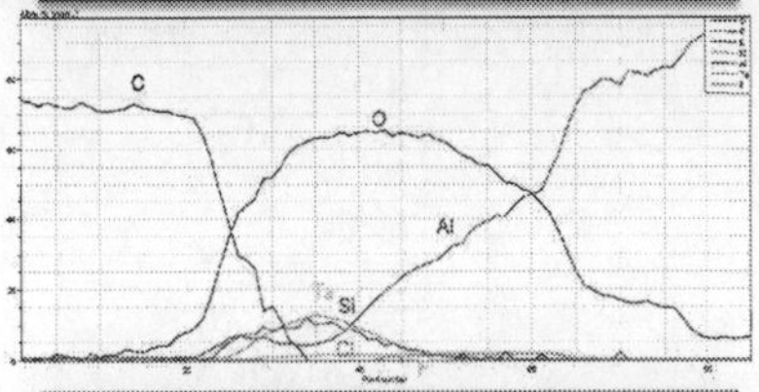

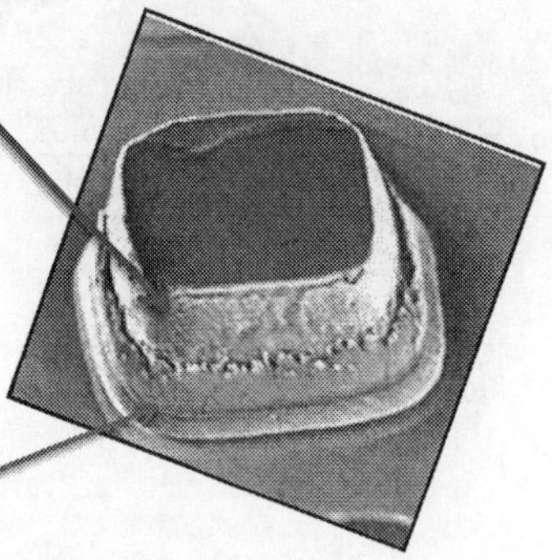

Ta rich Residues
→ barely attacked by conventional chemicals

Removing Ta/Zr/Hf based PER requires HF-based system
BUT
Need to control metal, oxide, and dielectric attack in HF

The Challenge is to tailor and control HF activity:
to concurrently address amorphous HK dissolution and prevent Al attack

Genesis of the Solution
--HfO_2 dissolution in HF--

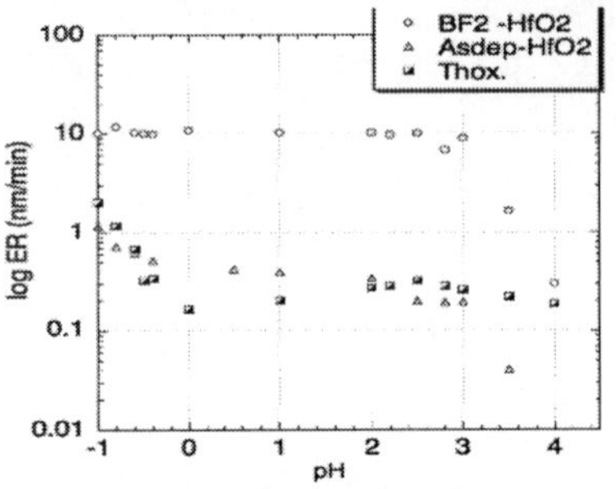

ER measurements in 0.05M, 25°C of HfO_2 (as deposited) and thermal SiO_2 (after dry etch)

Solvent	Thox ER (nm/min)	Asdep-HfO_2 ER (nm/min)
H_2O	0.472	0.3
DMSO	0.02	0.09
DMF	0.14	0
CH_3CN	0.156	0.15
CH_3OH	0.05	0.13
ETHANOL	0	0.08
2-PROPANOL	0	0.11
ACETONE	0	0.2

T 25°C [HF] at 0.05M

HfO_2 etch rate depending on:

- pH solution
- Crystallography
  - (Amorphous, post annealed, post dry etch)
- Material selectivity depending on co-solvent

Source: Solid State Phenomena Vols. 103-104 (2005) pp 97-102

Genesis of the Solution
--HfO_2 dissolution in HF--

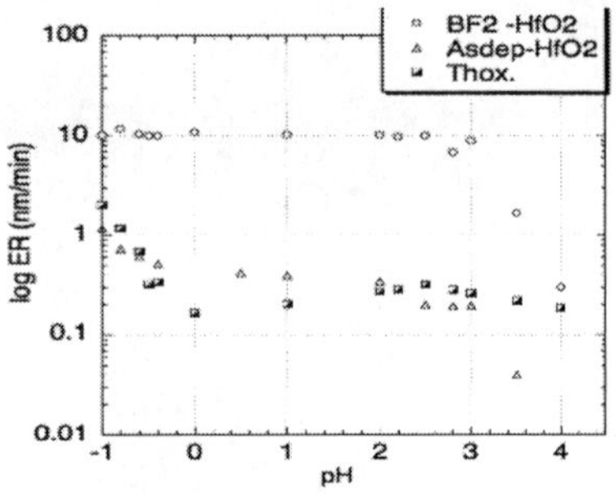

ER measurements in 0.05M, 25°C of HfO_2 (as deposited) and thermal SiO_2 (after dry etch)

Solvent	Thox ER (nm/min)	Asdep-HfO_2 ER (nm/min)
H_2O	0.472	0.3
DMSO	0.02	0.09
DMF	0.14	0
CH_3CN	0.156	0.15
CH_3OH	0.05	0.13
ETHANOL	0	0.08
2-PROPANOL	0	0.11
ACETONE	0	0.2

T 25°C [HF] at 0.05M

Amorphous HfO_2 = test vehicle for Ta / Zr / Hf etch polymers study

TECHNIC

Source: Solid State Phenomena Vols. 103-104 (2005) pp 97-102

Genesis of the Solution
--Al and AlF_x salt solubility in highly acidic media--

$2Al+6HF= 2AlF_3 +3H_2$

Aluminum is corroded in usual aqueous HF, but it is inhibited in anhydrous acid (80%-95%)

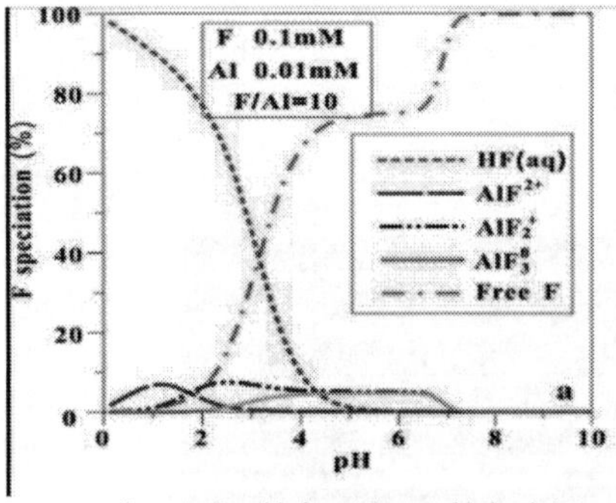

Source: Geochimica et Cosmochimica Acta
Volume 75, Issue 16, 15 August 2011, Pages 4476-4489

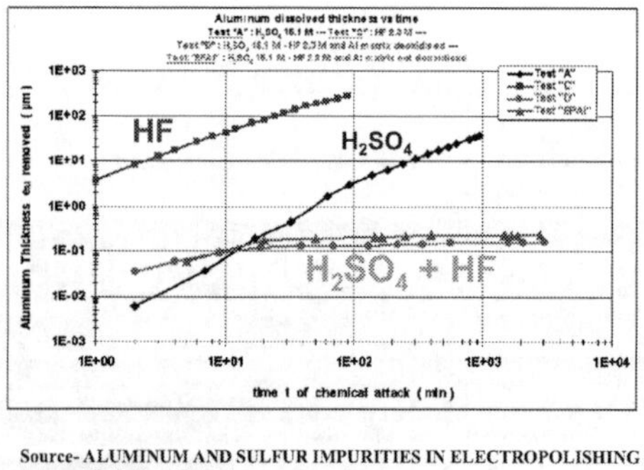

Source- ALUMINUM AND SULFUR IMPURITIES IN ELECTROPOLISHING BATHS*
A. Aspart, F. Eozénou, C. Antoine
CEA-Saclay, DSM/DAPNIA/SACM - 91191 Gif/Yvette - FRANCE.

AlF_3, AlF_2^+, AlF^{2+} are poorly soluble in anhydrous media

TECHNIC

HF-based formulation development:

Aqueous HF Solutions

A series of aqueous hydrofluoric acid solutions of 1% initial concentration across a pH range ~1-6 by addition of HCl and NH_3.

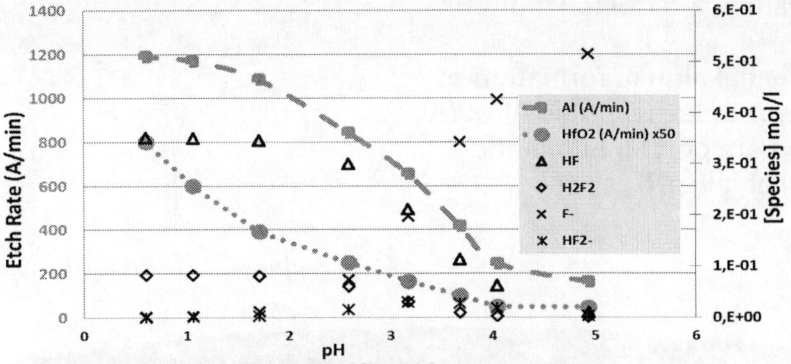

The etch rates of Al and HfO_2 vs. pH and [fluoride]

Al etch increases with aqueous [HF]

Al & High k etch rate:

highly influenced by pH

= f (non ionic species HF & H_2F_2 concentrations)

TECHNIC

HF-based formulation development:

Strong acidified aqueous HF Solutions

Increasing up to 10 mol/L the proton concentration in the HF solution (pH -1), through the use of a strong concentrated acid (pKa <<0), yields interesting and useful behavior on Al and high K materials

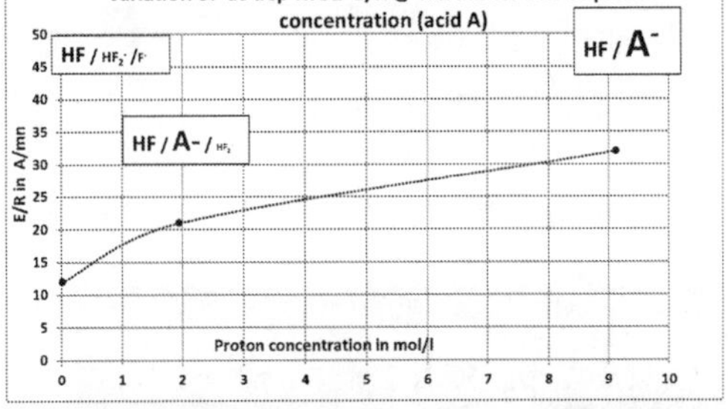

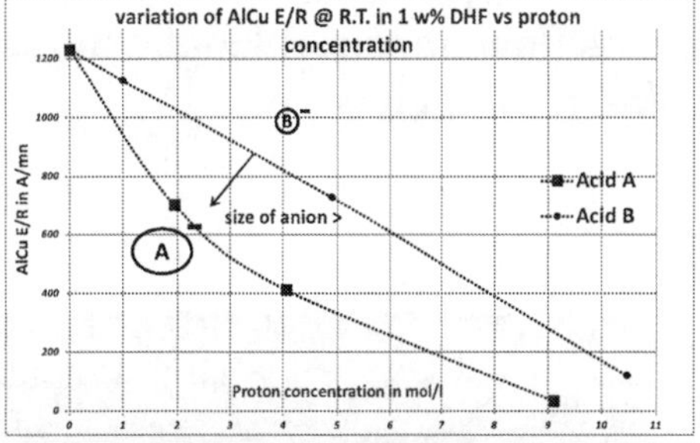

ER (HfO_2) increases

ER(Al) drastically reduces

- ***Strong acid counter-ions are also playing a role in AlFx salt dissolution mechanism at negative pH***

TECHNIC

HF-based formulation development:
Strong acidified aqueous HF Solutions

- Anodic branch -> diffusion limited

Potential sign of formation of protective layer or less soluble $AlFx^{+}/A^{-}$ species (low transport kinetic in the double layer) at low pH.

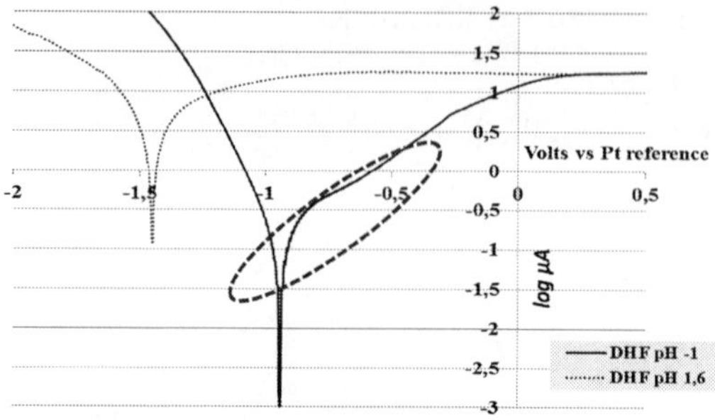

Tafel plot of pure Al in 1% HF at 2 pH levels

Better Al protection at high [proton] concentration

TECHNIC

HF-based formulation development:
Strong acidified aqueous HF Solutions

- Anodic branch -> diffusion limited

Potential sign of formation of protective layer or less soluble $AlFx^{+}/A^{-}$ species (low transport kinetic in the double layer) at low pH.

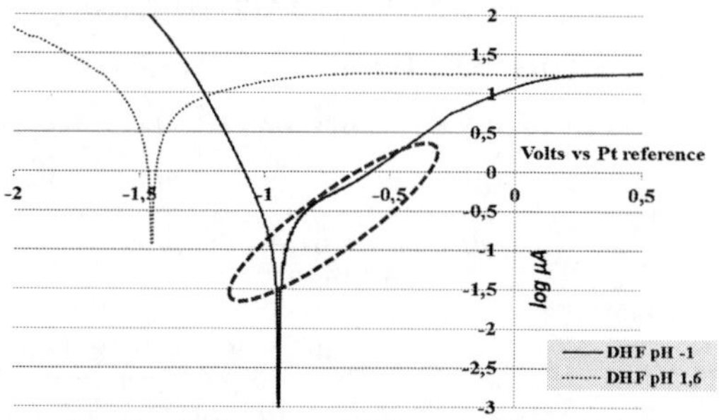

Tafel plot of pure Al in 1% HF at 2 pH levels

Developed a strong acidic blend addressing:

- ➔ Low heat generation during mixing and water rinse
- ➔ Good surface wetting attributes and rinsing
- ➔ Capability to be mixed with organic solvents/additives

TECHNIC

Co-solvent to optimize water ratio and limit strong acid content

Standard approach | **TechniClean IK73 alternative**

The co-solvent need to prevent HF dissociation

Co-solvent to optimize water ratio and limit strong acid content

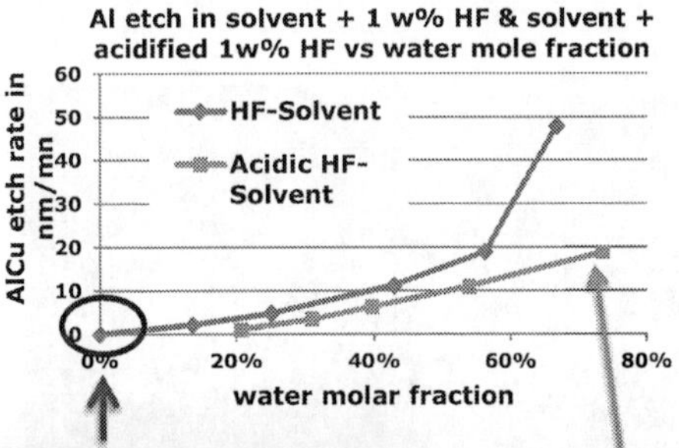

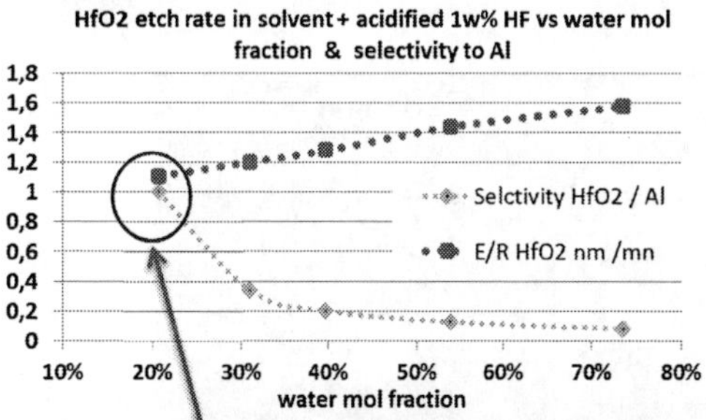

Anhydrous solution → no etch

Surface hydration required to initiate dissolution

- low ER in acidified HF + solvent < 20nm/min
- Aqueous HF (no solvent) ER120 nm/min !

At low water content, HfO_2 / Al selectivity markedly increases

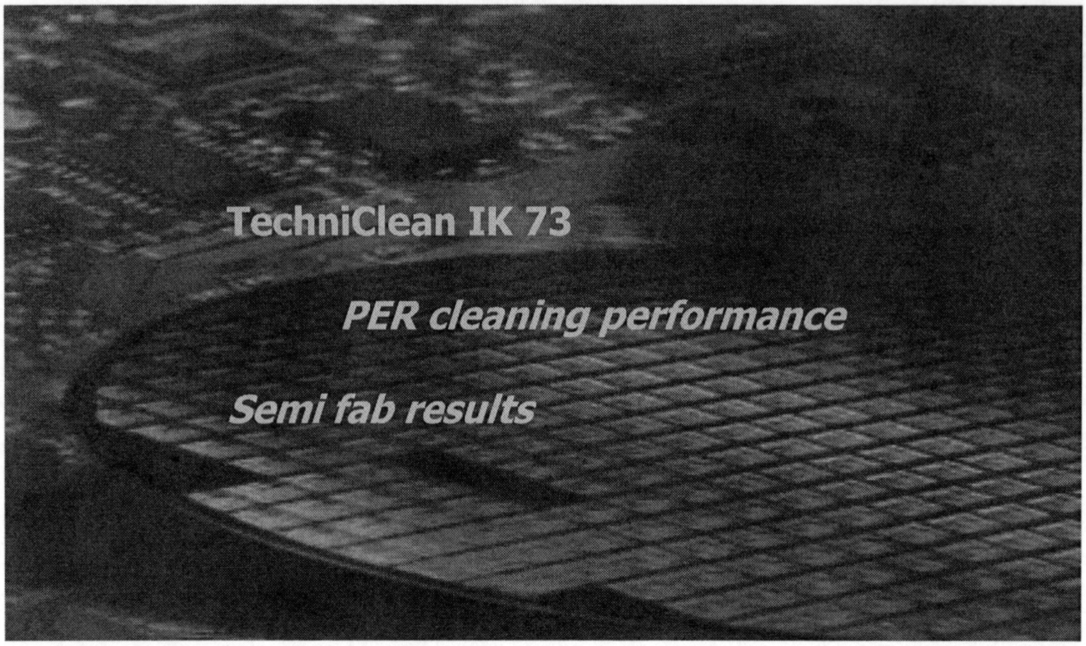

TECHNIC

TechniClean IK 73

Al PAD on TaN barrier

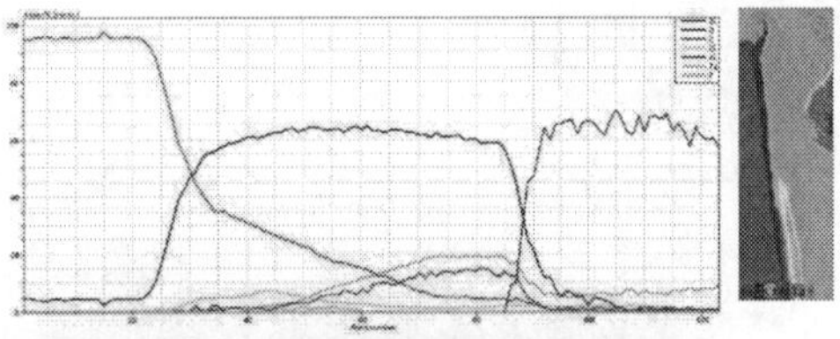

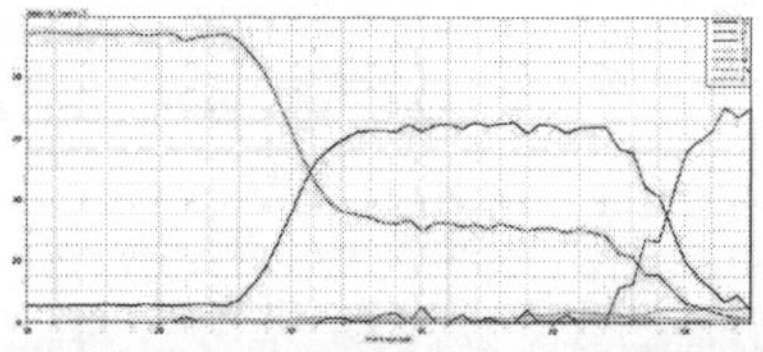

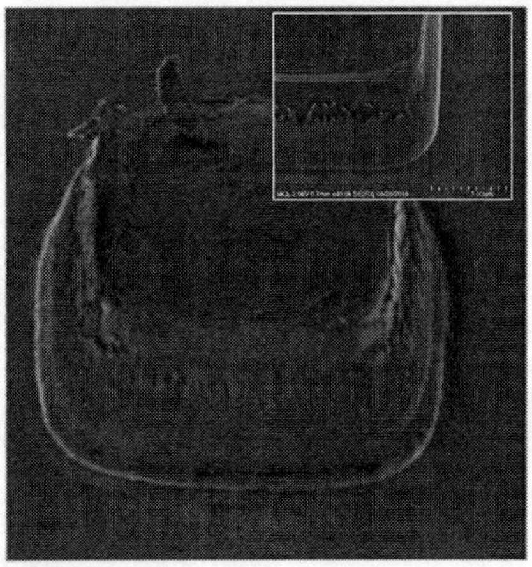

Before cleaning

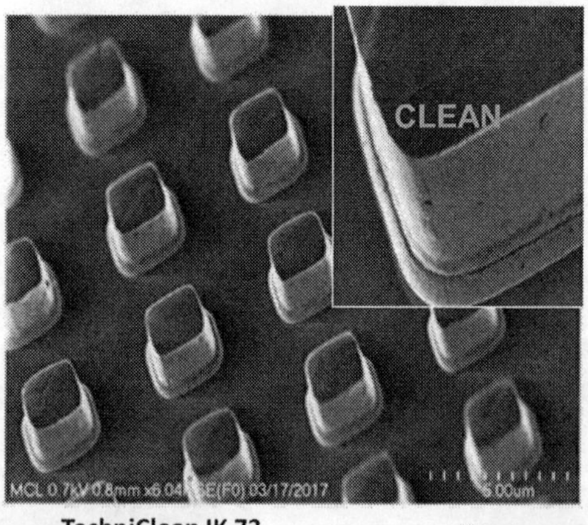

TechniClean IK 73
1min30 @ 20C

TECHNIC

TechniClean IK 73

MIM capacitors Ta_2O_5 high K

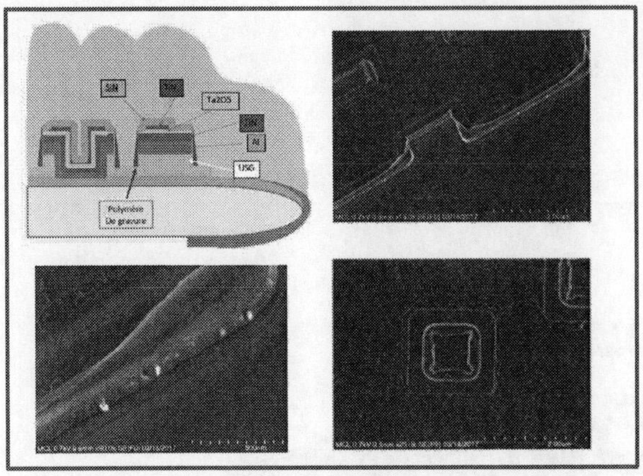

After Dry etch

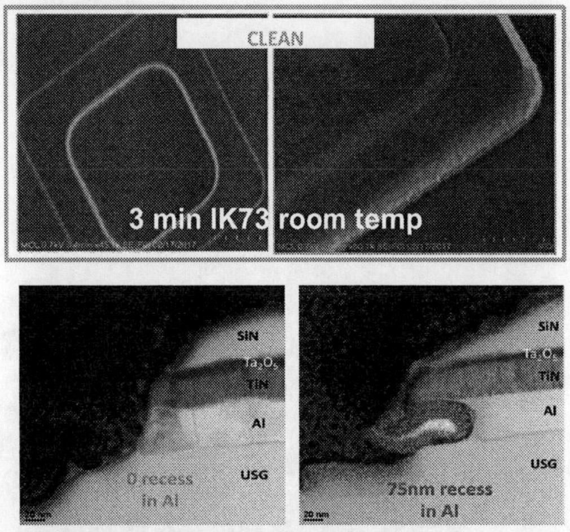

3 min IK 73 room temp | HF BASED POR

TechniClean IK 73

Validated on ZrO_2, HfO_2, Ta_2O_5

TECHNIC

TechniClean IK 73

20

FEOL/MEOL Manufacturing Compatible MRE

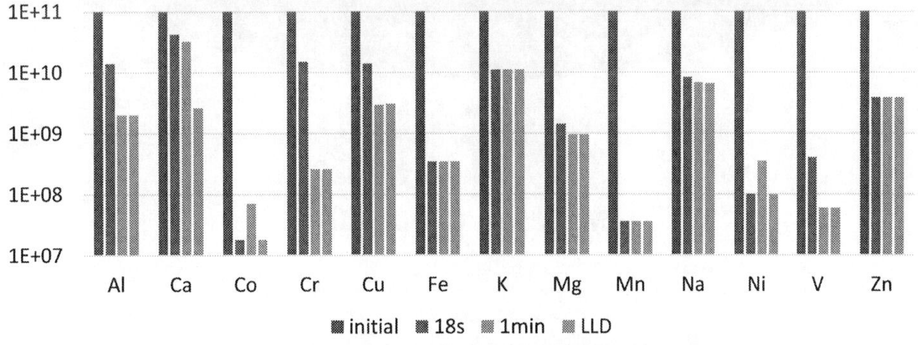

Surface Cleanliness level (TXRF)
- No metal deposition on bare Si → excellent complexation
  - Surface metal cleaning properties
- Aligned with manufacturing specifications

TECHNIC

Conclusion

- A ready-to-use solution for a large variety of PER
  - MIM cap with High K (Ta, Hf, Zr,…)
- Easy to work with, reclaimable, RT, water rinse, not flammable, not CMR…
- Complementary investigations on Al/Cu, Cu, Si P-RIE

life.augmented

Dr. L. Broussous, P. Garnier

TECHNIC

M. Cazes, C. Pizzetti

leti

L. Gabette, P. Besson

DRY REMOVAL OF A SURFACE FUNCTIONALIZATION CHEMISTRY USED FOR PATTERN COLLAPSE PREVENTION

GUY VEREECKE[1]*, KANA KOMORI[2], YUTA NAKANO[2], FARID SEBAAI[1], XIUMEI XU[1], YUSUKE ONIKI[1], AND FRANK HOLSTEYNS[1]

[1]IMEC, KAPELDREEF 75, 3001 LEUVEN, BELGIUM *guy.vereecke@imec.be

[2]SCREEN SEMICONDUCTOR SOLUTIONS CO., LTD., JAPAN

PUBLIC

INTRODUCTION

2

DOMINANT FORCES IN PATTERN COLLAPSE FORMATION

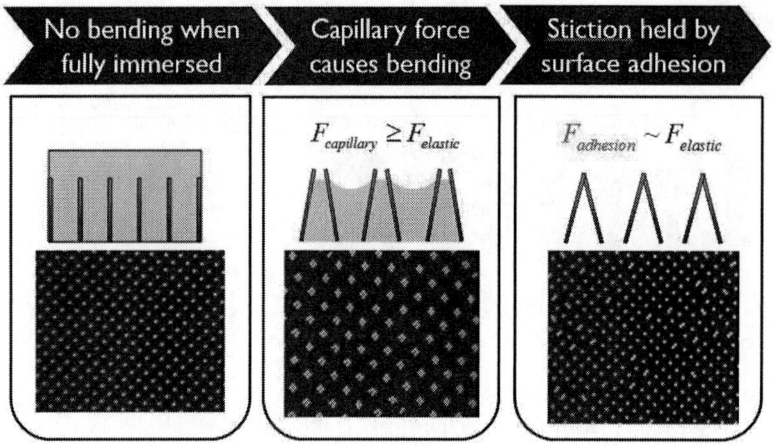

- Pattern collapse occurs at the end of drying phase
- Different forces dominate at different stages

Si micropillars:
- Pitch: 2 um;
- D: 0.8 um;
- H: 45 um;

In-situ observation by optical microscopy

- The elastic restoration force determines the requirements for preventive measures

imec

PREVENTION OF PATTERN COLLAPSE

- Reduce capillary force
  - Low surface tension liquid: e.g. IPA drying, but max 3x reduction in force
  - Surface modifications: effective, capillary force can be reduced more than 10x
- Reduce adhesion forces using organic coatings

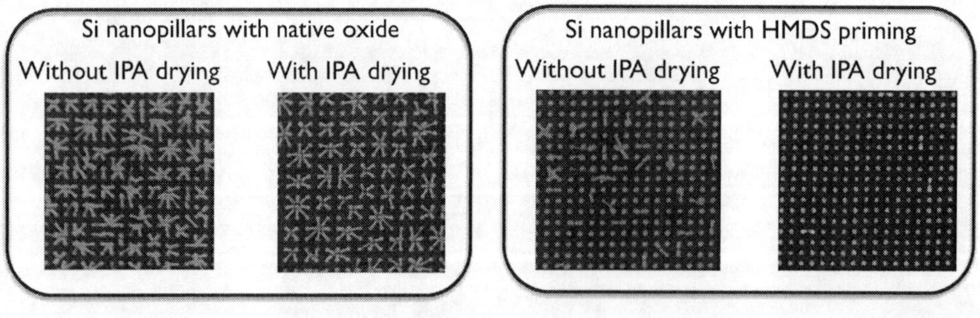

imec

EFFICIENCY OF SURFACE FUNCTIONALIZATION CHEMISTRY IN FIN POST-ETCH CLEAN

CLEAN / RINSE / SFC / RINSE / SPIN-DRY

- SFC → contact angle increased to 85±2°
- Pattern collapse prevention on fins with CD ~ 9 nm and pitch = 25 nm

Si fin height: 120 nm, total AR ~ 15

Si fin height: 160 nm, total AR ~ 20

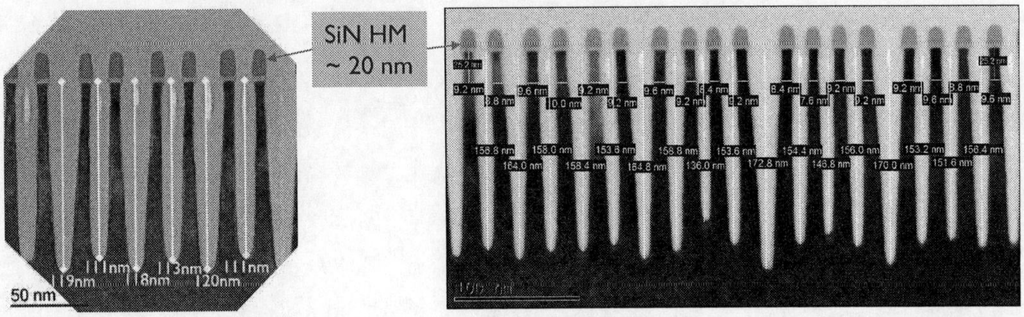

imec

5

NEED TO REMOVE SFC

POSSIBLE LEAKAGE ISSUES AT DUMMY OX/FIN INTERFACE IF SFC NOT REMOVED

Core	I/O
SRAM: leakage current though STI/Fin interface → N/P isolation fail	Dummy oxide used as gate oxide → high Dit and Jg expected

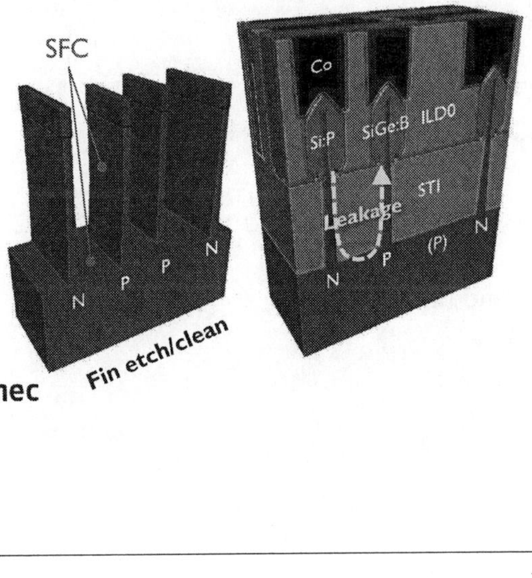

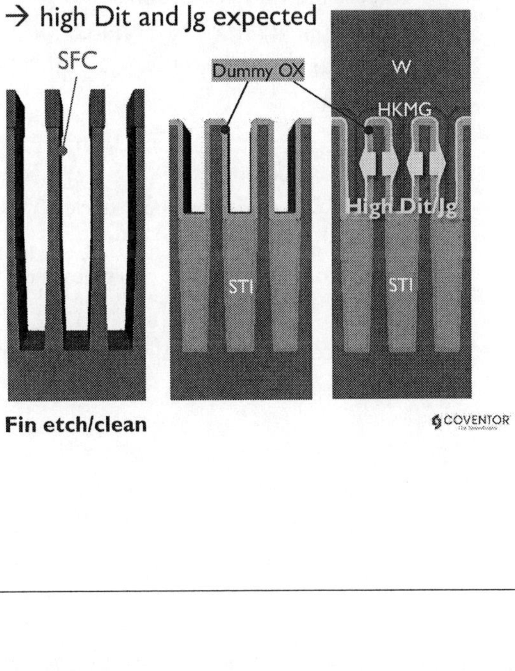

imec

RESULTS & DISCUSSION

SFC REMOVAL METHODS TESTED

- Microwave plasma strip
  - Axcelis Integra MW chamber
  - Chemistry: N_2/H_2 or $N_2/H_2/N_2O$
  - Chuck temperature: 245 / 200 / 150°C
  - Isotropic
- UV strip
  - VUV (λ < 200 nm)
    - SCREEN coater developer track, SOKUDO DUO
  - Atmosphere: N_2
  - Irradiation not isotropic → removal on fin sidewall?

Covalently bounded organic layers on Si are resistant to FEOL thermal budget

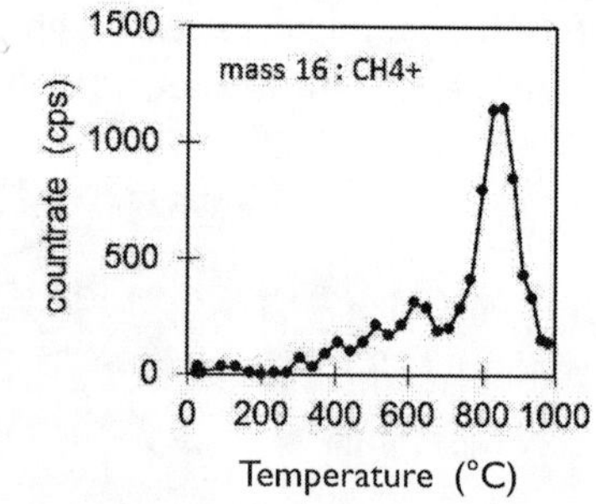

- Boundary conditions: ideally no oxide growth and no roughening

8

CHARACTERIZATION OF SFC REMOVAL

- Water contact angle

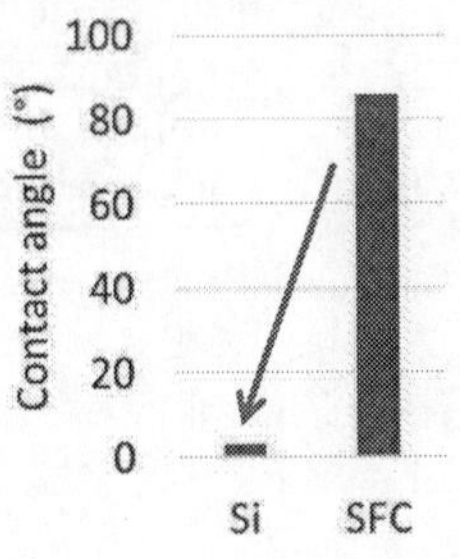

Water contact angle very sensitive to the removal of organic surface layers

- TOF-SIMS not applicable

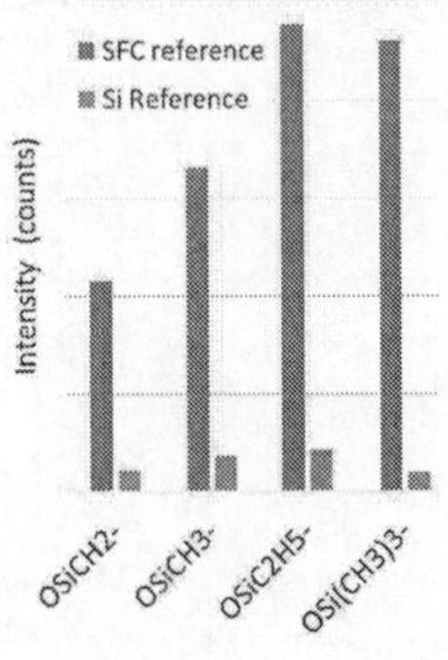

Background too high for accurate determination of removal efficiency

- GATR-FTIR

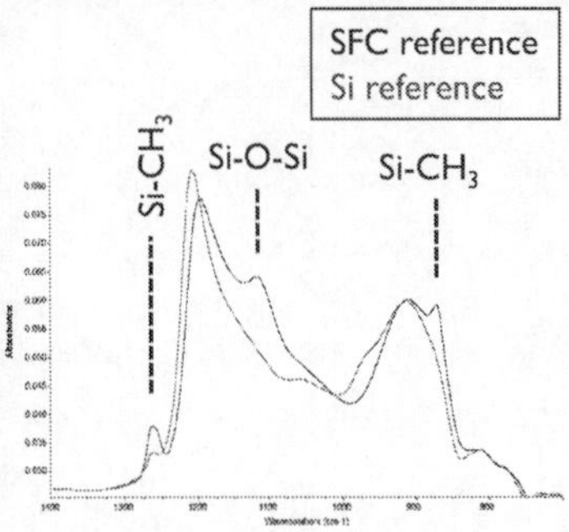

Detection of characteristic SFC-related peaks

imec

SFC REMOVAL BY **REDUCING PLASMA STRIP**

- Contact angle

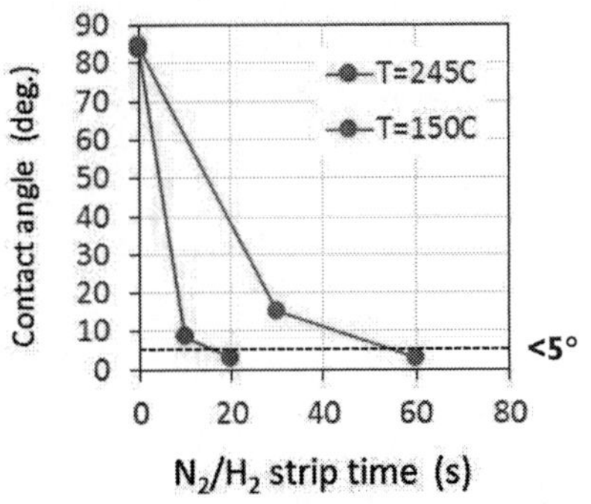

- Longer strip time needed at lower chuck temperature

- GATR-FTIR

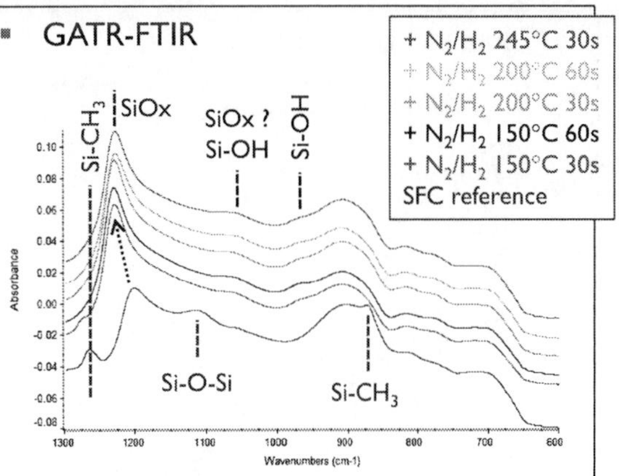

- Disappearance of SFC peaks in agreement with contact angle
- Plasma strip accompanied by oxide growth

imec

SFC REMOVAL BY **SLIGHTLY OXIDIZING PLASMA STRIP**

- Contact angle < 5° in all conditions
- Complete removal in all conditions tested

- GATR-FTIR

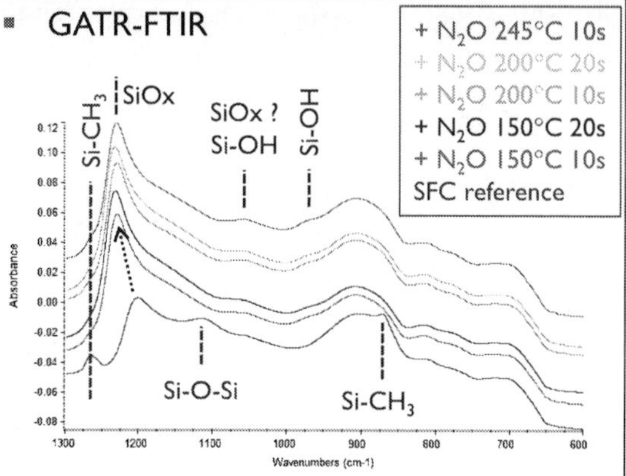

- Disappearance of SFC peaks in agreement with contact angle
- Plasma strip accompanied by oxide growth

SFC REMOVAL BY **UV STRIP**

- Contact angle

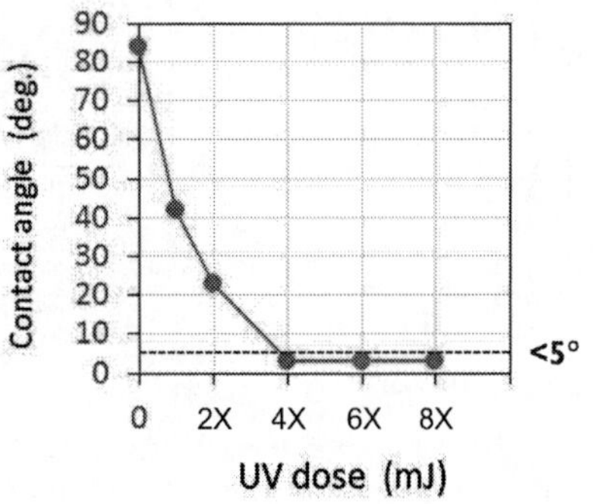

- Apparent complete removal with dose ≥ 4X

- GATR-FTIR

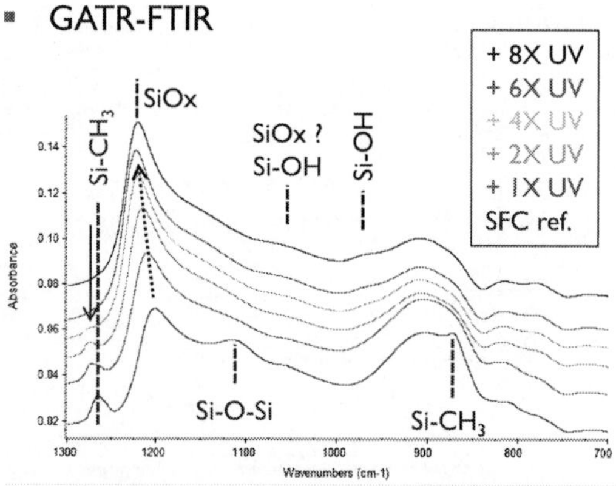

- Disappearance of SFC peaks in agreement with contact angle, but for a small residual peak at 1275 cm^{-1} at a dose of 4X
- UV strip accompanied by oxide growth

SURFACE ROUGHENING BY SFC STRIPS

- Plasma strips

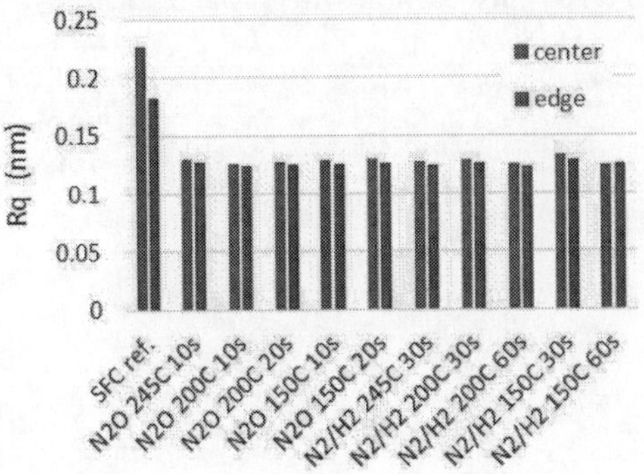

- UV strip

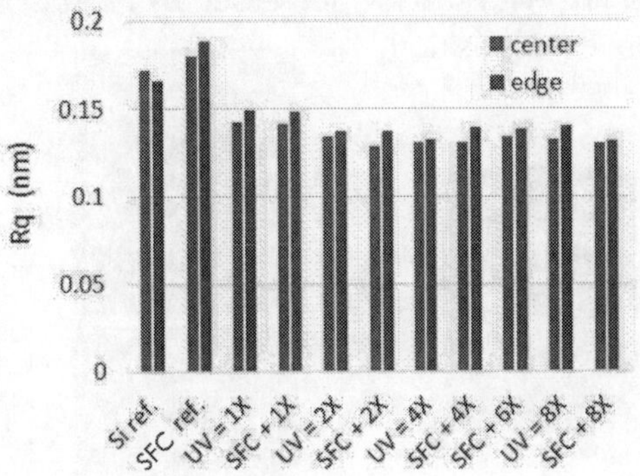

- No significant roughening by strips

imec

13

STRIPS DIFFERENTIATOR : OXIDE GROWTH

- Plasma strips

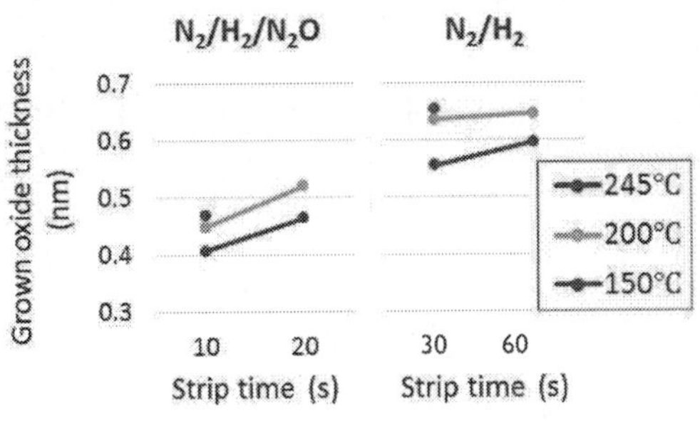

- Limited impact of lower T chuck

- UV strip

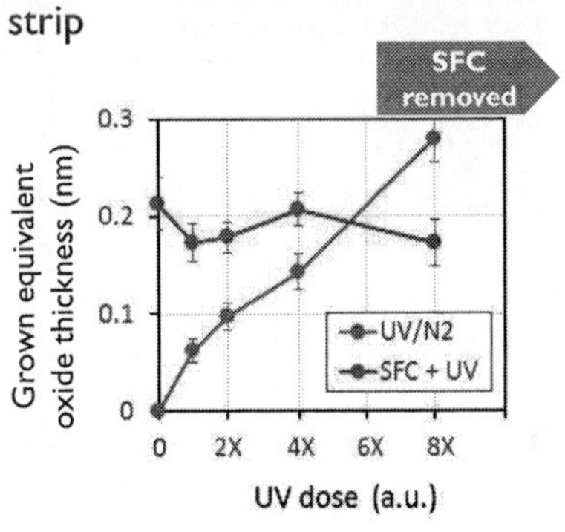

- SFC stripping lowers/delays Si oxidation

- Oxide growth by UV (~0.2nm) < oxidative strip (0.4-0.5nm) < N_2/H_2 strip (0.6-0.7nm)

imec

14

EFFICIENCY OF UV REMOVAL ON FINS
MATERIALS & METHODS

- Si fins with spacing decreased by PEALD oxide deposition

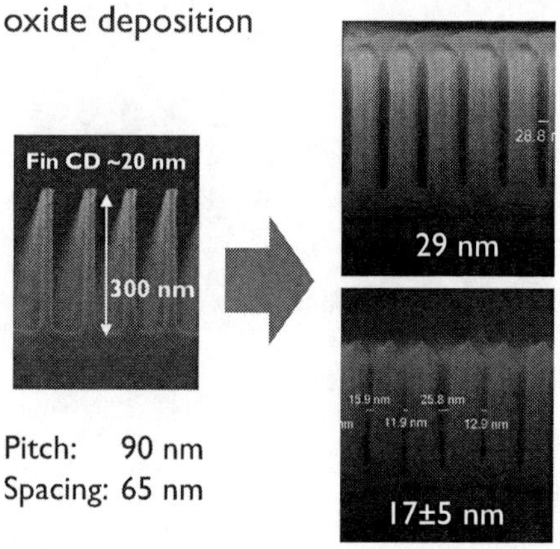

Pitch: 90 nm
Spacing: 65 nm

- GATR-FTIR on fins with polarization

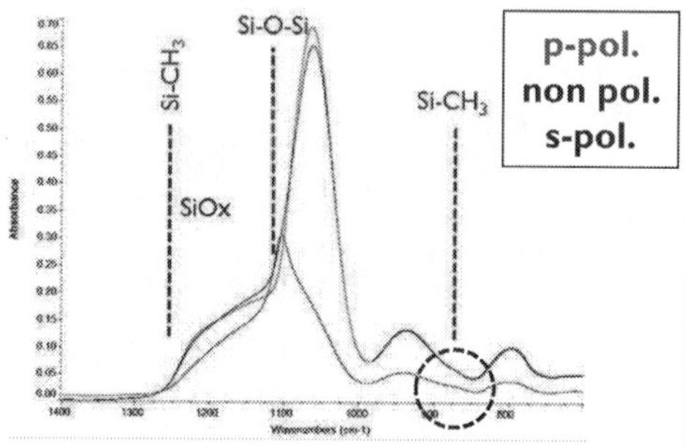

- Highest sensitivity for the SFC peak at 870 cm^{-1} with **s-polarization**
- Other SFC peaks not detected

imec

15

EFFICIENCY OF UV/N$_2$ REMOVAL ON FINS

17 nm spacing	29 nm spacing

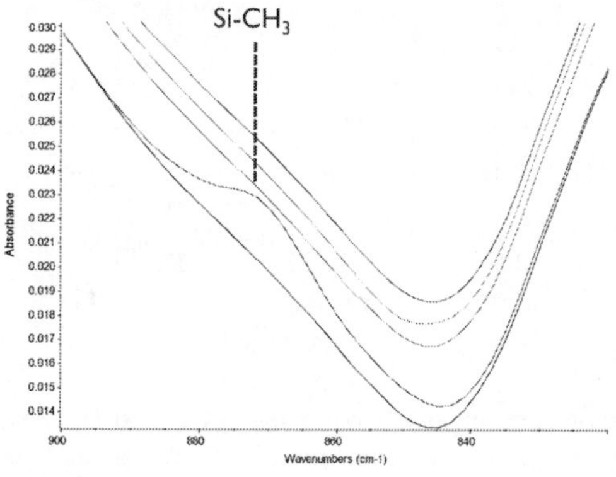

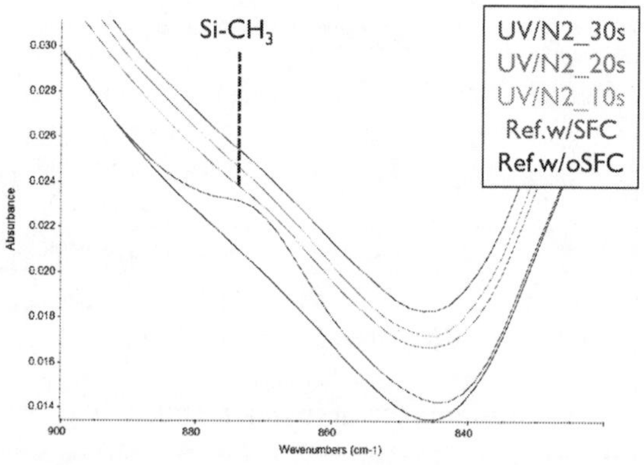

- Complete removal in all spacings with 10 s irradiation

imec

16

SUMMARY

- Efficient removal of SFC by plasma and UV strips
    - $N_2/H_2/N_2O$ strip ~ UV/N_2 strip > N_2/H_2 strip
- No significant surface roughening observed
- UV/N_2 strip outperforms plasma strips wrt Si surface oxidation
- UV/N_2 strip can remove SFC on fins sidewall

imec 17

imec

embracing a better life

imec PUBLIC

Cleaning Surfaces from Nanoparticles with Polymer Film: Impact of the Polymer Stripping

A. LALLART [1,2,3,4], P. GARNIER [1], E. LORENCEAU [2], A. CARTELLIER [3], E. CHARLAIX [2]

[1] STMicroelectronics, Crolles, France
[2] LIPHY, Université Grenoble Alpes, Saint Martin d'Hères, France
[3] LEGI, Université Grenoble Alpes, Saint Martin d'Hères, France
[4] CEA Leti, Grenoble, France

life.augmented

Outline

- Motivations
- Experimental setup
- Results
- Hypothesis
- Conclusion

life.augmented

Miniaturization & particle impact on yield

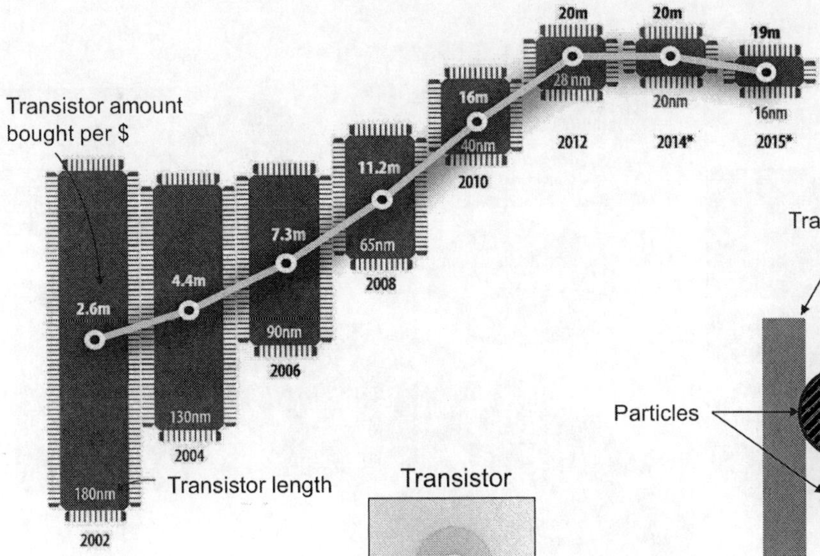

www.economist.com/news

Transistor

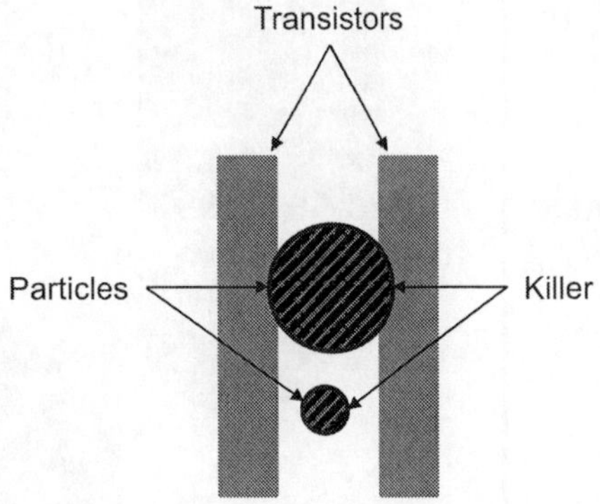

Miniaturization ➔ Critical particle size decrease

Challenge: Particle removal / No damage

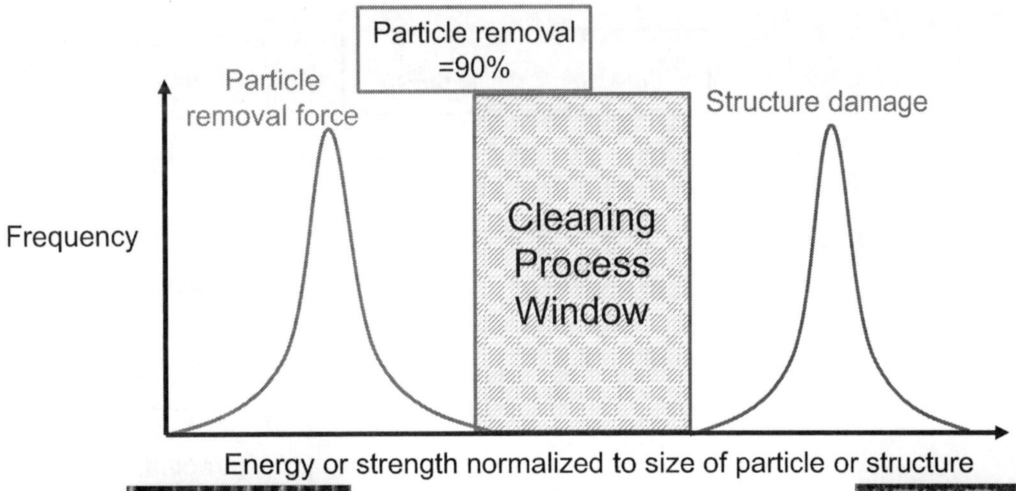

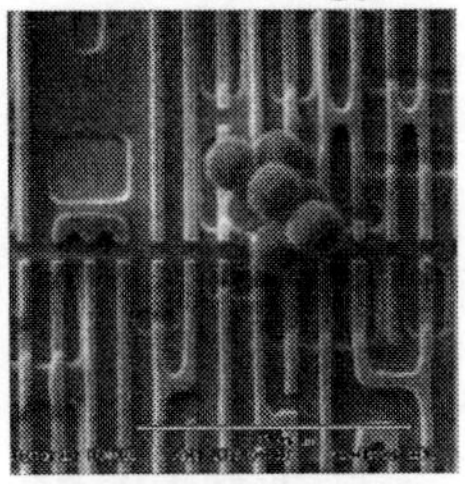

Particles on transistors

Structure damage

Challenge: Particle Removal / No Damage

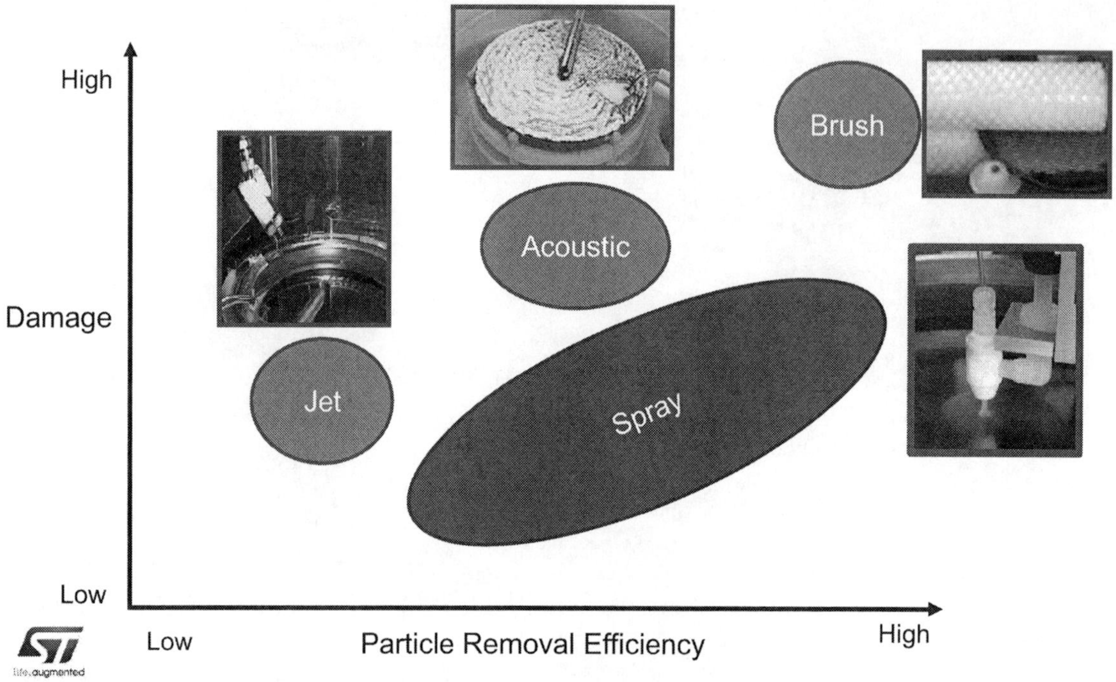

A. LALLART | Cleaning Surfaces from Nanoparticles with Polymer Film | SPCC 2018 Boston | April 10 -11

Particle Removal Condition

$$F_{adhesion} < F_{external}$$

Parameters influencing $F_{adhesion}$:

$$F_{adhesion}(\tau_{aging}) = \gamma d \frac{1}{\ln(\frac{P_{sat}}{P_v})} \ln(\frac{\tau_{aging}}{\tau_0})$$

Aging (τ_{aging})

System (γ, P_{sat}/P_v, τ_0)

Particle Characteristic (d)

γ liquid surface tension
d distance taking into acount the geometrical characteristics of the contact
P_{sat} saturated water pressure
P_v vapour pressure
τ_{aging} time since contamination
τ_0 time needed to condense one liquid layer

Bocquet, L., & *al*, Nature (1998)

Spray still efficient enough ? 7

$$F_{adhesion}(\tau_{aging}) = \gamma d \frac{1}{\ln(\frac{P_{sat}}{P_v})} \ln(\frac{\tau_{aging}}{\tau_0})$$

Bocquet, L., & *al*, Nature (1998)

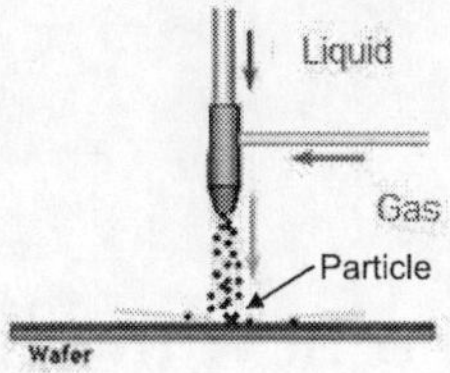

Spray cleaning efficiency on silicon wafer depending on particle size

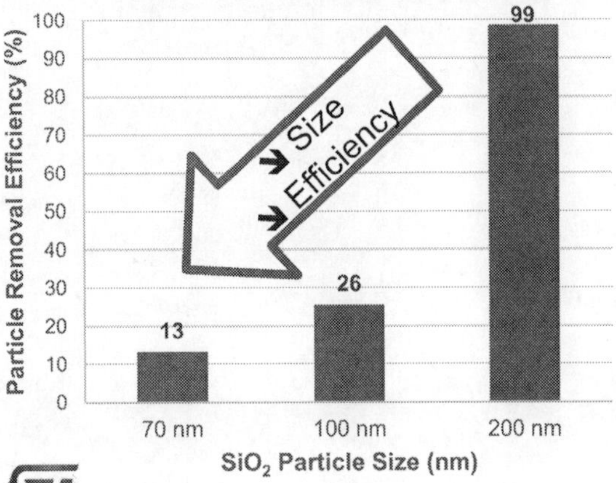

Aibara, M., & *al*, ECS conf. (2017)

Spray cleaning efficiency on silicon wafer depending on aging time

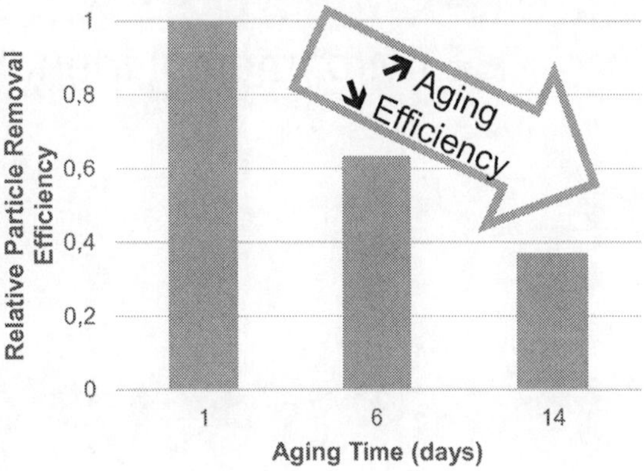

Spray still efficient enough ? 8

- Strong particle size dependence on removal efficiency
- Strong aging time dependence on removal efficiency
- Droplet size limitation

➔ New solution required

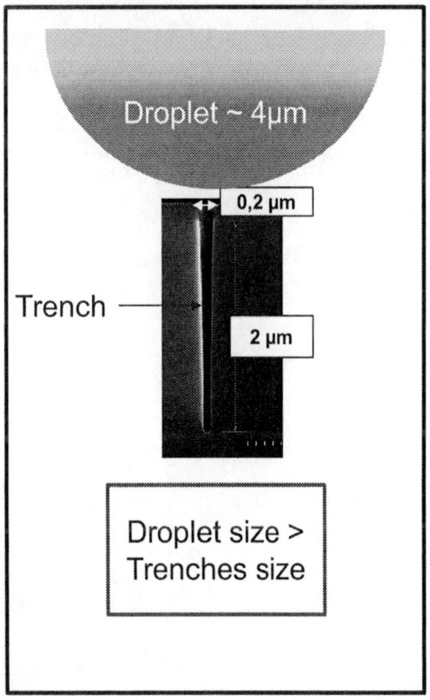

Motivations

Find a new method to remove nanoparticles :

- Without structure damage
- Environmental friendly

New solution based on polymer removal

$$F_{adhesion}^{\ polymer/particle} > F_{adhesion}^{\ surface/particle}$$

$$\text{Contact area}^{\ polymer/particle} > \text{Contact area}^{\ surface/particle}$$

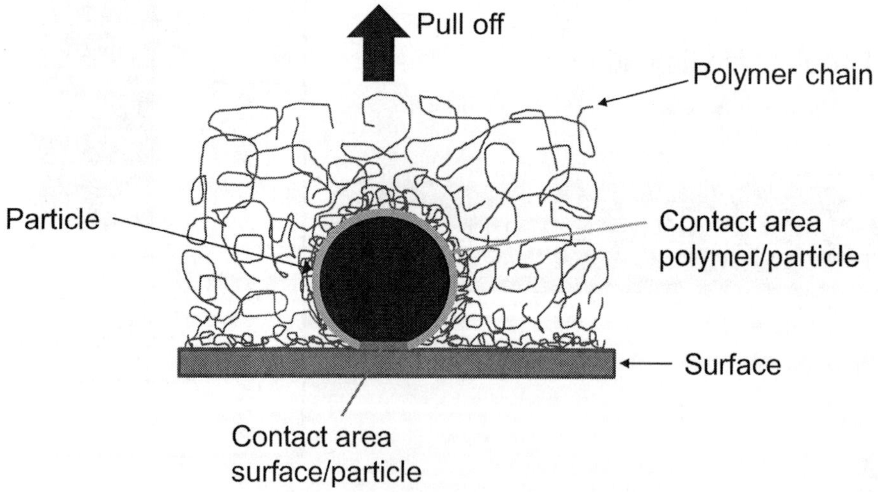

2 postulated ways to remove the polymer

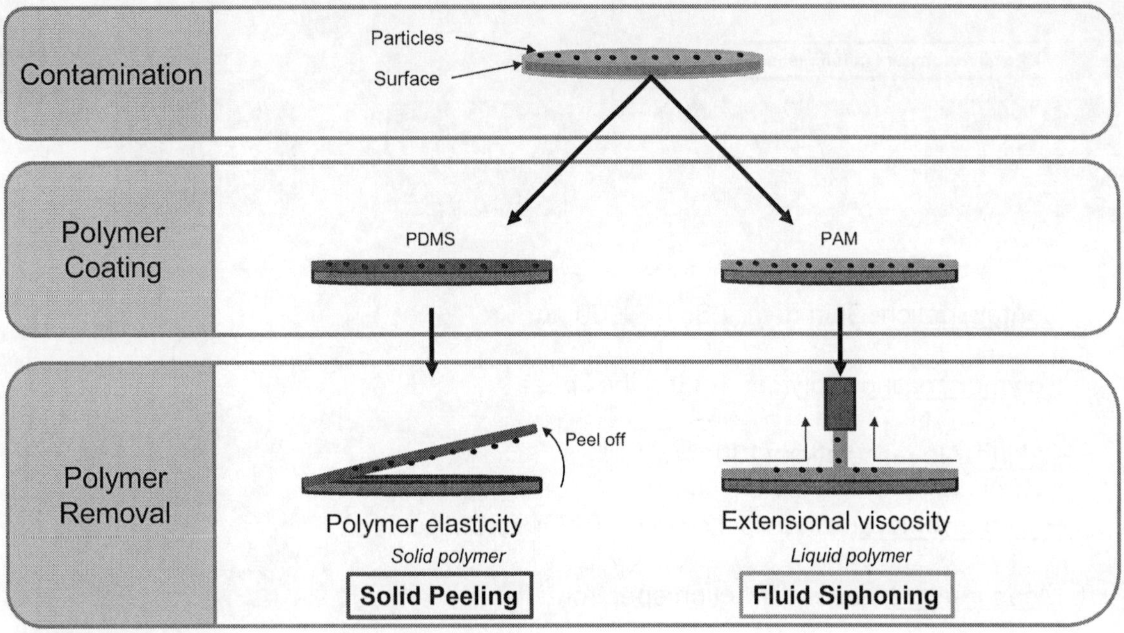

Lin, P., & *al*, Nanotechnology (2013)

Walker, T.W., & *al*, JoR (2014)

Outline

- Motivations
- Experimental setup
  - Process step
  - Polymer removal
- Results
- Hypothesis
- Conclusion

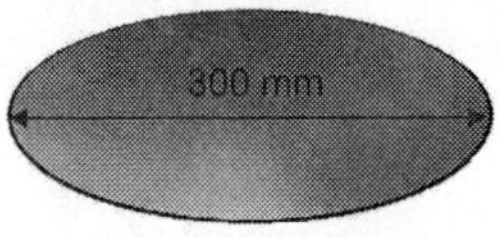

Silicon wafer

Method - Process Steps

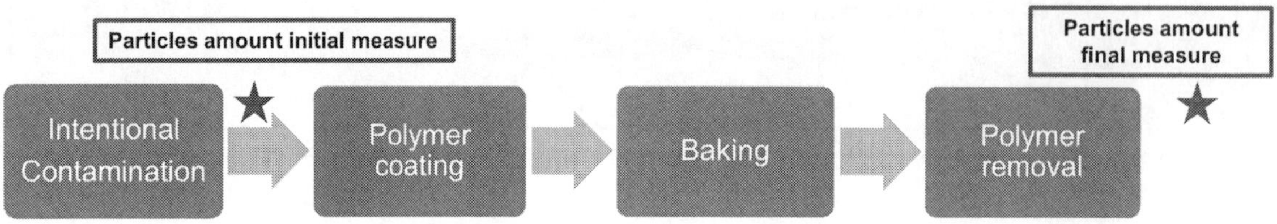

Contamination: Spin dryer / SiO_2 ➔ 60 nm

Polymer coating: Polymer 1,8 µm thickness

Baking: No reticulation (130°C)

Polymer removal: Chemical solution (SPM / SC1)

Measurement: Laser diffraction spectrometer

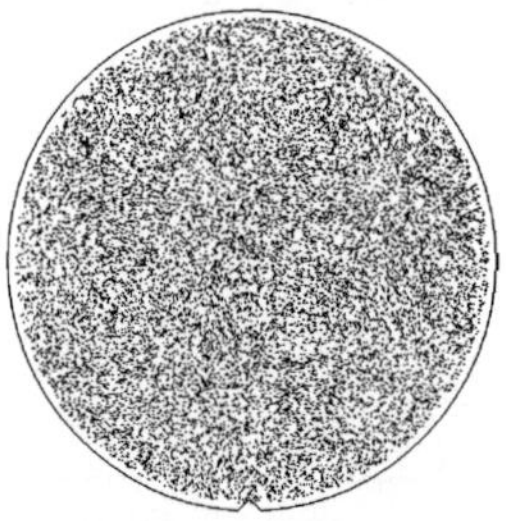

Wafer defect mapping – Initial contamination

life.augmented

Method – Polymer Removal

Wet Bench

Chemical removal only

Spray Batch

Chemical & **Physical** removal

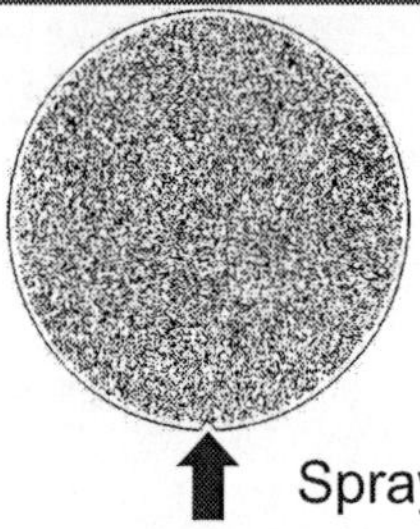

Spray direction

Side view

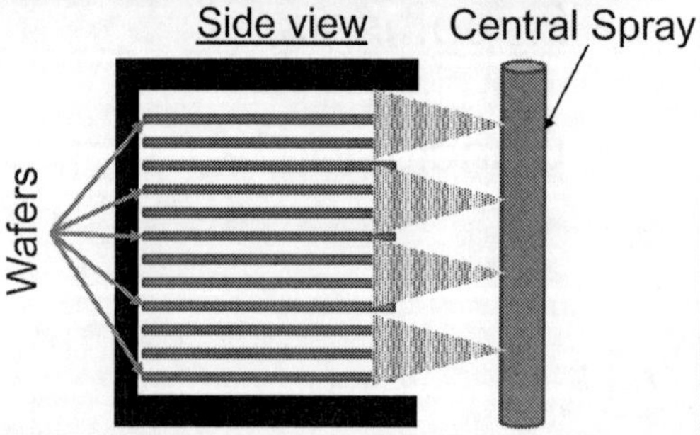

life.augmented

Outline

- Motivations
- Experimental setup
- Results

Polymer removal:
- Wet Bench vs Spray Batch

Polymer coating

Cleaning efficiency

Surface

Contamination:
- Particle Nature (SiO_2, Si_3N_4,…)
- Particle Size
- Aging time

- Hypothesis
- Conclusion

life.augmented

Wet Bench versus Spray Batch

	Polymer	
	Chemical solution (SPM / SC1)	
	Wet Bench	Spray Batch
Initial Particles Mapping	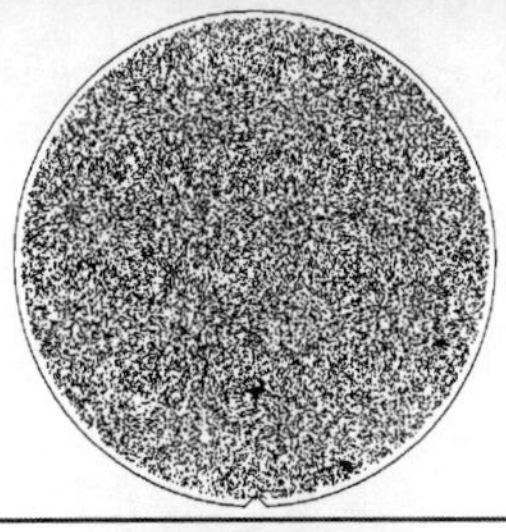	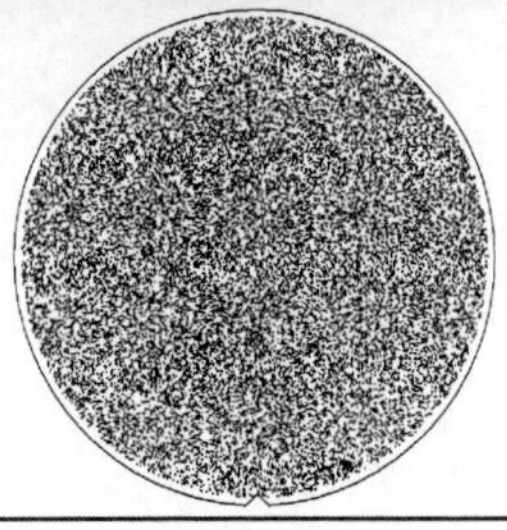
Final Particles Mapping	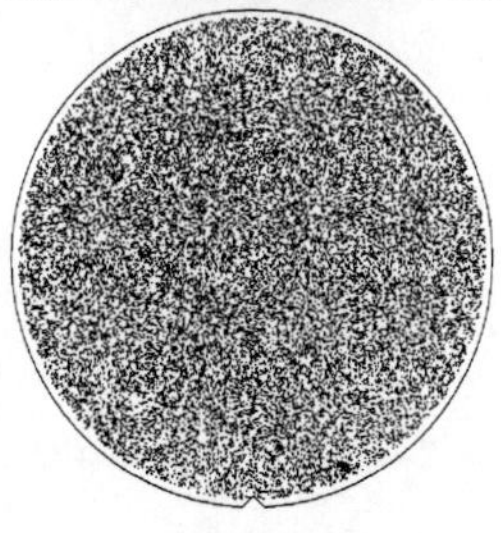	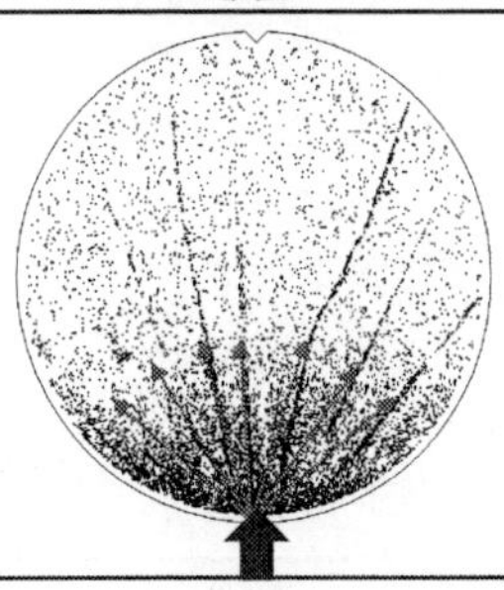
Removal efficiency (%)	0 %	87 %

- ➢ Physical action mandatory
- ➢ Non uniform removal
  Particles remain along the spray flow lines

Polymer Presence 17

	No Polymer	Polymer
	Chemical solution (SPM / SC1) - Spray Batch	
Initial Particles Mapping	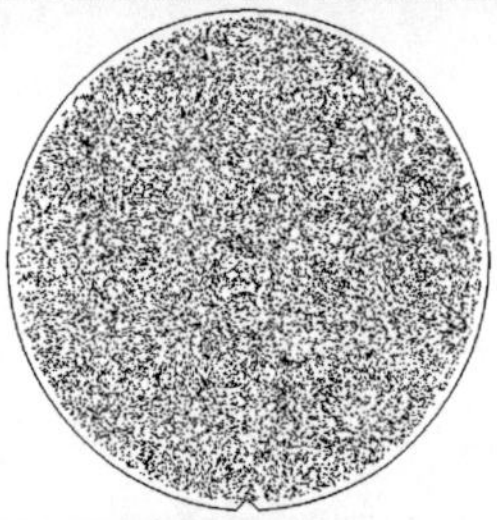	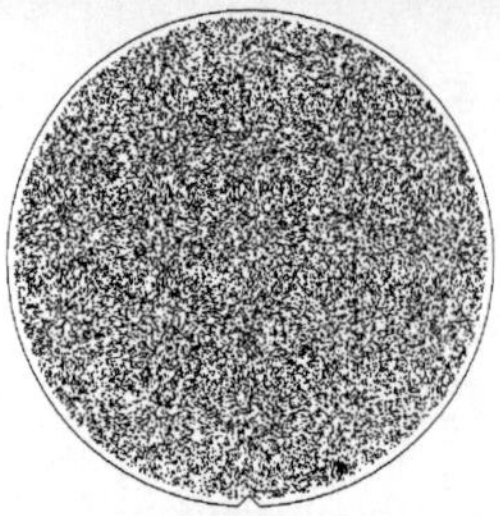
Final Particles Mapping	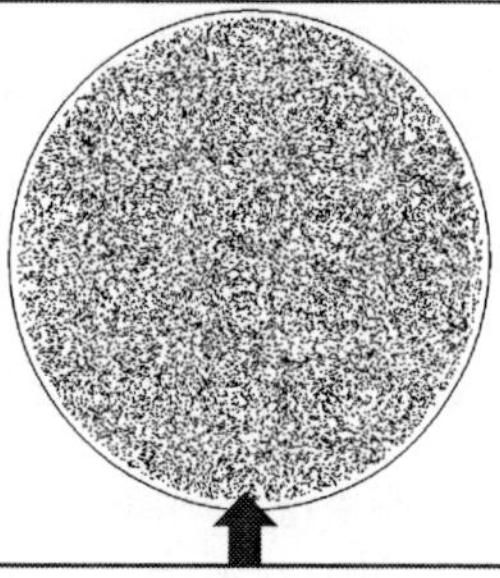	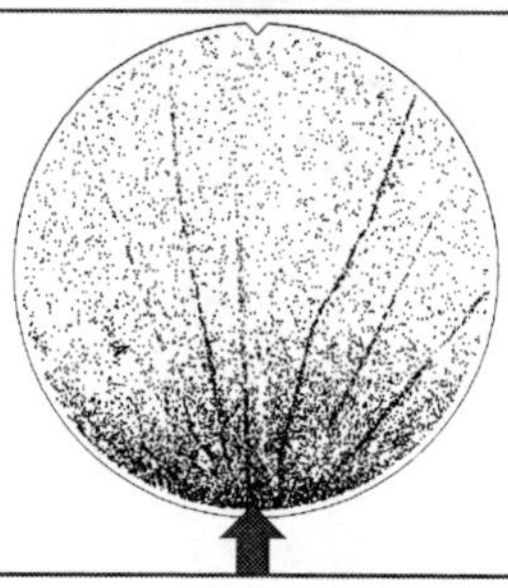
Removal efficiency (%)	0 %	87 %

- Polymer needed for particles removal

A. LALLART | Cleaning Surfaces from Nanoparticles with Polymer Film | SPCC 2018 Boston | April 10 -11

Aging Time 18

SiO_2 Monodisperse Contamination → Polymer Coating Baking → Polymer Removal Chemical Solution (SPM / SC1) – Spray Batch

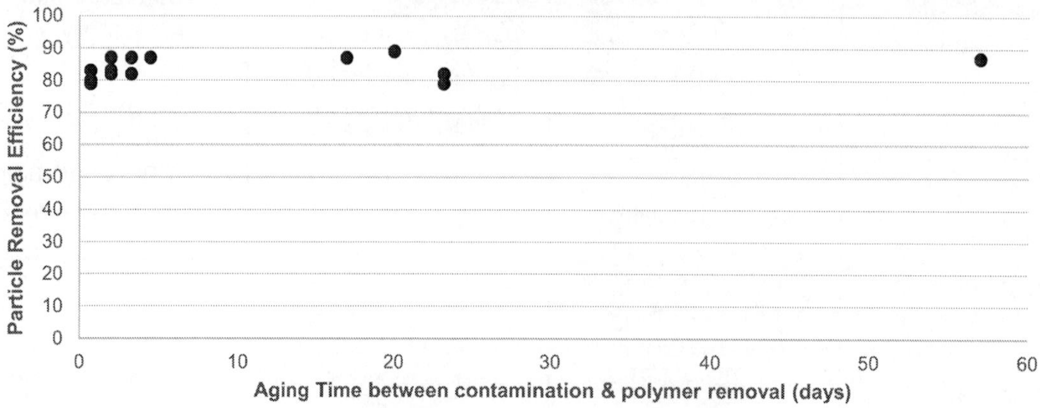

- No aging time dependence on Particle Removal Efficiency

life.augmented

Particles Size & Nature

19

	SiO_2 monodisperse	Si_3N_4 polydisperse
	Polymer	
	Chemical solution (SPM / SC1) - Spray Batch	
Initial Particles Mapping	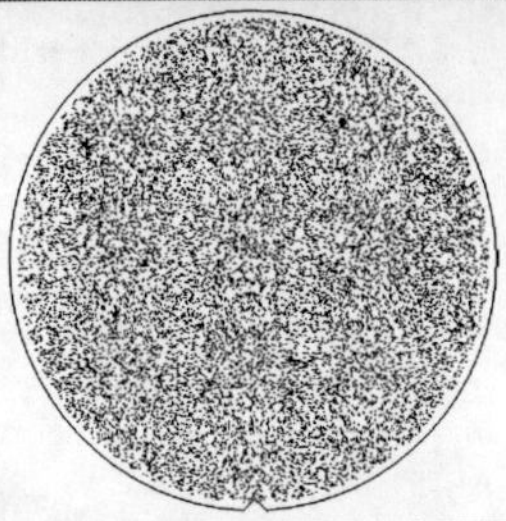	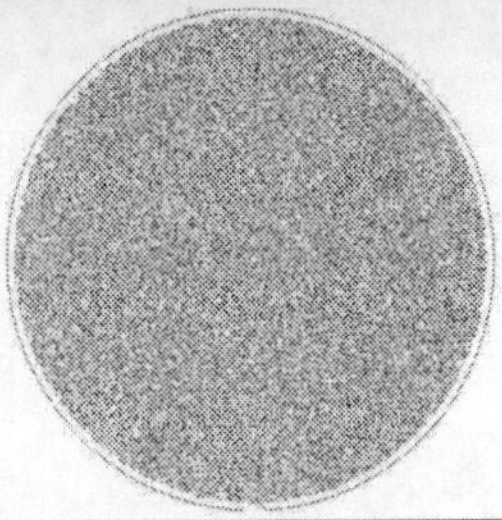
Final Particles Mapping	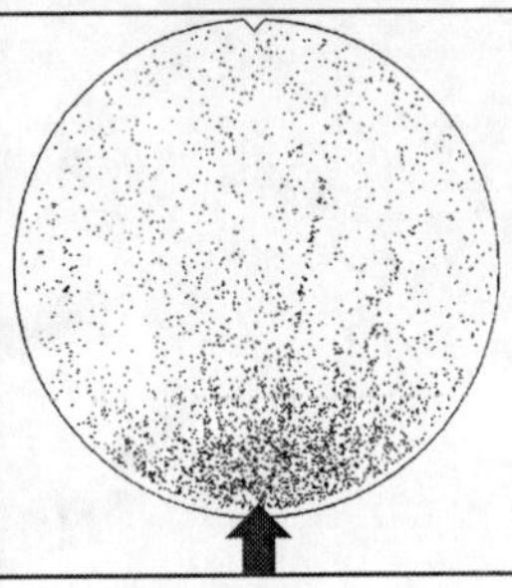	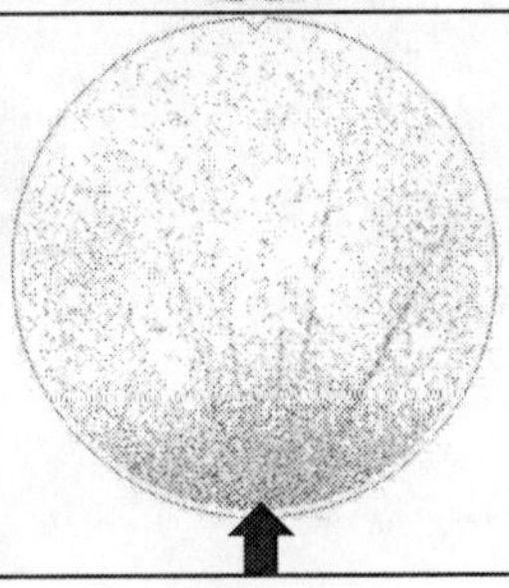
Removal efficiency (%)	85 %	85 %

➢ Efficient removal for both particles nature

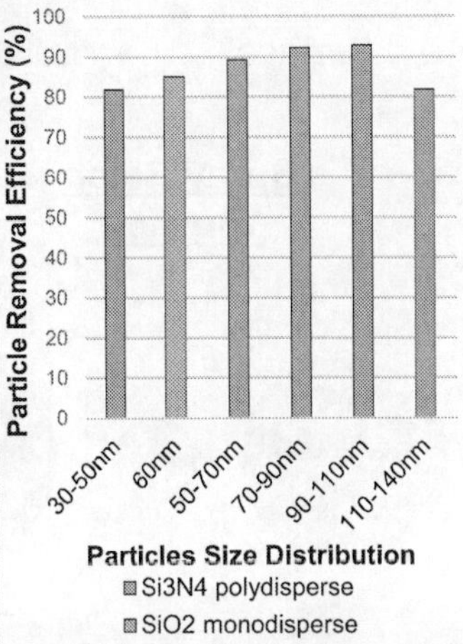

➢ No size influence on Removal Efficiency

A. LALLART | Cleaning Surfaces from Nanoparticles with Polymer Film | SPCC 2018 Boston | April 10 -11

Summary

20

- **<u>No</u> aging time** dependence on removal efficiency
- **<u>No</u> particle nature** dependence on removal efficiency
- **<u>No</u> particle size** dependence on removal efficiency

➔ No influence of the $F_{adhesion}$ ➚ on removal efficiency

- Hypothesis ?

life.augmented

Hypothesis : Peeling Zero Degrees

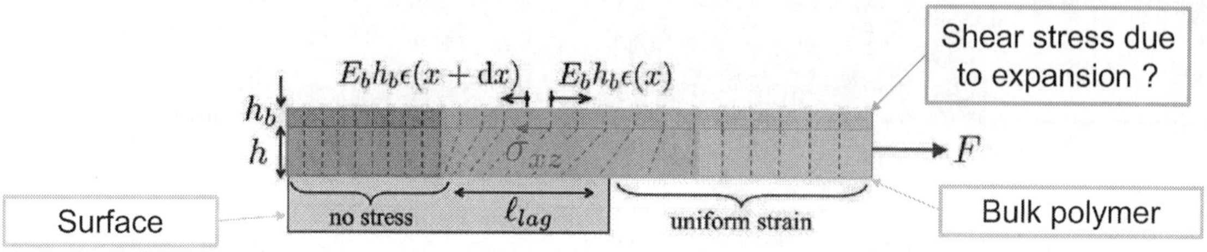

E_b Young Modulus of the hard layer

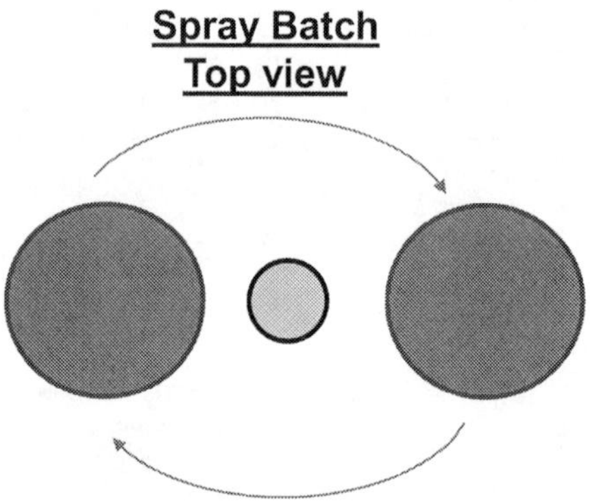

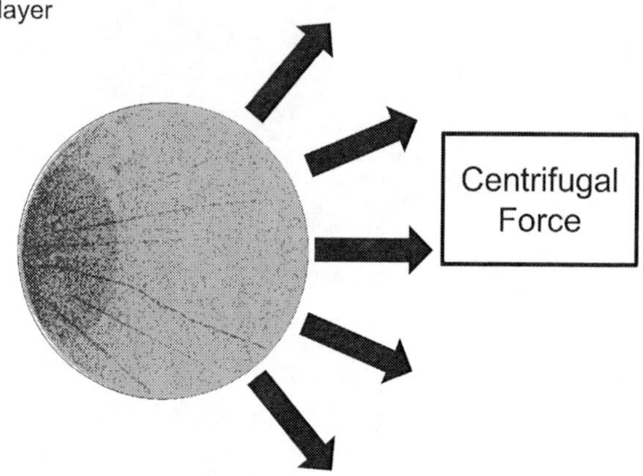

Rotation 80 RPM

Ponce, S., & *al*, Soft Matter (2015)

A. LALLART | Cleaning Surfaces from Nanoparticles with Polymer Film | SPCC 2018 Boston | April 10 -11

Hypothesis : Peeling Zero Degrees

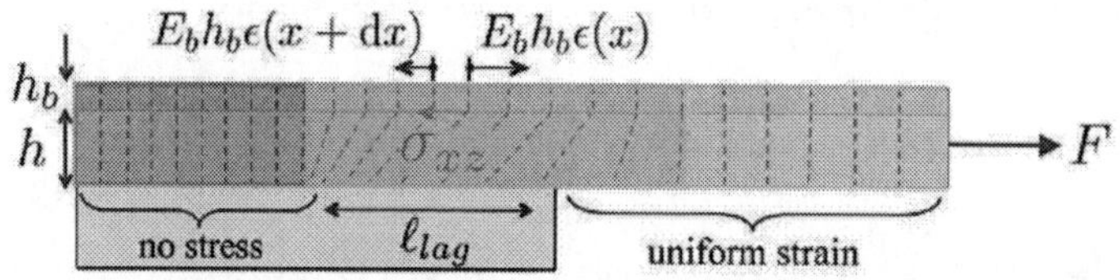

E_b Young Modulus of the hard layer

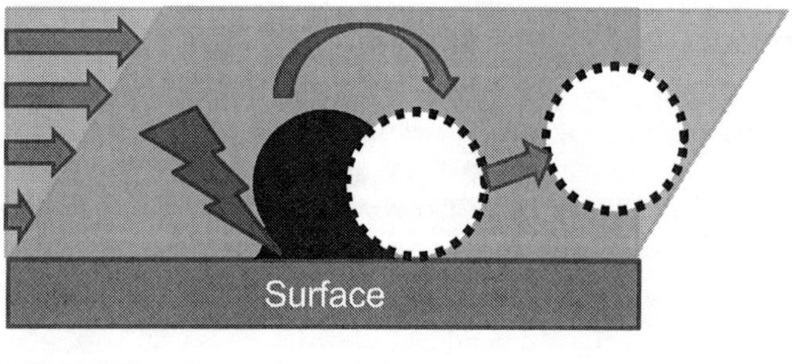

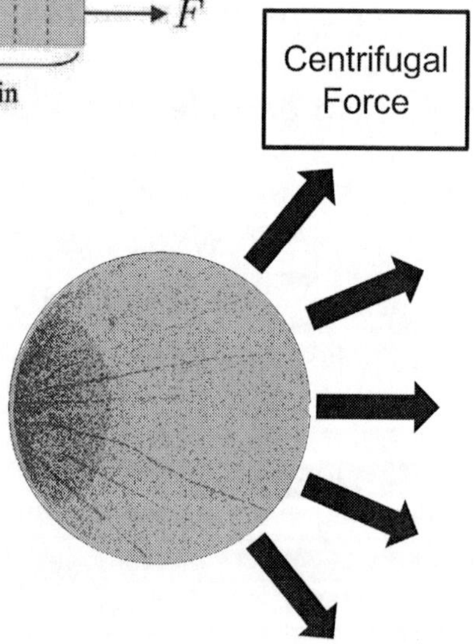

Ponce, S., & *al*, Soft Matter (2015)

Conclusions & Perspectives

Conclusions

- New solution to remove nanometric particles

➔ Aging independent

➔ Particle nature/size independent

➔ Without pattern damage

Perspectives

- To fully understand the mechanism of polymer removal
- Verify the peeling zero degrees hypothesis

➔ Adhesion & rheological studies

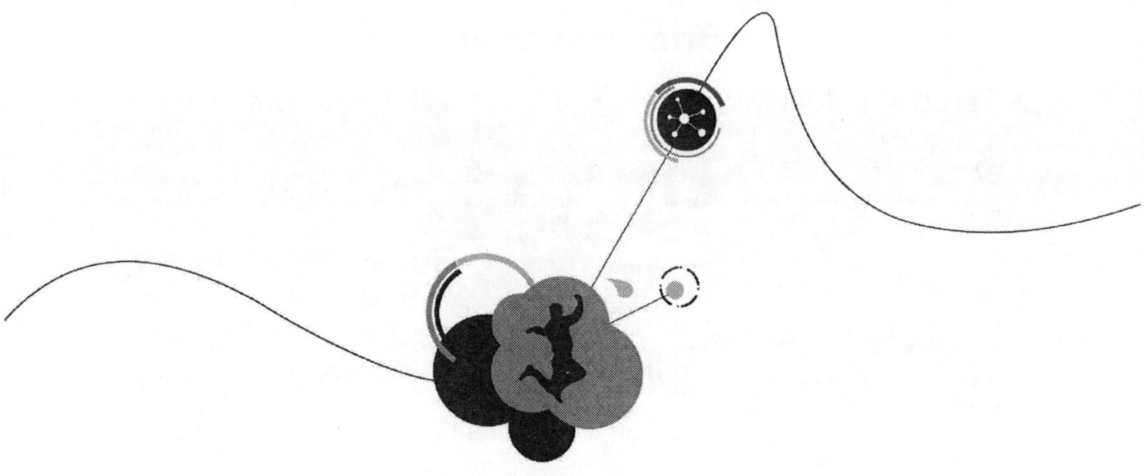

Thank you for your attention

TOSHIBA
Leading Innovation >>>

SPCC 2018 (2018 4/9,10,11 Cambridge, MA)

Thank you Mr. Chairman. Good morning, everybody.

Development of NIL-Template dry-cleaning process
- VUV cleaning technology -

Nobuyoshi Sato*, Masayuki Hatano, Takuya Kono, and Tetsuro Nakasugi

nobuyoshi.sato@toshiba.co.jp
Institute of Memory Technology Research & Development, Toshiba Memory Corporation
8, Shinsugita-Cho, Isogo-Ku, Yokohama 235-8522, Japan
+81-45-776-5882

Brief overview

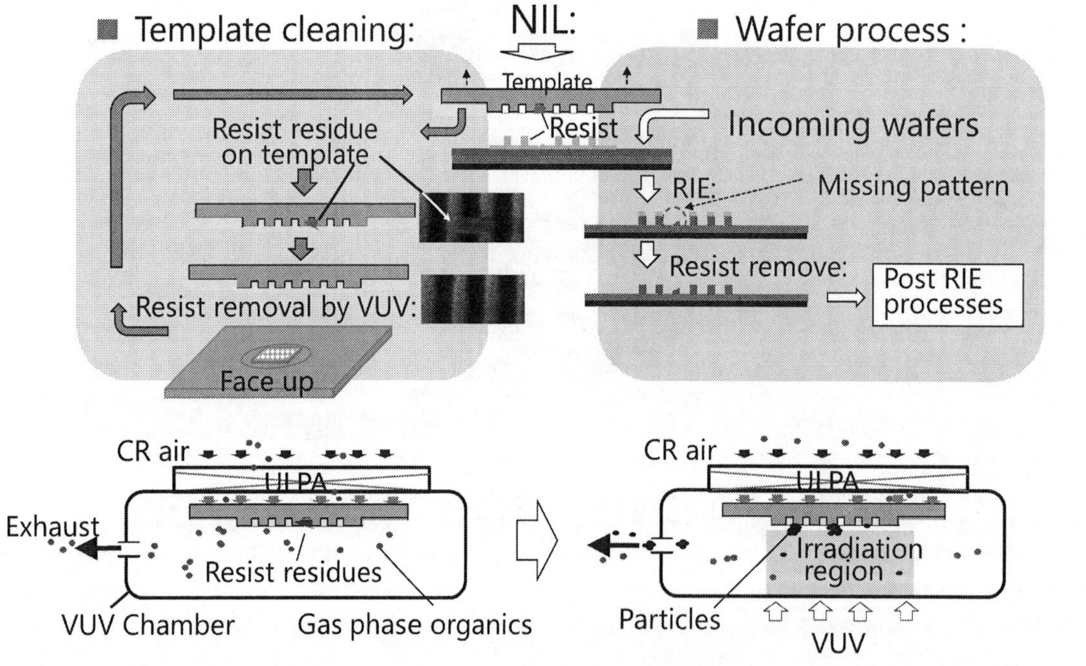

Particle adsorption onto NIL-Template by Gas-phase organics excited during VUV cleaning

Contents

- Introduction
  - Photolithography on Memory Device
  - Nano-Imprint Lithography (NIL) applications
  - Resist residue and its removal on NIL template
  - Vacuum Ultraviolet (VUV) cleaning performance
- Experiments and Results
  - Analysis of VUV damage and Particle on NIL template
  - Analysis of Organic contamination on NIL template
  - Model of particle adsorption on NIL template
- Summary

Conventional lithography vs. NIL on Memory Device

	Conventional lithography	Nano imprint lithography (NIL)
Cost	High	Low ☺
Process	Non-contact	Contact
Cleaning margin	Wide (Qz mask)	Narrow (Qz template) ☹

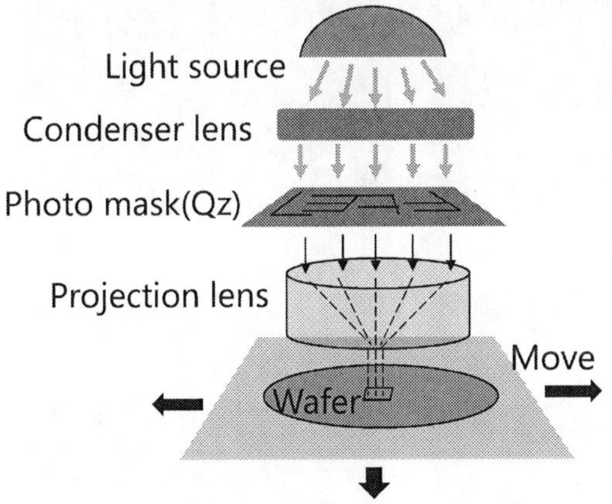

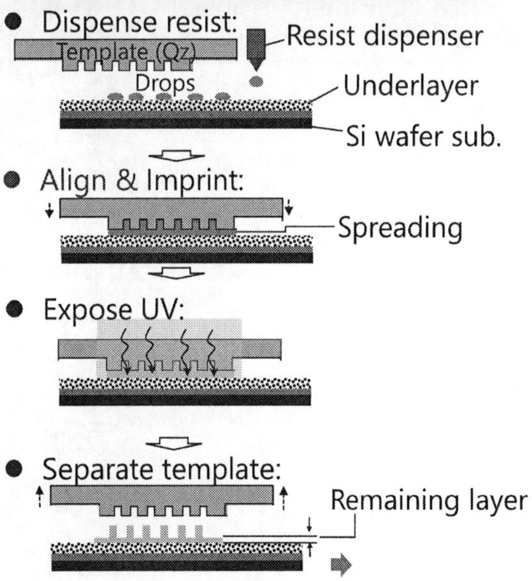

Photo mask vs. NIL Template on cleaning margin

- Conceptual diagram -

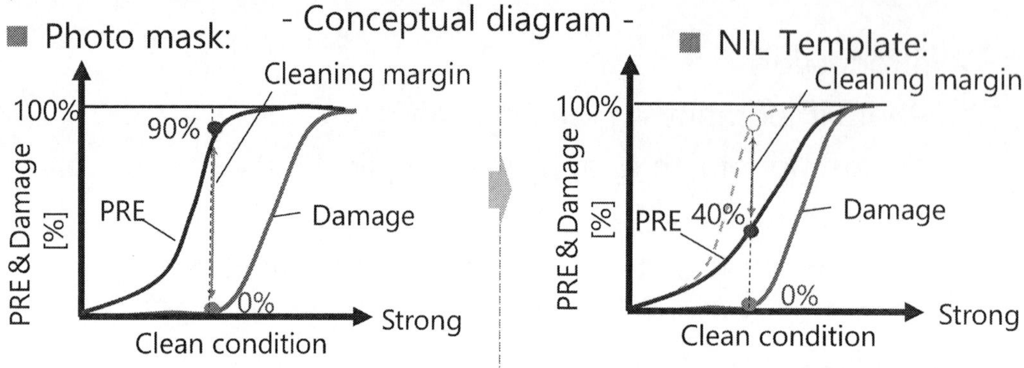

PRE: Particle removal efficiency

Damage: Physical/chemical Pattern damage

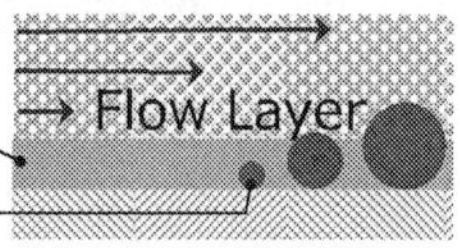

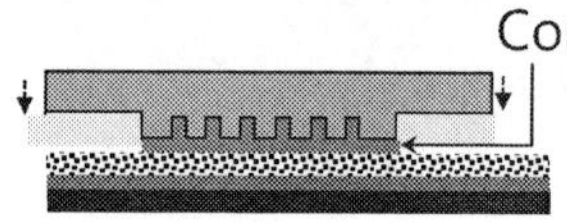

The cleaning margin on NIL template is getting much narrower due to contact process as well as smaller pattern size.

Various applications on NIL

Characteristics of NIL
- Direct Pattern Transfer
- High Resolution

Hp1x-nm line;

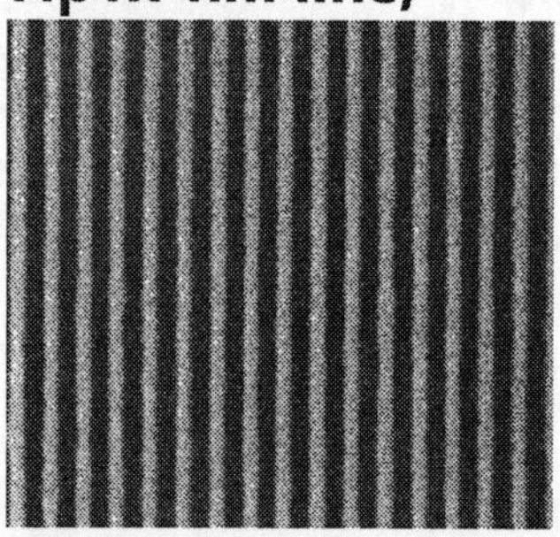

Diagonal line;

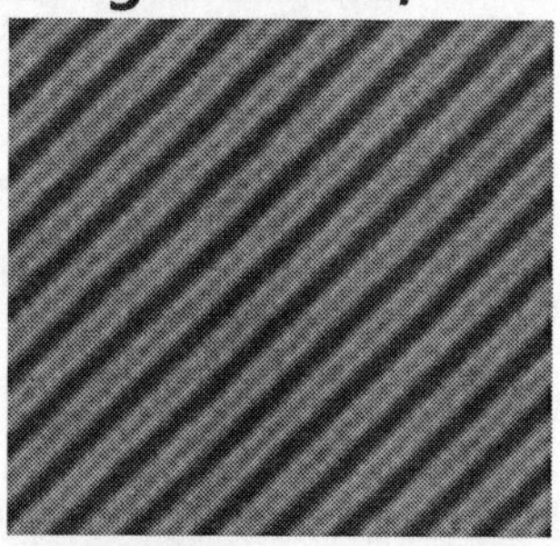

2x-nm hole;

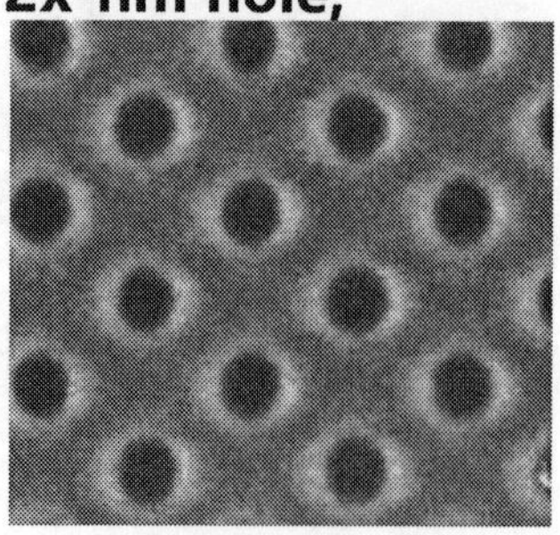

Pillar;

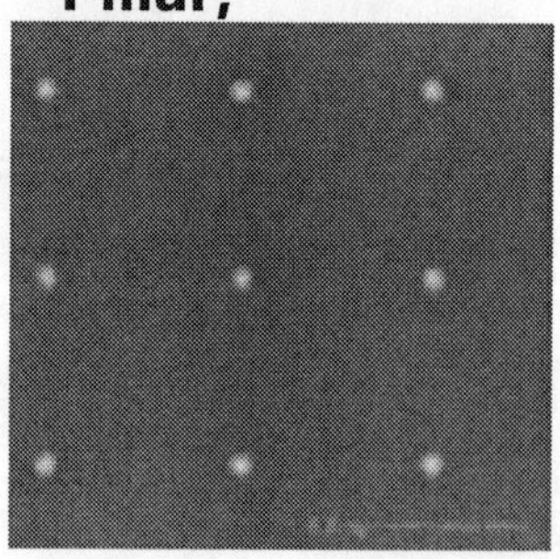

Resist residue & cleaning on NIL template

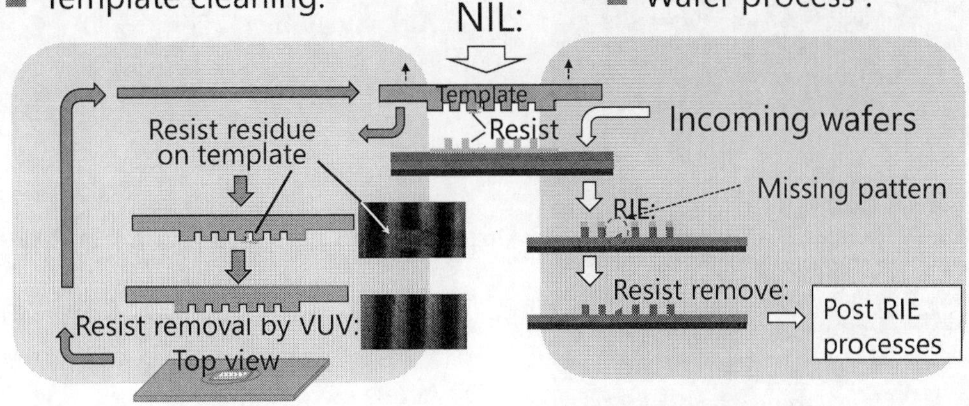

Resist residue happens on the template. It should be removed.

Cleaning processes for NIL template

Process step:	Removal material:	Throughput (a.u.):	Concerns:
APM (Wet) NH_4OH:H_2O_2	Inorganic	1.0	show these concerns Pattern damage
SPM (Wet) H_2SO_4:H_2O_2	Organics	1.0	Sulfur residue
SPM/AMP (Wet)	Organics/Inorganic	0.5	Pattern damage, Low throughput
DIO_3 (Wet)	Organics	1.5	Poor removability
O_2- Asher (Dry)	Organics	3.0	Metal mark damage
VUV-O_3 (Dry)	Organics	2.5	Metal mark damage?

SPM/APM clean is current process (throughput: 0.5)

→How does VUV remove organics (e.g. resist, carbon contamination)?

Photochemical reaction on VUV cleaning

■ Components of organics:

Components	Binding energy (kcal/mol)
C-C	84
C=C	141
C-H	98
C-F	115
C-N	64
C-O	76
C=O	190
O-O	118
O-H	109
Si-O	105
Si-C	70
Chemical bonds are disconnected by VUV.	

Hishinuma, et al., Current technology of VUV Excimer lamp, Kougaku, 30, 790 (2001).

■ Organics are removed as followings:

O_2 (in Air) + VUV hν (172nm) → O^* + O (^{3}P)

O (^{3}P) + O_2 → O_3

O_3 + hν (172nm) → O^* + O_2

Organics: $C_mH_nO_k$ + hν (172nm) → $C_{m'}H_{n'}O_{k'}$

$C_{m'}H_{n'}O_{k'}$ + O^* → H_2O, CO, CO_2, C_mH_n ↑ (Gas)

O^*: Active oxygen [Radical]

O (^{3}P): Ground state oxygen

Basic removal performance on VUV cleaning

■ Resist film removal:

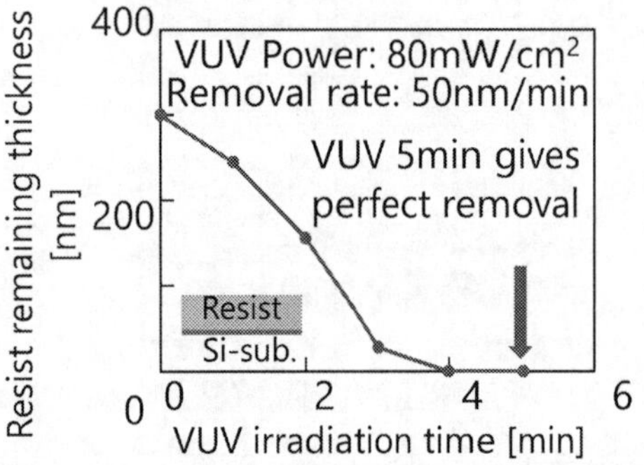

■ XPS analysis:

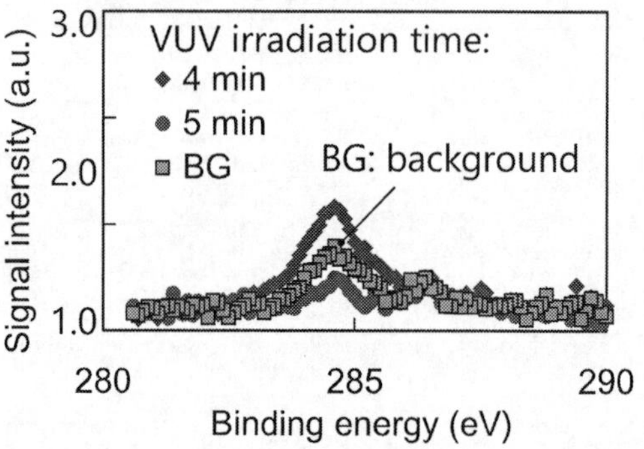

■ Residual resist removal:

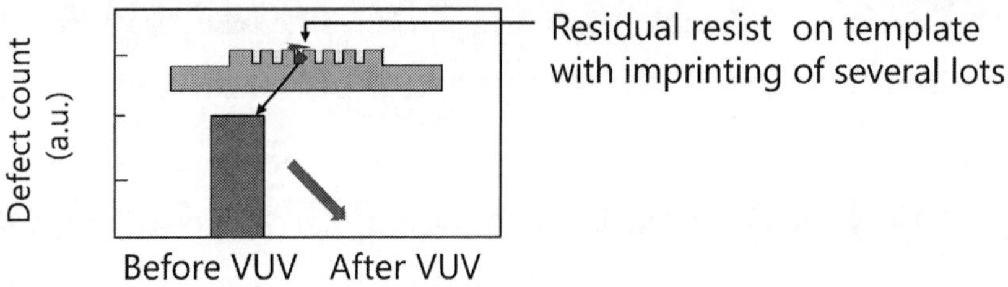

VUV clean perfectly removes resist. How about VUV damage?

Test template on VUV damage evaluation

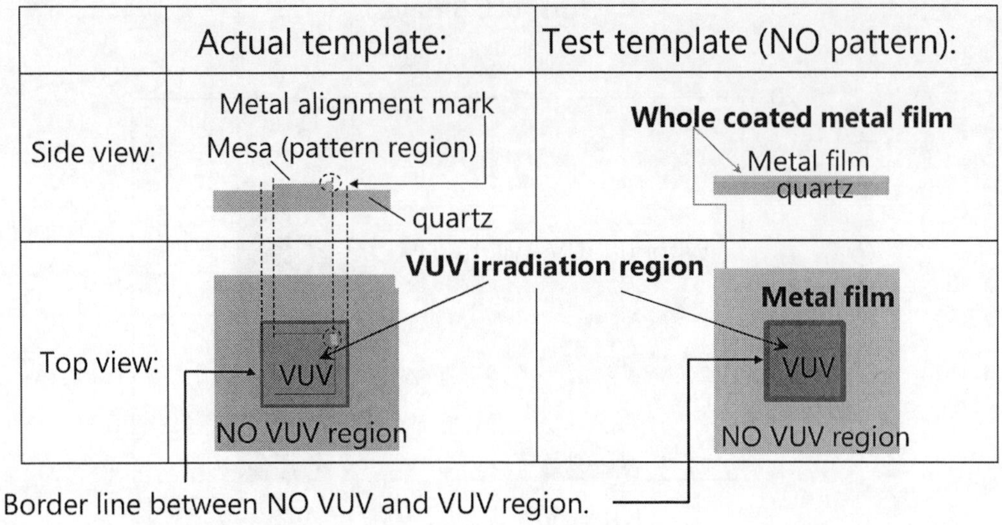

Test template wholly coated with metal film is evaluated.

Evaluation flow: VUV damage & Chamber cleanness

■ Evaluation flow:

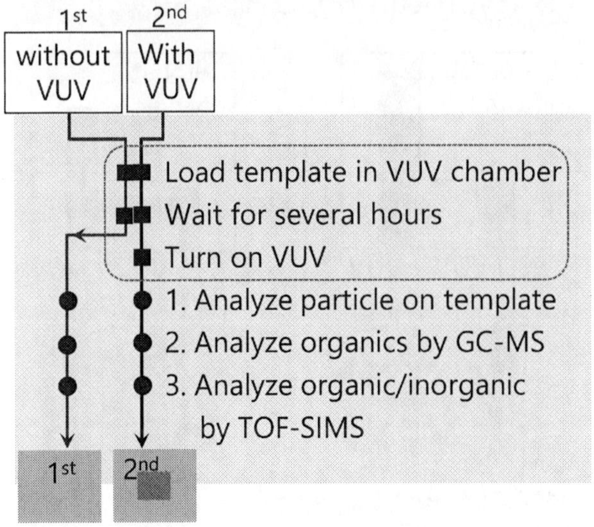

■ VUV irradiation:

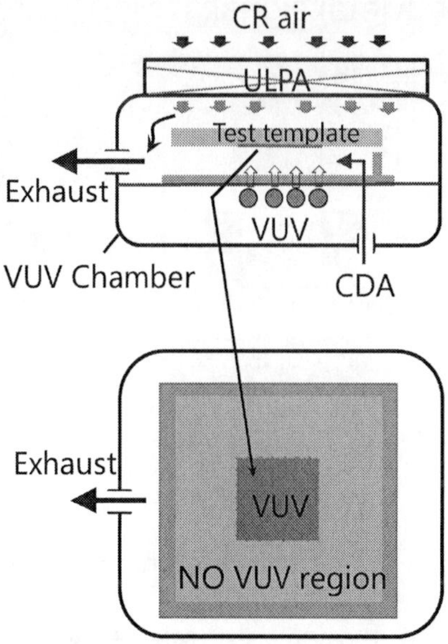

- Both templates are placed in SIMF during transportation.
- VUV is turned off for 5 days before loading the both templates.
- But, only 2nd template is irradiated with VUV.

TOSHIBA
Leading Innovation >>>
TOSHIBA MEMORY CORPORATION
Institute of Memory Technology Research & Development
© 2018 Toshiba Memory Corporation

Result: Particle inspection on the template

Turn off: 5 days

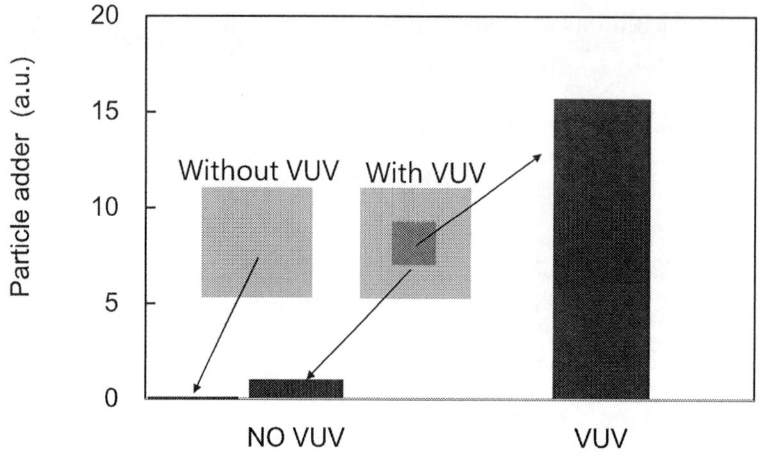

➢ Particles generated at VUV region and also near the border line.

Result: VUV damage evaluation - TOF-SIMS -

Turn off: 5 days

■ Metal signal: ■ Hydrocarbon, $C_xH_y^+$:

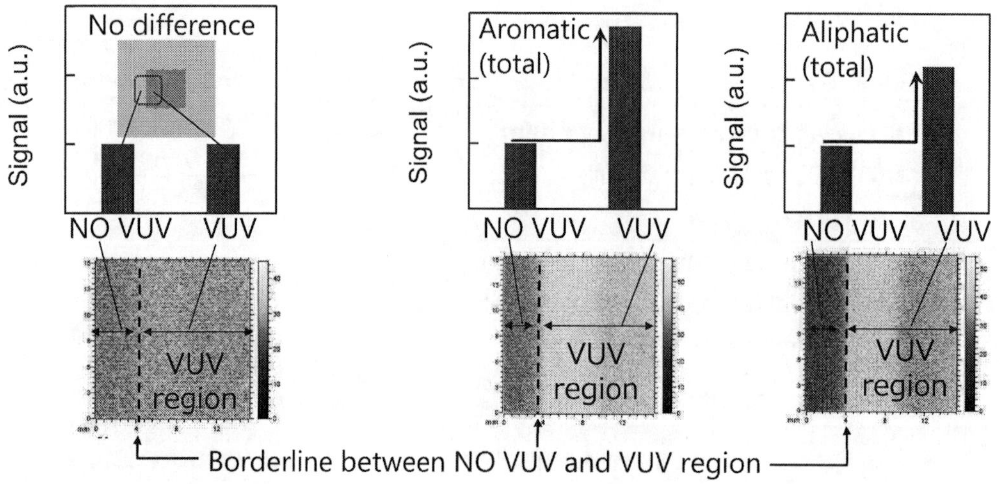

➢ Since there is no difference at both regions on metal signal, it is concluded that VUV does not damage metal film.

➢ However, Hydrocarbon ($C_xH_y^+$) increased on VUV region where particles generated.

Result: Si+ and SiO_3- Signal on VUV region - TOF-SIMS -

Turn off: 5 days

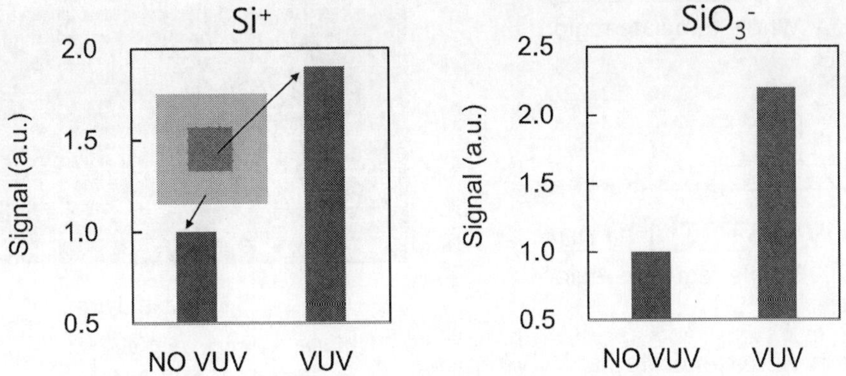

- TOF-SIMS also detected Si+ and SiO_3- at the VUV region.

Results: VUV turn off term vs. Particle increase

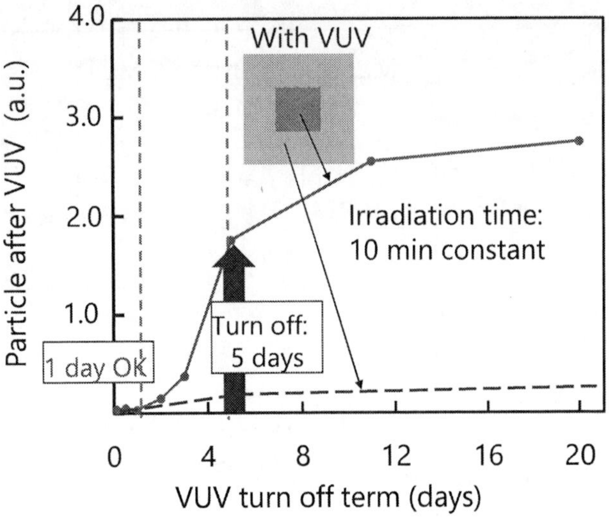

- Particles increased with increasing VUV turn off term.
- Organics could be accumulated in the chamber with off term.

Results: GC-MS analysis on whole template region

Turn off: 5 days

■ without VUV, TOC: 310 ng

■ with VUV, TOC: 110 ng

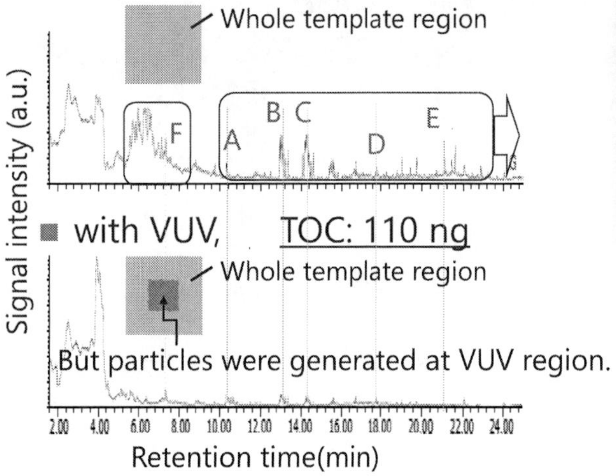

■ Organics (Aliphatic) :

	Material	Weight [g/mol]	Formula
A	1-Dodecene	168	$C_{10}H_{21}CH=CH_2$
B	7-tetradecene	196	$C_{14}H_{28}$
C	Cyclopentadecanone	210	$C_{15}H_{30}$
D	Octadecane	254	$C_{18}H_{38}$
E	Heneicosane	290	$C_{21}H_{44}$
F	Organosiloxane	296	$4[(CH_3)_2SiO]$

Structure:

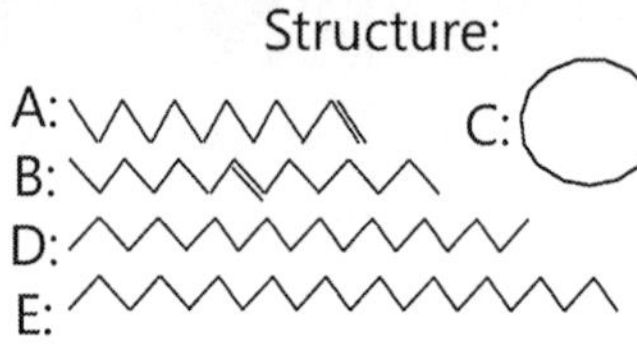

- C_xH_y and Organosiloxane were found on the whole template without VUV.
- These organics were dramatically decreased by VUV irradiation (one third), however particles were generated at VUV irradiation region.

Results: - Organosiloxane - GC-MS analysis

Turn off: 5 days

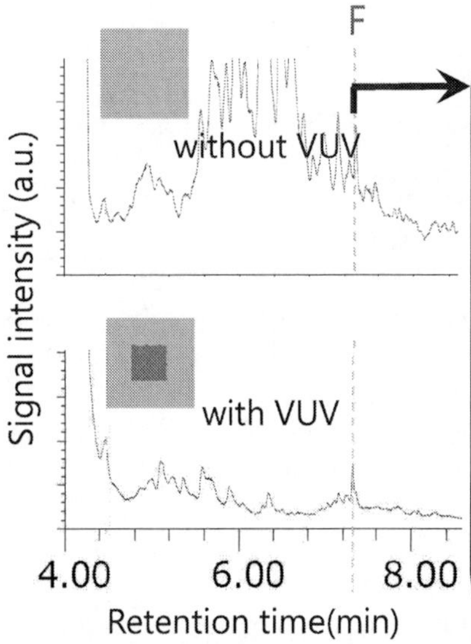

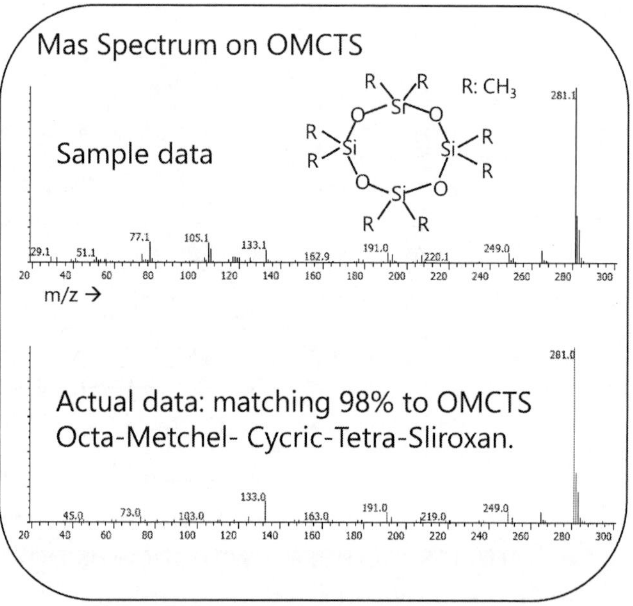

Model of Particle generation & adsorption

- On production template:

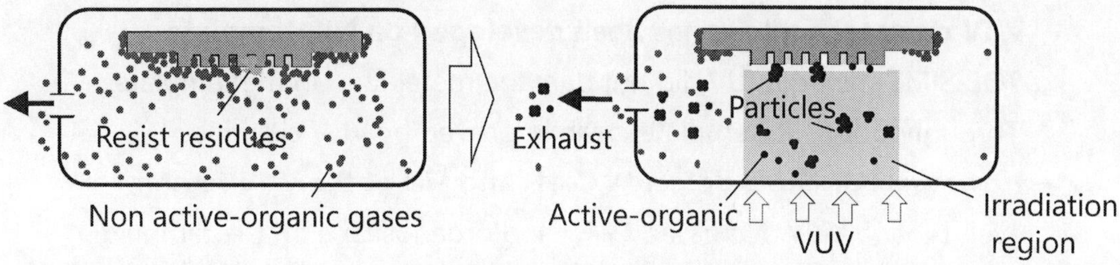

- ➢ Organic gases are accumulated in the chamber.
- ➢ Concentration of the Organic gases goes over the threshold.
- ➢ Most organic gases are activated by VUV on the irradiation region.
- ➢ Active-organics condense each other, then grow as particles on both gas phase and template surface. Most organics condensed by VUV could be exhausted out of chamber, according to GC-MS analysis.

It is important to manage the turn off term how long not over the threshold of the organic gases' concentration.

Chemical reaction related to SiO_2 particle generation:

Generation of radicals:

O_2 + VUV hν (172nm) → O* + O (^3P)

O (^3P) + O_2 → O_3, O_3 + hν → O* + O_2

H_2O + VUV hν (172nm) → *OH + H*

VUV activates and/or degrades OMCTS.

OMCTS:

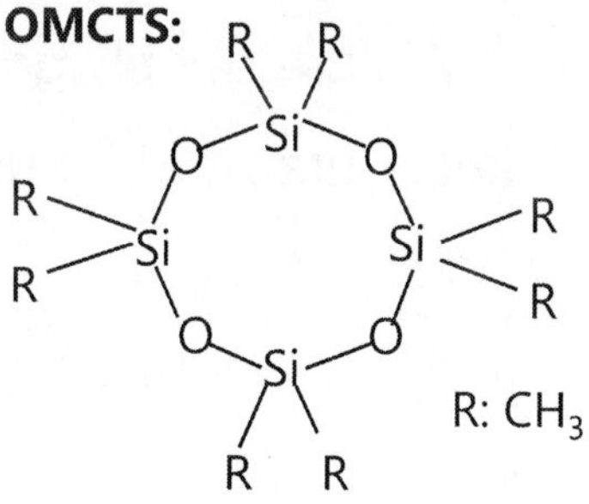

Overall reaction 1:

Oxygen radicals oxidize OMCTS.

$4[SiO_{,}(CH_{3,})_{2,}]$ + 4O* → $4SiO_2$ + $4C_2H_6$ ↑(Ethane gas)

Overall reaction 2:

Hydroxyl and Hydride radicals react OMCTS.

$4[SiO_{,}(CH_{3,})_{2,}]$ + 4 (*OH, H*)→ $4SiO_2$ + $8CH_4$ ↑(Methane gas)

TOSHIBA Leading Innovation >>>
TOSHIBA MEMORY CORPORATION
Institute of Memory Technology Research & Development
© 2018 Toshiba Memory Corporation

Summary

- **VUV clean technology has been developed on NIL-template.**
  - TOF-SIMS showed VUV did not damage metal film on the template.
  - Throughput of NIL-template clean improved by VUV cleaning.
- **However, TOF-SIMS detected C_mH_n and SiO at the VUV region**, even though VUV decreased C_mH_n and organosilane on the template.
  - Cluster particles were also found at the VUV region.
  - Particles increased with increasing VUV turn-off term.
- **Model of particle generation is proposed**:
  - 1st step, organic gases are accumulated into the VUV chamber,
  - 2nd step, the organics, O_2, and H_2O gases are activated by VUV.
  - 3rd step, the active gases condense each other, then grow as particles
- **Solution of particle generation is proposed**:
  - Managing the turn off term on VUV irradiation.
  - VUV irradiation on the appropriate timing.

Acknowledgments

We are deeply grateful to Masaya Kamiya and Satoshi Nakamura of Shibaura Mechatronics Corporation, in addition, they have been greatly tolerant and supportive for this experiments. They supported to gather particle data, and setting up of VUV tool in this experiment.

We also wish to thank Kim Young Ho and Jung Woo Yeoung of SK Hynix Inc., Keita Yoshihara of Ushio electronics, Hideaki Sakurai, and Kashiwagi hiroyuki of Toshiba memory Corpration for their suggestion on this experiment.

TOSHIBA Leading Innovation >>> TOSHIBA MEMORY CORPORATION Institute of Memory Technology Research & Development
© 2018 Toshiba Memory Corporation

ETCHING OF SILICON NITRIDE WITH HIGH TEMPERATURE WATER AND DEUTERIUM OXIDE

Joshua Barclay, Jesse Smith, Rick Reidy

Department of Materials Science and Engineering

University of North Texas, Denton TX

JoshuaBarclay@my.unt.edu | (979)-549-6129

Outline

- *Rationale*
- Deuterium oxide
- Dissociation of water and D_2O
- Experimental conditions
- Results
  - Ellipsometry
  - SEM
  - Molybdate blue UV-Vis (dissolved silica)
  - ATR D_2O post etch liquid
- Etch rate comparison
- Summary
- Future work

Rationale

- SiN is commonly etched using ≈160°C (85%) H_3PO_4--it is widely believed that the principal etchant is water
  - *At this temperature, surface Si-N are protonated followed by water attack creating Si-OH (soluble groups)**
- Previous work showed high temperature water (HTW) is an effective etchant for SiN
- Compare HTW and HT deuterium oxide (HTD_2O) to understand the etching mechanism
  - *Because dissociation of D_2O is ≈10x less than H_2O*

*Martin Knotter, Handbook for Cleaning for Semiconductor Manufacturing: Fundamentals and Applications, Chapter 3, Scrivener, 2011

Deuterium Oxide (D_2O)

H_2O

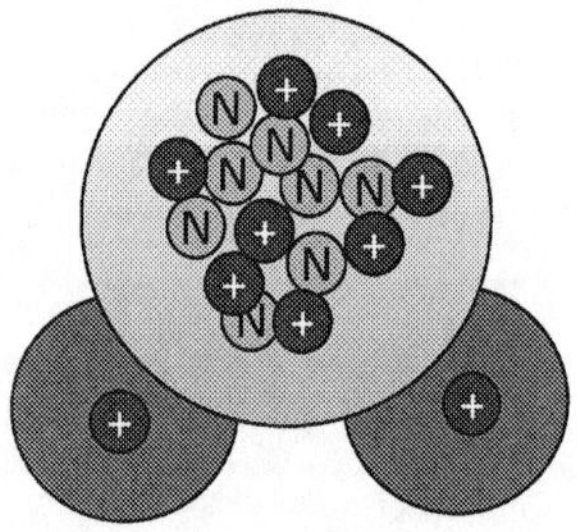

- 18.01 g/mol
- Boiling Point 100°C
- Vapor Pressure 23.8 torr

D_2O

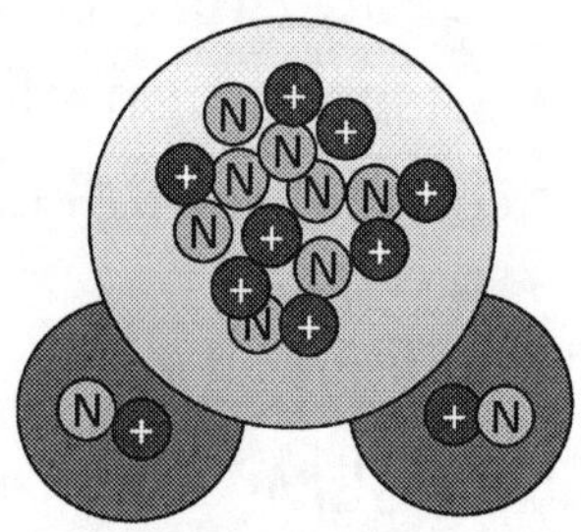

- 20.02 g/mol
- Boiling Point 101.4°C
- Vapor Pressure 20.6 torr

UNT EST. 1890 | DEPARTMENT OF MATERIALS SCIENCE & ENGINEERING College of Engineering

Dissociation

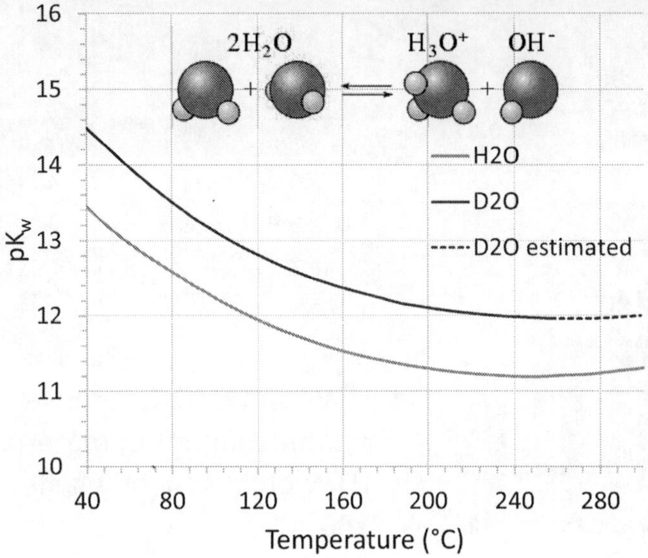

- H_2O 140 °C pK_w =11.71
- D_2O 140 °C pK_w =12.56
- H_2O 160 °C pK_w =11.53
- D_2O 160 °C pK_w =12.35
- Order of magnitude difference in reactive specie concentration between H_2O and D_2O

AV Bandura, AN Lvov, J. Phys and Chem Ref Data 35 (1) 15–30, 2006
DW Shoesmith, W Lee, Can J. Chem, 54 ,3553-3558, 1976

UNT EST. 1890 | DEPARTMENT OF MATERIALS SCIENCE & ENGINEERING College of Engineering

Experimental Conditions

- 30ml water and D_2O
- Heated to 140, 160°C
  - ramp time: 120 mins
- Hold times of 0,10,20 min at temp
- 3 samples for each condition
- $P_{equil,}$ H_2O (140°C) = 3.5 atm
- $P_{equil\,,}$ H_2O (160°C) = 6.1 atm
- Reactor is quickly cooled with water
- Post etch water is collected for characterization

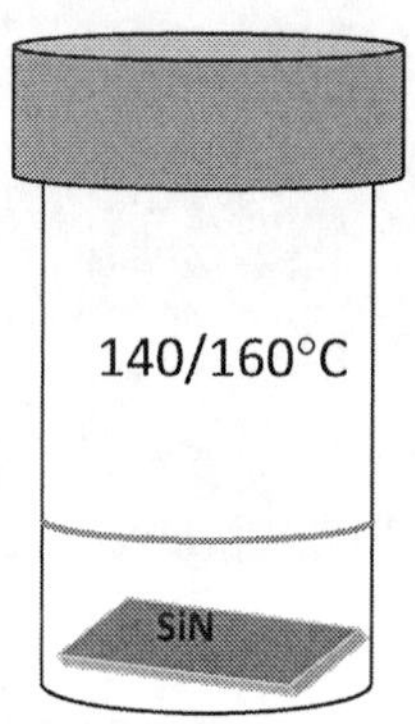

UNT EST. 1890 | DEPARTMENT OF MATERIALS SCIENCE & ENGINEERING College of Engineering

Thickness Losses

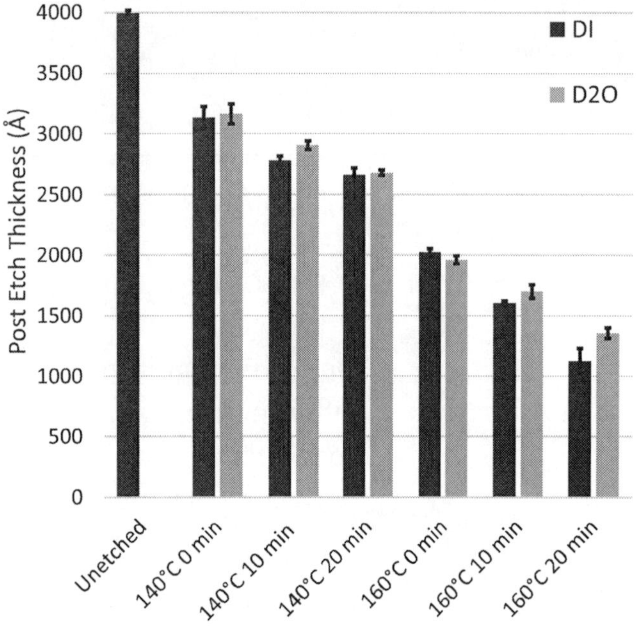

Ellipsometry Results

Samples	Etch Rate (Å /min)
D_2O 140°C	24±8
D_2O 160°C	30±8
DI 140°C	25±12
DI 160°C	45±9

- HTW and HTD_2O remove ≈ same amount of material
- HTW etches slightly faster at 160 °C
- **<u>Etching more dependent on temperature</u> than reactive species concentration**

UNT EST. 1890 | DEPARTMENT OF MATERIALS SCIENCE & ENGINEERING College of Engineering

SEM 140°C

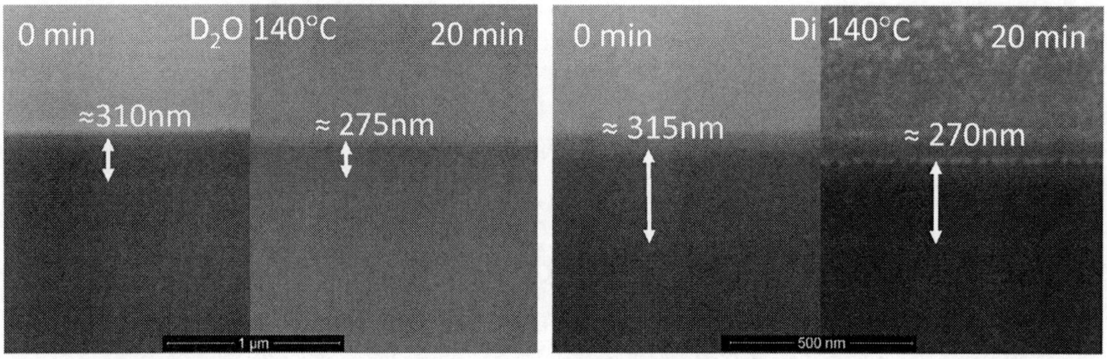

- SEM images of 140°C samples agree with ellipsometry results
- **Given the dissociation behavior, expected HTW to etch more (faster) than D_2O**

UNT EST. 1890 | DEPARTMENT OF MATERIALS SCIENCE & ENGINEERING College of Engineering

Thickness Loss vs Dissolved SiO_2

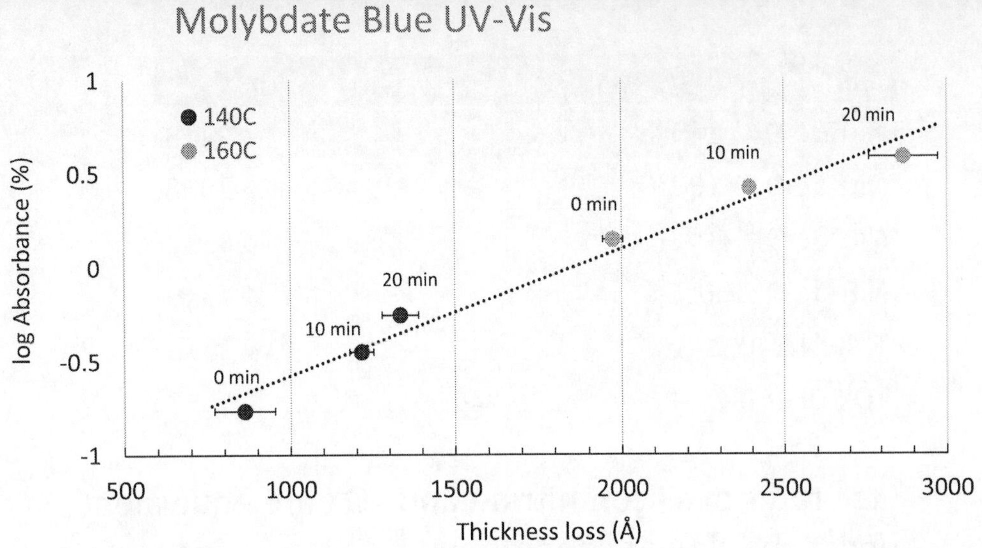

- Increasing dissolved silica concentration with increased etching
- Molybdate blue method did not work in D_2O—need a different method

UNT EST. 1890 | DEPARTMENT OF MATERIALS SCIENCE & ENGINEERING College of Engineering

ATR Post 160°C Etch D_2O

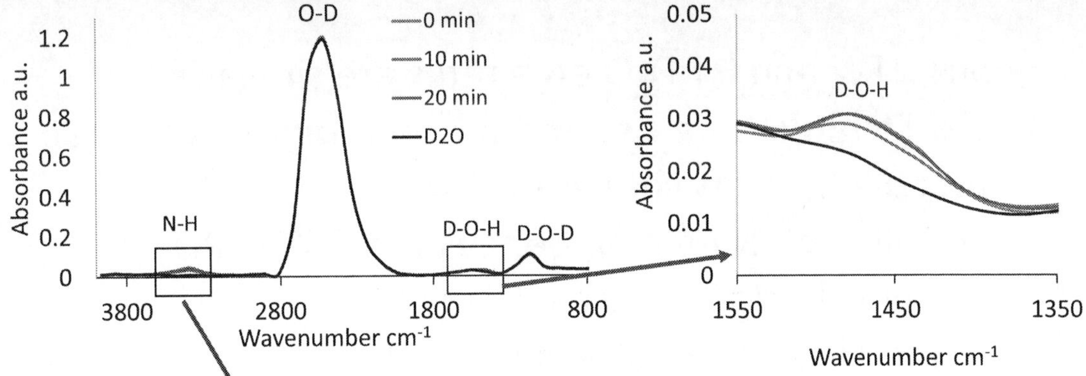

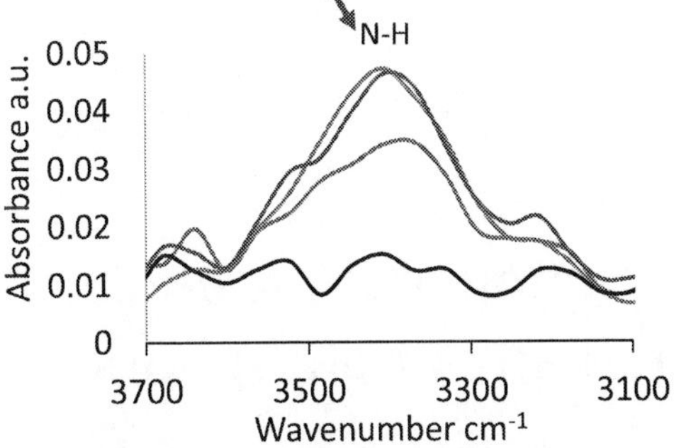

- N-H: Ammonia formed during reaction
- D-O-H: Exchange of dissociated D_2O with surface OH

UNT EST. 1890 | DEPARTMENT OF MATERIALS SCIENCE & ENGINEERING College of Engineering

Etch Rate Comparison

Samples	Etch Rate (Å /min)
Nitride D_2O 140°C	24±8
Nitride D_2O 160°C	30±8
Nitride DI 140°C	25±12
Nitride DI 160°C	45±9
TOX DI 140°C	18±3
TOX DI 160°C	45±6

- Etch rates of silicon nitride and TOX are equivalent under the similar conditions
- Poor selectivity between TOX and SiN without spiking

UNT EST. 1890 | DEPARTMENT OF MATERIALS SCIENCE & ENGINEERING College of Engineering

Summary

- SiN HTW and HT D_2O etch rates are similar
- **SiN Etching more dependent on temperature than reactive species concentration**
  - Follows SiN etching behavior in 85%H_3PO_4→ phosphoric increases [H^+] but does not show increased etch rates over HTW
- HTW is a viable etchant for silicon nitride
  - However with poor selectivity over silica without spiking
- HTW is potentially more environmentally friendly and cost effective SiN etchant than hot phosphoric

UNT EST. 1890 | DEPARTMENT OF MATERIALS SCIENCE & ENGINEERING College of Engineering

Future Work

- Dilute hot phosphoric under the same conditions
- Substitute HT DI water with ultra pure water (UPW) to determine if other ions impact etch rates.
- Determine concentration and identity dissolved siliceus species in post etch water
- Silica spiking and other methods to gain selectivity between SiN and SiO_2
- Investigate etching of bare silicon
  - Significant etching of underlying silicon

UNT EST. 1890 | DEPARTMENT OF MATERIALS SCIENCE & ENGINEERING College of Engineering

Acknowledgements

- University of North Texas
- UNT Materials Research Facility
- Qorvo: Nitride Wafers

UNT EST. 1890 | DEPARTMENT OF MATERIALS SCIENCE & ENGINEERING College of Engineering

Thank You

Questions?

SELECTIVE WET REMOVAL OF THE SiN CONTACT ETCH STOP LAYER PRIOR TO S/D CONTACT FORMATION.

A PACCO[A], F. HOLSTEYNS[A], M. SCHAEKERS[A], J-L. EVERAERT[A], D. DICTUS[B], D. CUYPERS[B]

[A] IMEC, 75 KAPELDREEF, 3001 LEUVEN, BELGIUM

[B] LAM RESEARCH, 75 KAPELDREEF, 3001 LEUVEN, BELGIUM

PUBLIC

OVERVIEW

- **Benefits** of a Contact Etch Stop Layer (CESL)
- Requirements and options for its removal
  → Ultra diluted HF at high temperature as an alternative solution

- **Results** of SiN etch in HT udHF and it's selectivity towards SiO_2
- Selectivity of HT udHF towards the contact materials

- MR CTLM *(multi ring circular transmission line model)* as a test vehicle for contact resistance evaluation
  - Contact resistance results for SiO_2 liner removal (reference)
  - Contact resistance results for SiN liner removal (comparison)
  - Effect of a post-annealing step on the contact resistance

- Conclusions and prospect

2

SELECTIVE CONTACT ETCH STOP LAYER (CESL) REMOVAL

- **Potential benefits** for the use of a SiN liner (CESL) and the advantage for the **wet over dry** removal of this SiN CESL.

- No/(or less) oxidation of the S/D areas during the post deposition anneal of the FCVD ILD0 material
- No plasma damage of the S/D areas during the replacement material pull
  → keeps EPI diamond shapes intact
- No/(or less) residues after contact etch

Contact etch | CESL removal | Contact metallization

imec Lam

OPTIONS & REQUIREMENTS FOR SELECTIVE SiN CESL REMOVAL

Requirements:

- SiN CESL has to be removed selectively towards S/D EPI contact materials (SiGe:B and Si:P).
- High selectivity towards surrounding ILD0 and/or spacer SiO_2
- Limited lateral etch of the SiN CESL

Options:

H_3PO_4	HF	dry removal	formulated chemistries
+ commodity chemistry	+ commodity chemistry	+ selectivity towards SiOx	+ selectivity
− post wet etch residues	− selectivity towards SiOx	− • post etch residues • selectivity towards SiGe	− cost of ownership
	↑ this study		

imec Lam

ULTRA DILUTED HF AT HIGH TEMPERATURE
STRATEGY FOR IMPROVEMENT OF SELECTIVITY TOWARDS SiO_2

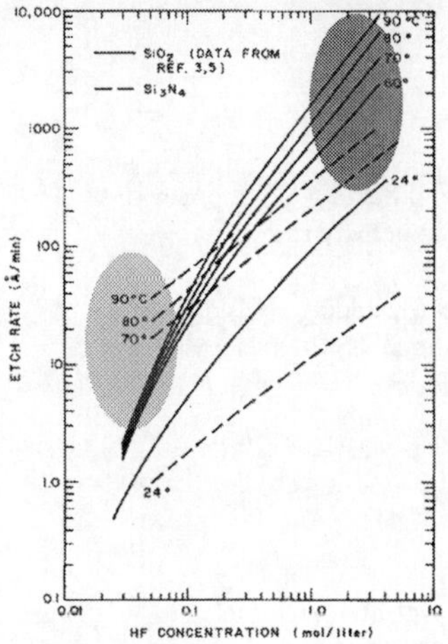

$Si_3N_4/SiO_2 < 1$

↓

$Si_3N_4/SiO_2 > 1$

SHIFT in selectivity going towards high temperature (> 70°C) and ultra diluted solutions (< 0.1 wt.%)

imec Lam RESEARCH C.A. Deckert, JECS 127, 2433 (1980) 5

ULTRA DILUTED HF AT HIGH TEMPERATURE
STRATEGY FOR IMPROVEMENT OF SELECTIVITY TOWARDS SiO_2

difference in etching mechanism* can be utilized to obtain better selectivity towards SiO_2 :

Difluorides are the main active species for etching of SiO_2 (HF_2^- with $SiOH_2^+$) →

Monofluorides are the main active species for etching of SiN (F^- with $SiNH_3^+$) - - - →

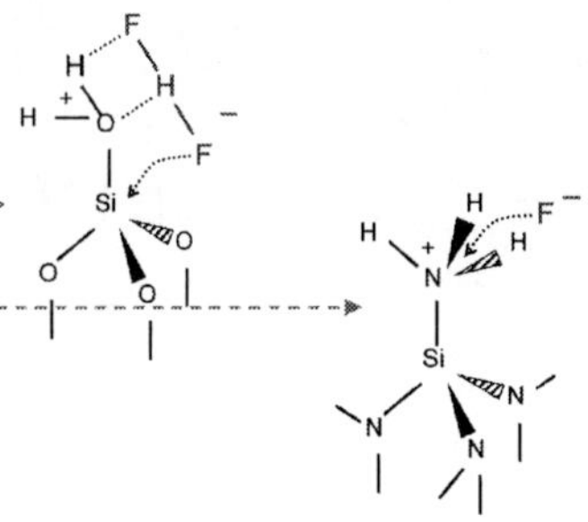

H_2F_2/ HF_2^- —dilution→ $2\ HF / F^-$

H_2F_2/ HF_2^- —heating→ $2\ HF / F^-$

imec Lam RESEARCH 6

* D.M. Knotter, JECS 2001

RESULTS
SiN / SiO_2 SELECTIVITY

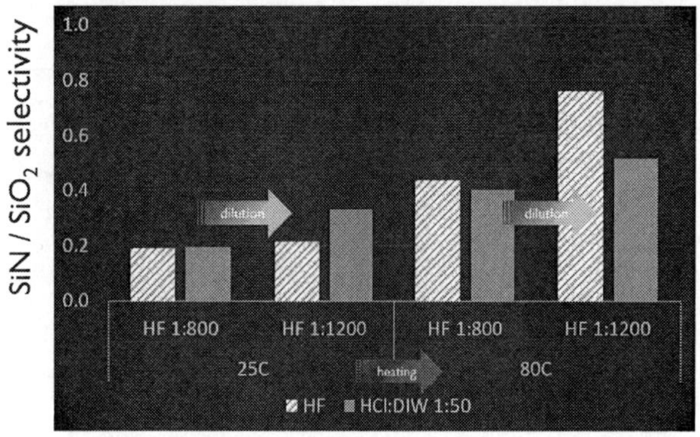

- Dilution and heating did improve the SiN/SiO_2 selectivity ratios.
- Lower pH (by adding HCl) did not improve the SiN/SiO_2 selectivity ratios at high dilution.

At 80°C and for HF dilution ratios around 1:1500 we obtained SiN/SiO_2 wet etch ratios around 1.

imec Lam Research

ETCHING OF SiN ON Si:P
IMPACT OF PRE CLEAN

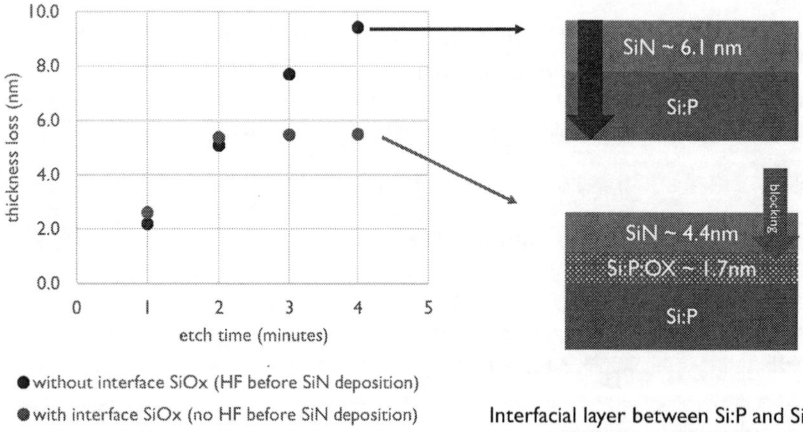

Start of etching of the Si:P substrate? "protective?" native Si:P:O_X removed for condition **with HF** clean before SiN deposition

For condition **without HF** clean before SiN deposition: **blocking** of further etching due to native Si:P:O_X?

Interfacial layer between Si:P and SiN delays the complete removal of the SiN layer and the over-etch into the Si:P substrate

imec Lam Research

HT udHF SELECTIVITY TOWARDS CONTACT MATERIALS
ETCH RATE OF Si:P IN HT udHF

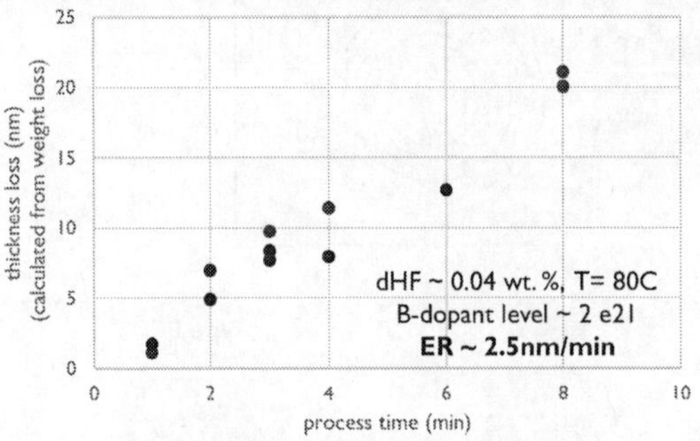

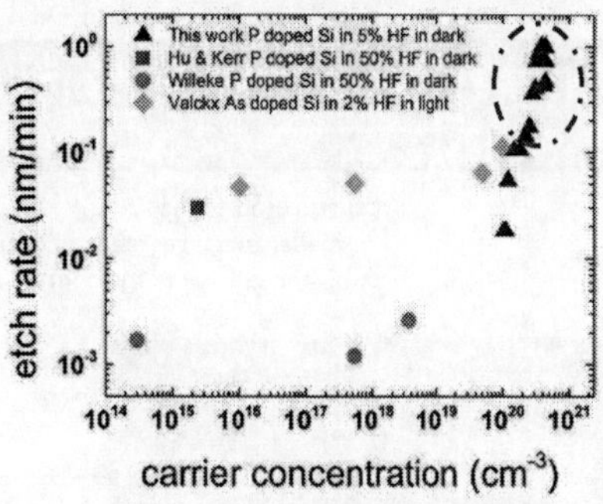

At 80°C and for HF dilution ratios around 1:1500, the Si:P loss is significant

HF ~ 5 wt. %, T= 40C
@ B-dopant level ~1 e20
ER ~ 1nm/min

**Unexpectedly High Etching Rate of Highly Doped n-Type Crystalline Silicon in Hydrofluoric Acid (Liu, 2013)*

imec Lam RESEARCH

HT udHF SELECTIVITY TOWARDS CONTACT MATERIALS
ETCH RATE OF Si:P vs. $Si_{25}Ge_{75}$:B IN HT udHF

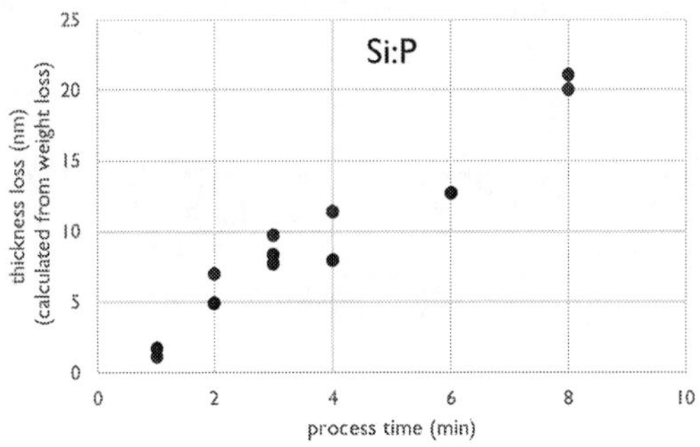

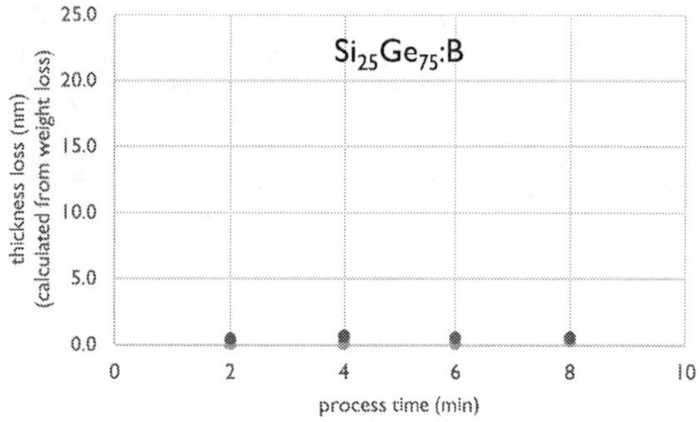

At 80°C and for HF dilution ratios around 1:1500 the Si:P loss is significant

At 80°C and for HF dilution ratios around 1:1500 the Si:Ge:B loss is negligible

imec Lam RESEARCH

10

MR CTLM *(MULTI RING CIRCULAR TRANSMISSION LINE MODEL)*

Test vehicle for the quantification of the contact resistance

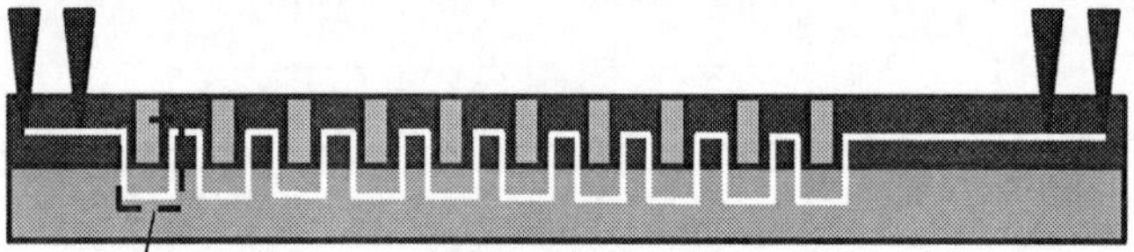

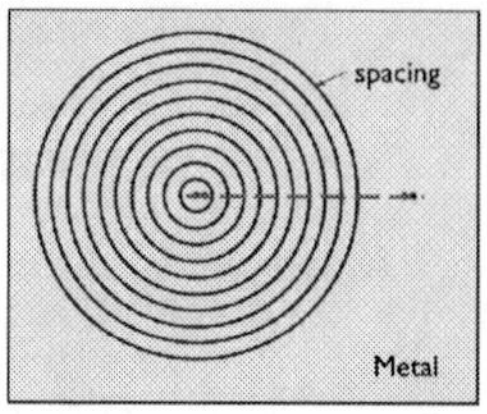

structures in series
→ smaller metal resistance contribution
→ higher sensitivity for contact resistance

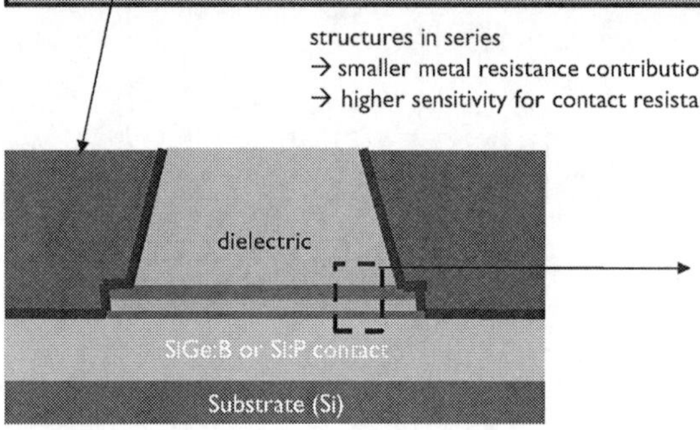

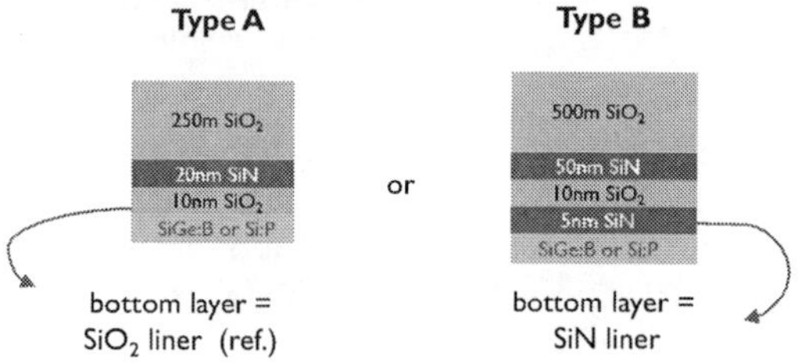

imec Lam RESEARCH

11

MR CTLM PROCESS SEQUENCE

SiO_2 vs. SiN CESL REMOVAL

MR CTLM structures **TYPE "A"**:

Objectives: selective SiOx removal
test the sensitivity of the structures to clean and surface preparation
reference / baseline for type "B" structures

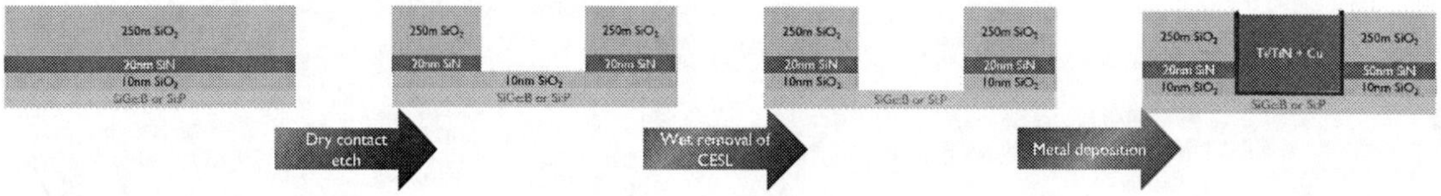

MR CTLM structures **TYPE "B"**:

Objectives: selective SiN removal
comparison with type "A"

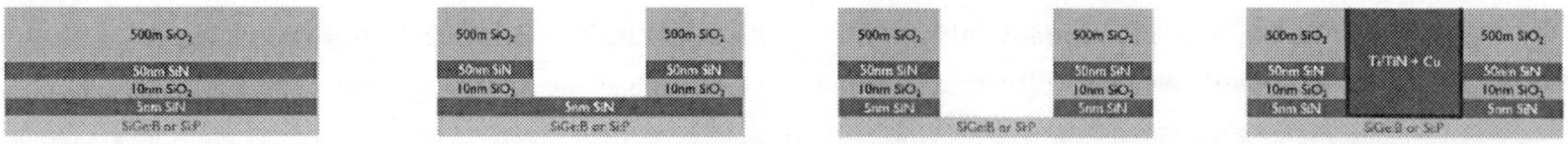

imec Lam RESEARCH

12

RESULTS SiO_2 LINER REMOVAL WITH RT dHF (0.7 wt. %, 20°C)

MR CTLM TYPE "A"

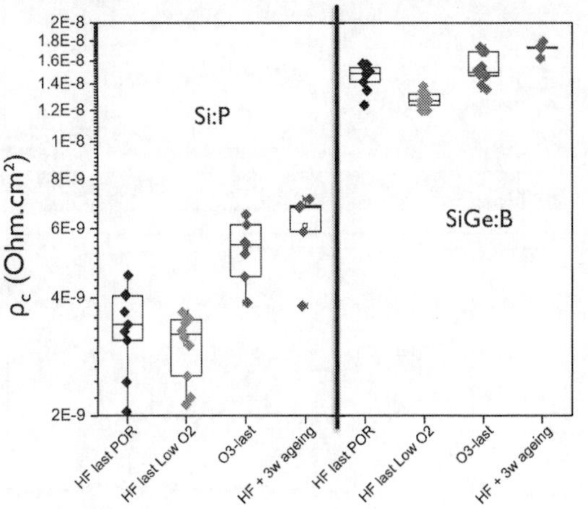

- Difference between Si:P and SiGe:B due to intrinsic doping (less relevant here)
- Worst ρ_c for SAW's with exaggerated ageing (3 weeks)
- O_3-grown oxide has a negative impact on ρ_c (for both Si:P and SiGe)
- Low vs. high O_2 has marginal influence on ρ_c.
- Queue times to metal (1h to 5h) does not have an impact on ρ_c (not shown here).

MR CTLM short loop test structures are sensitive to interface termination and cleans.

imec

13

EDS ANALYSIS OF THE METAL CONTACTS

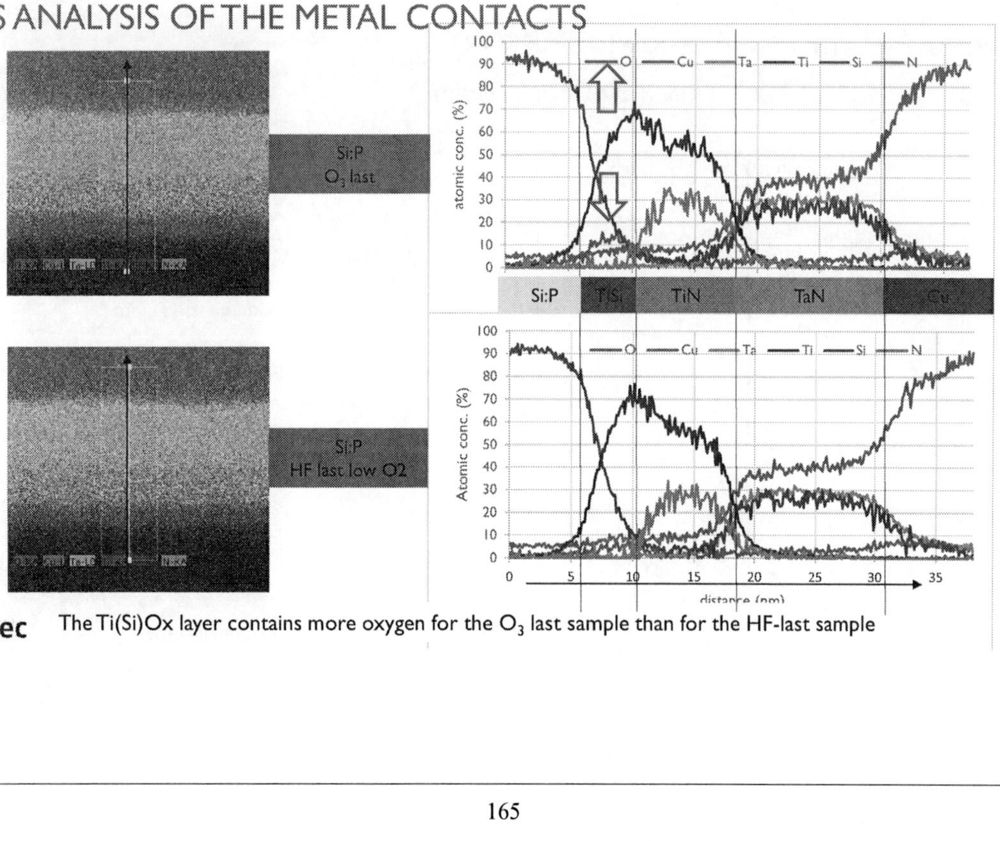

imec The Ti(Si)Ox layer contains more oxygen for the O_3 last sample than for the HF-last sample

EDS ANALYSIS OF THE METAL CONTACTS

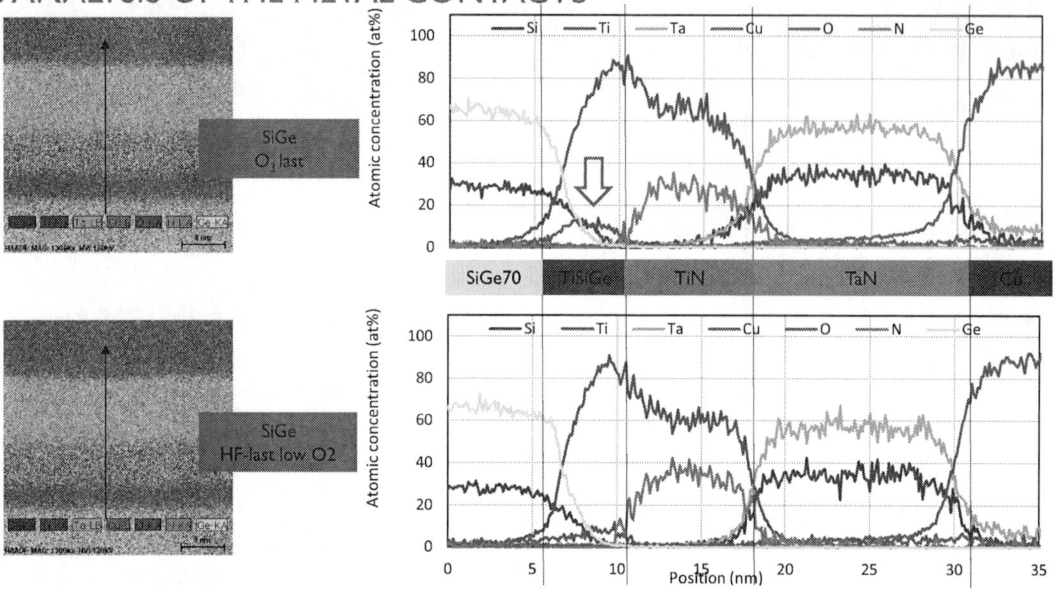

imec The Ti(Si)Ox layer contains more oxygen for the O_3 last sample than for the HF-last sample

EFFECT OF AN ADDITIONAL POST-ANNEALING STEP

SiGe:B 450°C, N_2, 1 MIN Si:P 500°C, N_2, 1MIN

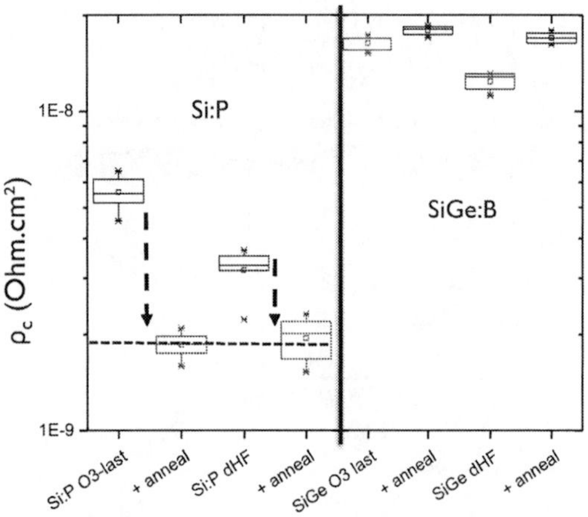

- A high T step is beneficial for ρ_c for the Si:P substrates. This is probably due to the scavenging of oxygen from the interfacial Si:P oxide by the Ti film.*
- ρ_c equals out for Si:P and SiGe: an oxide free interface preparation does not look critical when applying a high-T step.

* *Direct observation of both contact and remote oxygen scavenging of GeO_2 in a metal-oxide-semiconductor stack*
Journal of Applied Physics116, 164101 (Fadida, 2014)

imec Lam

16

RESULTS SiN LINER REMOVAL WITH HT udHF (0.04 wt. %, 80°C)

MR CTLM TYPE "A" vs. TYPE "B"

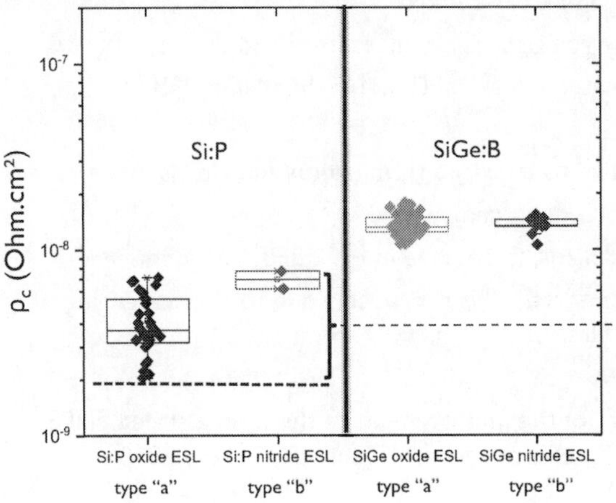

SiGe:B: ρ_c values comparable

Si:P: somewhat higher ρ_c values for SiN ESL compared to SiOx

imec Lam RESEARCH 17

EDS ANALYSIS OF THE DIELECTRIC STACK

- SiN is O-rich at the **Si:P** contact interface.

→Possibly Si:P:Ox
→Importance of a (in–situ) pre clean

- SiN is N-rich at the **SiGe** contact interface.

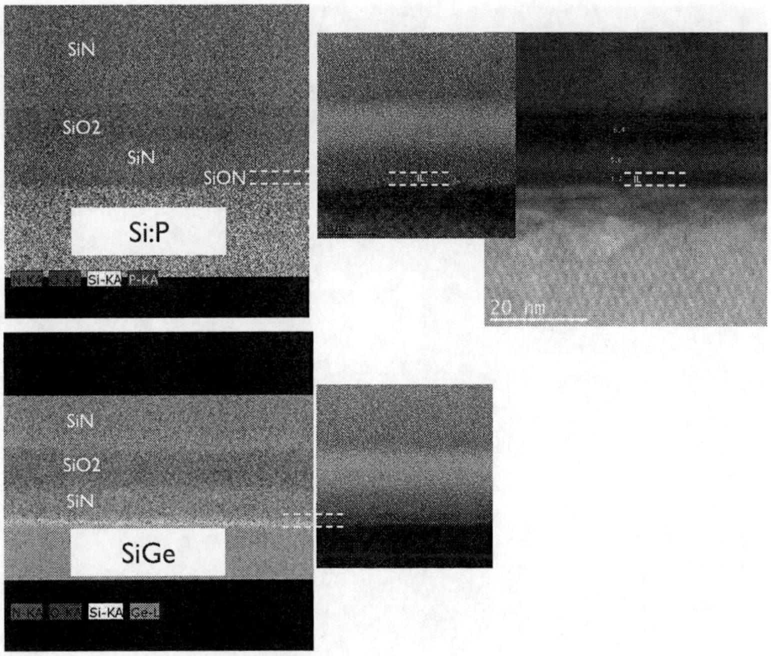

imec

CONCLUSIONS AND PROSPECT

- HF was evaluated as an alternative for selective SiN CESL removal.
- Selectivity of HF towards SiO_2 was improved by going to high temperature and dilution.
- Si:P as a contact material is more sensitive to material loss in HT udHF then $SiGe_{75\%}$:B

- MR CTLM short loop test structures are sensitive to interface termination and cleans.
- SiN CESL removal was benchmarked towards SiOx and comparable.
- An interfacial blocking SiON, which was more difficult to remove in HT udHF, was identified.
- An anneal step is beneficial for ρ_c for the Si:P contacts. This is probably due to the scavenging of oxygen from the interfacial Si:P oxide by the Ti film.

- Implementation of a wet CESL removal as a part of the improvement of the ρ_c in a scaled FinFET flow will be presented by imec @ International Interconnect Tech. Conf. (IITC), June 2018.

imec Lam RESEARCH 19

imec

embracing a better life

imec PUBLIC

Novel EHS-Friendly Ru Etch and SPM Alternatives for 5nm Applications

Chien-Pin Sherman Hsu
Avantor
Hsinchu, Taiwan
Sherman.hsu@avantorinc.com

April 11, 2018

avantor 1

Presentation Outline

1. ***Ruthenium (Ru) – Cu Interconnect Alternative***
2. **Ru Etch – Current Options**
3. **EHS-Friendly Ru Etch Development**
   - **RSE-1 Chemistry Platform Design**
   - **Selective Ru Etch with Broad Substrate Compatibility**
   - **Other Selective Etch Examples – WNx, W Selective Etch**
4. **Study of SPM Alternatives and Their Applications**
5. **Summary**

avantor 2

Ruthenium – Potential Cu Replacement in Interconnect

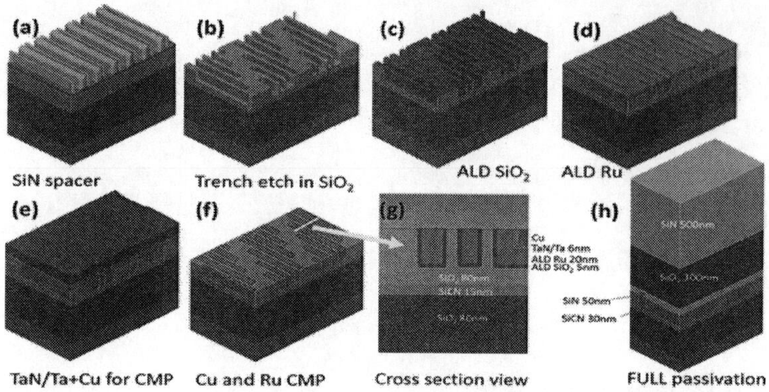

[Source: L.G. Wen et.al., ACS Appl. Mater. Interfaces, 2016, 8 (39), pp 26119–26125}

- **Ru – Good effective resistance; candidate as barrier-less metallization in interconnect; replacing Cu**
- **Ru lines – excellent EM behavior, no Ru ion drift into low-k**

Ru as Cu Interconnect Alternatives

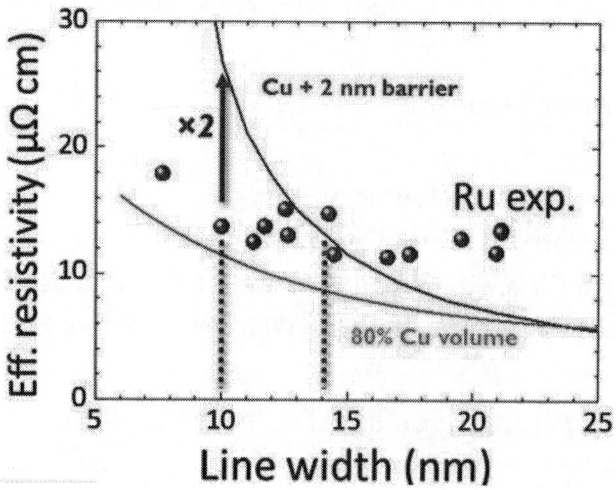

- **Effective resistivity as a function of line width for copper and ruthenium.**
- **At 10nm narrow lines, Ru interconnects has 2X better resistivity than Cu**

Source: courtesy of IMEC

Presentation Outline

1. Ruthenium (Ru) – Cu Interconnect Alternative

2. *Ru Etch – Current Options*

3. EHS-Friendly Ru Etch Development
   - RSE-1 Chemistry Platform Design
   - Selective Ru Etch with Broad Substrate Compatibility
   - Other Selective Etch Examples – WNx, W Selective Etch

4. Study of SPM Alternatives and Their Applications

5. Summary

avantor 5

Ru Wet Etch Chemistry Options

- Ru is a hard metal; resistant to common etch chemistries
- Ru can be oxidized to various oxides. RuO_2 is most common oxide of ruthenium
- Conventional etch chemistries do not provide enough Ru etch rates: SPM, HNO_3, Aqua Regia, HF, SC-1…

- Ru integration in advanced nodes has big challenges
  - Need high Ru etch selectivity vs. a number of substrates and metallization
  - Need better Ru CMP solutions

avantor 6

Ru Wet Etch Chemistry Options

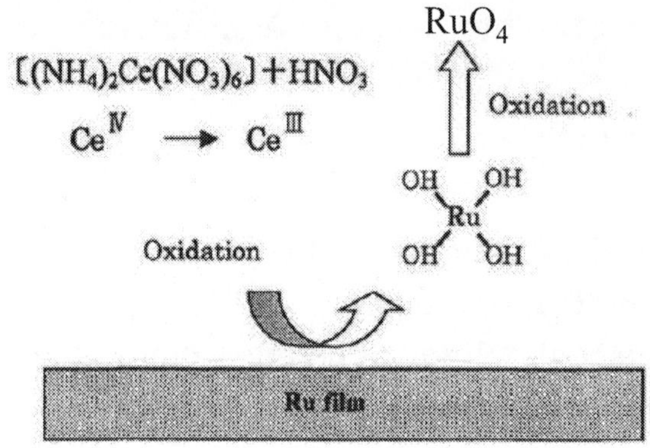

[H. Aoki et.al. Extended Abstracts of the 2001 International Conference on Solid State Devices and Materials. 2001]

- **Cerium ammonium nitrate (CAN) and nitric acid etch Ru.**
- **By-product: volatile RuO_4 (m.p. 25.4°C; b.p. 40°C; toxic); Formulation contains metal (Ce)**

avantor 7

Presentation Outline

1. **Ruthenium (Ru) – Cu Interconnect Alternative**
2. **Ru Etch – Current Options**
3. ***EHS-Friendly Ru Etch Development***
   - **RSE-1 Chemistry Platform Design**
   - **Selective Ru Etch with Broad Substrate Compatibility**
   - **Other Selective Etch Examples – WNx, W Selective Etch**
4. **Study of SPM Alternatives and Their Applications**
5. **Summary**

avantor 8

Ru Selective Etch by RSE-1 Chemistry

Ru + OH⁻ + Oxidant → RuO_4^- (major) + By-product X (reduced oxidant)

- **Aqueous, alkaline oxidative chemistry design**
- **Convert Ru to RuO_4^-; Fully dissolution mode. No toxic RuO_4 release**
- **Utilize a unique oxidant to boost Ru etch efficiency, but minimize alkaline bases' attack on low-k and TEOS**
- **Proposed mechanism – based on Pourbaix diagrams**

avantor 9

Ru Selective Etch by RSE-1
(with Ru – XPS analysis)

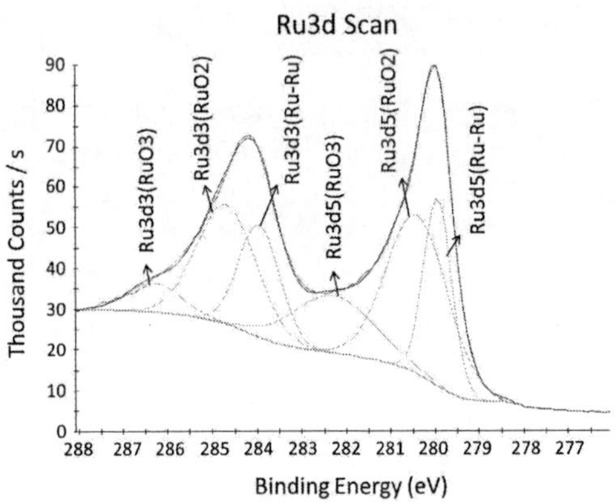

Untreated Ru film

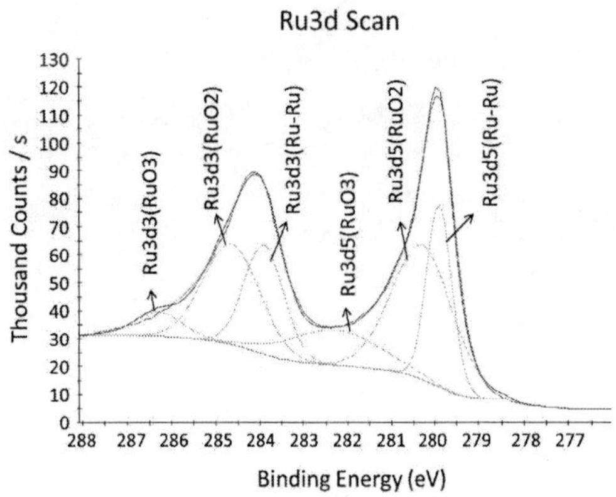

RSE-1, 65°C, 5m

*** Ru 3d XPS -**
Similar Ru-Ru, RuO_2 and RuO_3 are primary bonding in both Ru samples
* Ru oxide is also conductive

avantor 10

RSE-1 – Powerful PR/Residue Cleaner and Selective Ru Etch

RSE-1 etch rates at 65°C (Å/min)										
Cu	Ta	SiN	TEOS	SiC	Black Diamond II	SiOC	FSG	General CDO	Ru (PVD)	Ru (CVD/ALD)
<1	<1	<1	<1	<1	<1	<1	<1	<1	90-170	15-25
Co	Ti	TiN	TaN	WN	B-Doped Si		P-Doped Si	W	BARC	
<1	<1	<1	<1	>100	0		10	> 70	Compatibility Depends on BARC Types	

- RSE-1 is fully aqueous, but offers improved PR removal (high dose, highly cross linked polymer structure)
- RSE-1 is EHS friendly. No toxic RuO_4 gas release.
- RSE-1 has unique high Ru etch rates and broad compatibility with other materials ⇒ Expected applications: Selective Ru etch, Ru CMP & pCMP....
- Other Selective Etch examples: WNx etch selectivity vs. TiN, HfOx, LaOx -- - from >2500 to >10,000

avantor 11

RSE-1: Selective W, WNx Etch vs TiN

Etch rate comparison

Chemistry Description	Process Temp. (°C)	WNx (A/min)	W (A/min)	TiN (A/min)	WNx/TiN Selectivity	W/TiN Selectivity
10:1 H2O-(RSE-1)	35	> 100	-	0.04	> 2,500: 1	-
RSE-1	60	-	>100	0.8	-	> 125:1

- RSE-1 Chemistry platform (10:1 dilution): Excellent WNx Etch selectivity vs. TiN; Selectivity: >2,500:1
- RSE-1 Chemistry platform: Excellent W Etch selectivity vs. TiN; Selectivity: >125:1

avantor 12

RSE-1 Construction Material Compatibility

Compatible Materials		
Chemraz	PFE	PVC
Glass	Polyethylene	PVDF
Kalrez	Polypropylene	Quartz
PFA	PTFE	Telfon

* Viscosity: 1.30 mPa.s (30°C); 0.58 mPa.s (65°C)

Presentation Outline

1. Ruthenium (Ru) – Cu Interconnect Alternative
2. Ru Etch – Current Options
3. EHS-Friendly Ru Etch Development
   - RSE-1 Chemistry Platform Design
   - Selective Ru Etch with Broad Substrate Compatibility
   - Other Selective Etch Examples – WNx, W Selective Etch
4. *Study of SPM Alternatives and Their Applications*
5. Summary

RSE-1: Cu Compatible SPM Alternative

	Chemistry Description	Cu ER (Å/min)	TiN ER (Å/min)	Co ER (Å/min)	Ti ER (Å/min)
A	SPM, 90°C	>1000	>340	>80	>60
B	SC-1 (5:1), 65 °C	>25	>230	>15	>60
C	RSE-1, 65°C	<1	<1	<1	<1

Process Description	N5, Implanted BARC	Sub-20nm, PR/Organic BARC Bi-Layer
RSE-1, 65C, 5min	Implanted BARC film is removed	PR is removed; Organic BARC film remains
RSE-1, 65C; DHF (Multi-step)	---	PR and organic BARC film is removed

- RSE-1: Broad substrate compatibility (Cu, Co, Ti, TiN, TaN, TEOS, low-κ...)
- Fully aqueous, more environmentally friendly
- RSE-1 alone or RSE-1/DHF multi-step --- Effective bi-layer or tri-layer PR & BARC layers (100% removal, dissolution mode; including high dose, highly cross linked polymer structure)
- Suitable for typical FEOL/BEOL Single Spin Wafer Processors

avantor 15

Presentation Outline

1. Ruthenium (Ru) – Cu Interconnect Alternative
2. Ru Etch – Current Options
3. EHS-Friendly Ru Etch Development
    - RSE-1 Chemistry Platform Design
    - Selective Ru Etch with Broad Substrate Compatibility
    - Other Selective Etch Examples – WNx, W Selective Etch
4. Study of SPM Alternatives and Their Applications
5. ***Summary***

avantor 16

Summary

- Ru has been proposed as a potential Cu interconnect replacement. Ru selective etch and its CMP present integration challenges.
- Novel chemistry, RSE-1, delivers highly selective Ru etch and excellent compatibility with a broad range of substrates, such as Cu, Co, Ti, TiN, TaN, low-k and others. It also provides excellent W or WNx selective etch vs. TiN.
- Another application of RSE-1 chemistry platform is used for PR and BARC implant removal. It is fully aqueous, organic solvent-free, thus an EHS-friendly SPM alternative.

avantor 17

Thank you

avantor 18

Development of Wet-etch Chemistries for Tungsten Word-line Recess

CK Ge†, Wen Dar Liu and Yi-Chia Lee
Versum Materials Technology LLC., 2F, No.25, Lane 62, Chung Ho Street, Chupei City, Hsinchu County 30267, Taiwan

† Corresponding author: ck.ge@versummaterials.com

2018 Surface Preparation and Cleaning Conference
April 09-11, 2018 in Boston, MA

VERSUM MATERIALS

1

Agenda

✓ **Introduction**

- 3D NAND technical background
- 3D NAND technical challenge of W recess
  - Dry-etch methods
  - Conventional wet-etch methods
- Critical technical specification

✓ **Wet etchants for 3D NAND W recess**

- Design concept
- Etch rate performance

✓ **Summary**

VERSUM MATERIALS

2

Technical Background

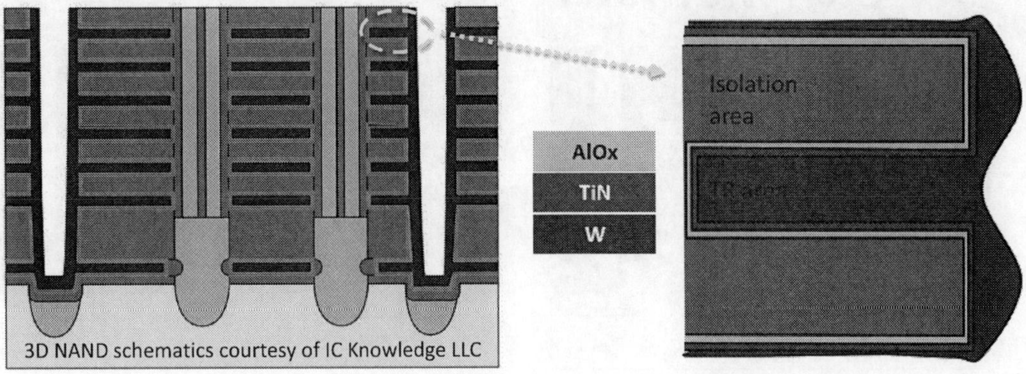

- **During fabrication of the 3D NAND memory device, tungsten(W) recess for word-line(WL) isolation is one of the key processes. Typically, high-k/metal gate are used for the connection of tungsten control gate.**

VERSUM MATERIALS

3

Technical Background

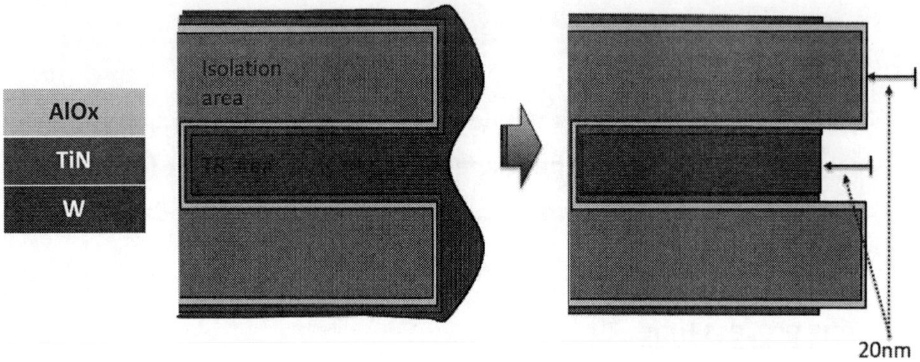

- **In the recessing process, TiN and W should be simultaneously etched with equal thickness**
- **AlOx is the protecting layer that should not be damaged**

VERSUM MATERIALS

4

Technical Challenge of W recess
- Dry-etch Methods

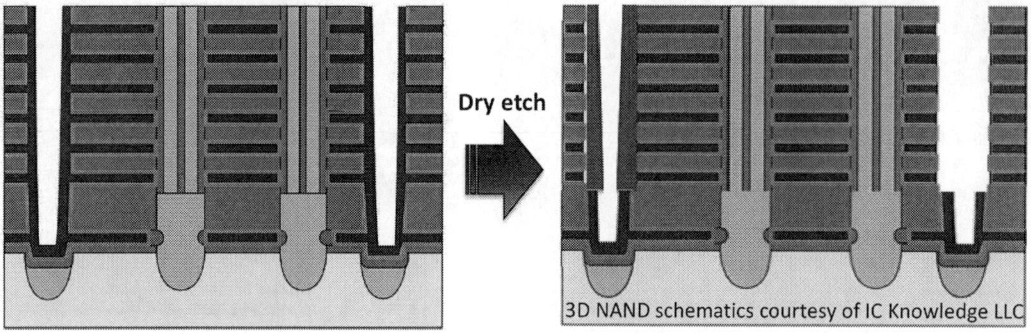

- As the number of layer increases, it's difficult to completely etch the bottom layer of W and TiN by dry etch-methods because the dry-etching byproduct from the top layer would remain in the trenches and restrict etching the bottom layer. Therefore, wet-etch method is proposed as an alternative for W recess.

VERSUM MATERIALS

5

Technical Challenge of W recess
- Conventional Wet-etch Methods

- <u>AlOx compatibility</u>

  Typical wet-etch chemicals would easily etch the AlOx and cause a recess in the side wall of the channel at the AlOx layer that forms an undesirable floating gate, and results in an on-current degradation for the NAND string.

- <u>Long process time</u>

  Conventional wet etchants show low TiN or W etch rates that result in the extremely long process time(over 1 hr). The long process time means the application needs to operate in batch type tool and restricts the SWT application.

VERSUM MATERIALS

6

Critical Technical Specification

	Target	Conventional Wet-etchant
TiN E/R (A/min)	>30	~2
W E/R (A/min)	>30	~2
TiN/W E/R selectivity	~1	~1
AlOx E/R (A/min)	<1	<1

- The wet etchant should perform equal TiN and W etch rates which means the TiN/W selectivity is close to 1
- AlOx is used as the protecting layer of transistors and barrier layer of plugs, so the chemical should be compatible with AlOx
- Conventional Wet-etchant shows low TiN and W etch rates, it needs extremely long time for processing
- Developing the etchants with higher TiN and W etch rates to shorten the process time is one of the important issues

VERSUM MATERIALS

7

Novel Wet Etchants for 3D NAND W Recess - Design Concept

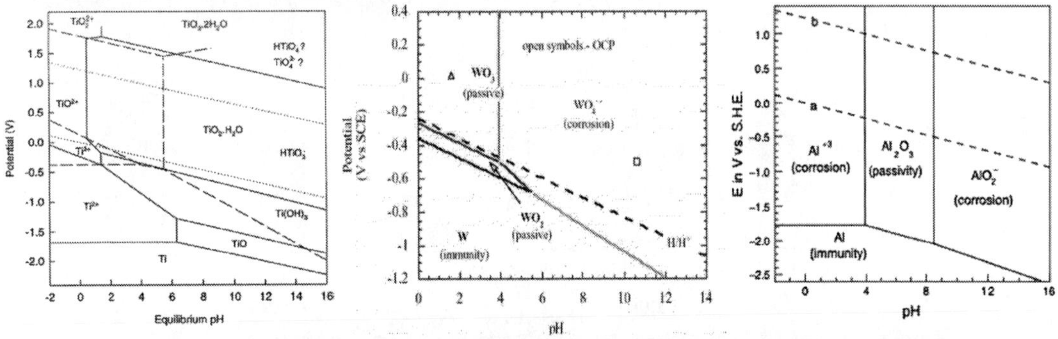

Key Components	Functions
A	Transform TiN to TiOx, and W to WOx
B	Adjust pH to control TiN, W and AlOx etch
C	Promote TiN etch rate, and improve TiN/W selectivity
D	Suppress TiN & W etch rates, and improve TiN/W selectivity

VERSUM MATERIALS

8

Novel Wet Etchants for 3D NAND W Recess
- Design Concept (pH Effect)

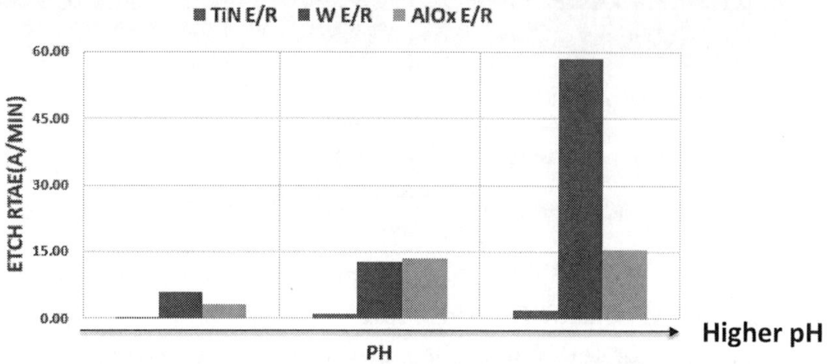

- TiN, W and AlOx etch rates increased while pH was higher. especially, the impact on W is much greater
- Need to control the pH to avoid AlOx being damaged
- Need additional component or method to increase TiN etch, and improve TiN/W selectivity

VERSUM MATERIALS

9

Novel Wet Etchants for 3D NAND W Recess
- Design Concept (TiN Promoter & Temp. Effect)

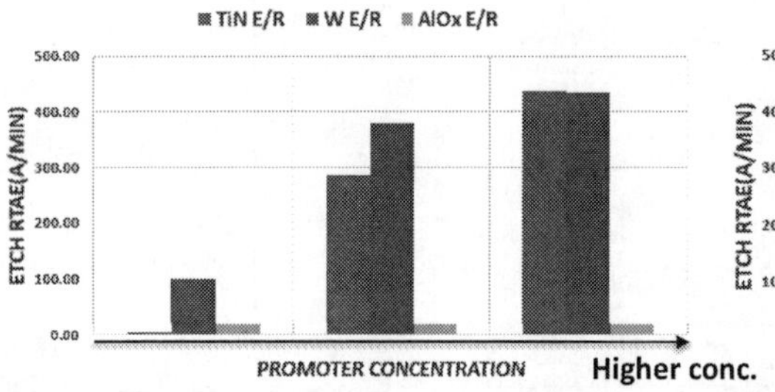

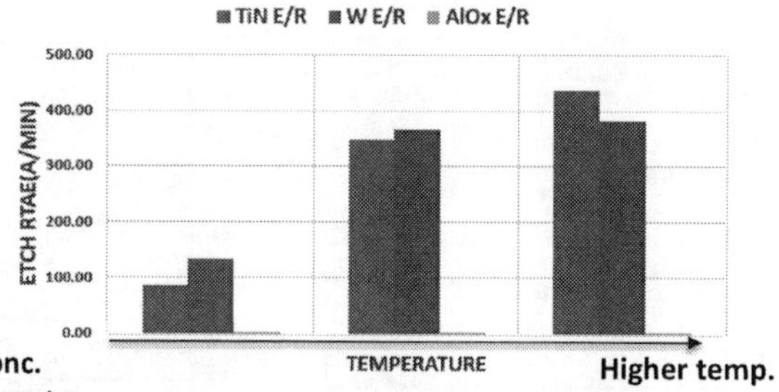

TiN/W selectivity	0.02	0.75	1.00

TiN/W selectivity	0.65	0.95	1.15

- TiN E/R can be boosted by two ways, adding promoter and adjusting process temperature. TiN/W selectivity can be also improved
- AlOx etch rate change was minimal

VERSUM MATERIALS

10

Novel Wet Etchants for 3D NAND W Recess

- Design Concept (Etch Rate Suppressor)

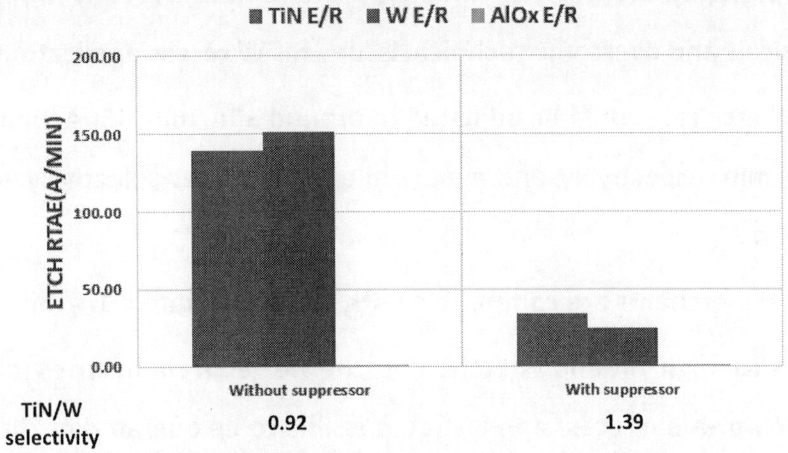

- Both TiN and W etch rates would decrease while suppressor is added, and the TiN/W selectivity also gets higher
- No significant influence on AlOx e/r

VERSUM MATERIALS

11

Novel Wet Etchants for 3D NAND W Recess

- Etch Rate Performance

	Target	Etchant 1	Etchant 2	Etchant 3
Process temp. (°C)	-	35	25	30
TiN E/R (A/min)	>30	>300	>150	>50
W E/R (A/min)	>30	>300	>150	>50
TiN/W E/R selectivity	~1	0.95	0.92	1.20
AlOx E/R (A/min)	<1	<1	<1	<1

- The W etch rate could be adjusted to around 50A/min, 150A/min and 300A/min respectively, and meanwhile the TiN to W selectivity are close to 1
- Additionally, all three etchants are compatible with AlOx

VERSUM MATERIALS

12

Summary

- We developed the wet-etch chemistries to replace the conventional chemicals and overcome technical issues for W recess application.
- The W etch rate could be adjusted to around 50A/min, 150A/min and 300A/min respectively, and meanwhile the TiN to W selectivity are close to 1.
- All of the etchants are compatible with AlOx (etch rate < 1A/min).
- With such high TiN and W etch rates, the wet-etch chemistries make the W word-line recess application possible to be operated on SWT.

VERSUM MATERIALS

13

Acknowledgments

✓ I would like to express my appreciation to all of the group members for supporting this research.

- *Yi-Chia Lee*
- *Wen Dar Liu*
- *Amy Chen*
- *Shawn Chen*
- *Tino Li*
- *Ryback Lo*
- *Lan Shing Lee*
- *Ginny Chung*
- *Cindy Chou*
- *Sheng-Wen Wu*
- *Allen Huang*
- *Aron Chen*
- *George Peng*
- *Alan Hsu*
- *Aiping Wu*
- *Hoon Song*
- *Steve Ryu*

- IC Knowledge LLC for providing the pictures of devices on pages 3 and 5

VERSUM MATERIALS

14

THANK YOU

VERSUM
MATERIALS

15

Study of cobalt etching speed controling by pH and oxidizer concentraion

Kurita Water Industries Ltd.

○Yuichi Ogawa, Nobuko Gan, Toru Masaoka, Minami Yoshimura, Hideaki Iino

imec vzw

Quoc Toan Le, Els Kesters , Frank Holsteyns

SCREEN Semiconductor Solutions Co., Ltd.

Yuya Akanishi, Akihisa Iwasaki

Agenda

1. Introduction
2. Coupon test
3. Surface analysis
4. Mechanism
5. 300mm wafer test
6. Summary

Introduction

KURITA

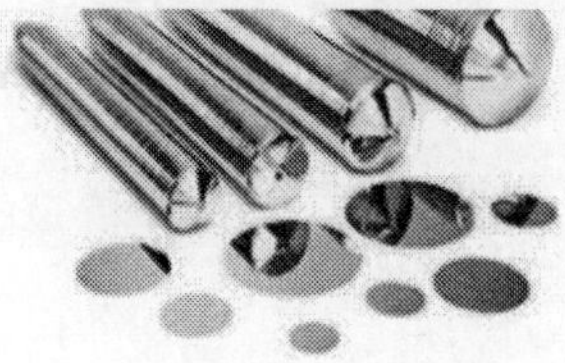

Scaling down

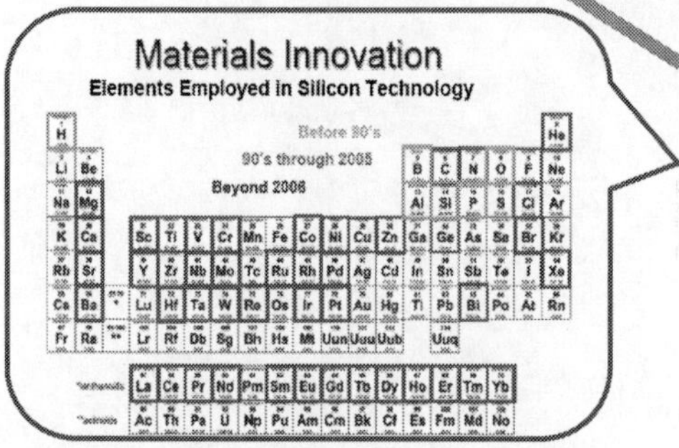

Process Innovation

- Fin
- STI (shallow trench isolation)
- Gate
- Spacer
- S/D (source/drain) EPI
- ILD0 (inter-layer dielectric)
- RMG (replacement metal gate)
- MOL (middle-of-line)
- BEOL (back-end-of-line)
- :

New processes and structures are used in device.

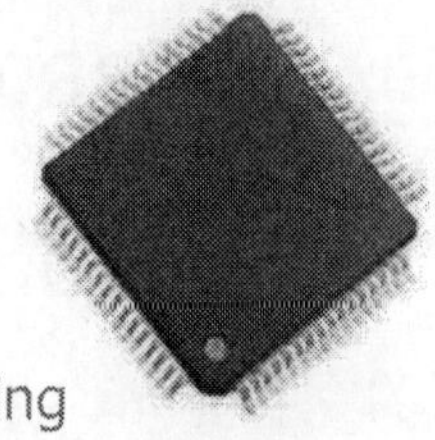

New materials are used in devices.
Some materials are corroded in water.

It is also necessary to develop the methods of cleaning new materials and new structures.

3

Introduction

KURITA

The transition of line width and material

Trend	iN8	iN7	iN5	iN5+	iN3
PP/MP/FP	56/40/30	42/32/24	42/21/21	42/21/21	36-42/16-21/16-21
Channel n/p Device architecture	Si/Si FF	Si/Si FF	Si/Si or SiGe FF	Si/Si or SiGe FF or GAA	Si/Si or SiGe GAA or CFET
BEOL Lines (Fill/Liner/Barrier) Via	Mx:Cu/TaNCo Vx:Cu Mint:**Co**/TiN Vint:**Co**	Mx:Cu/TaNCo Vx:Cu, Co Mint:Co/TiN Vint:**Co**	Mx:Ru, Cu/MnRu Vx:**Co** Mint:**Co**/TiN, Ru Vint:**Co**, Ru	Mx:Ru, Cu/MnRu Vx:**Co**, Ru Mint:**Co**/TiN, Ru Vint:**Co**, Ru	Mx:Ru, Cu/MnRu, Ru Vx:**Co**, Ru Mint:**Co**/TiN, Ru

Ref,:imec

Co will replace W or Cu in MOL and BEOL.

We have studied the new rinsing method for avoiding Co dissolution.

4

Agenda

KURITA

1. Introduction
2. Coupon test
3. Surface analysis
4. Mechanism
5. 300mm wafer test
6. Summary

Coupon test ① (pH) Experimental Procedure

KURITA

<Purpose>To study the behavior of Co dissolution concerning pH.

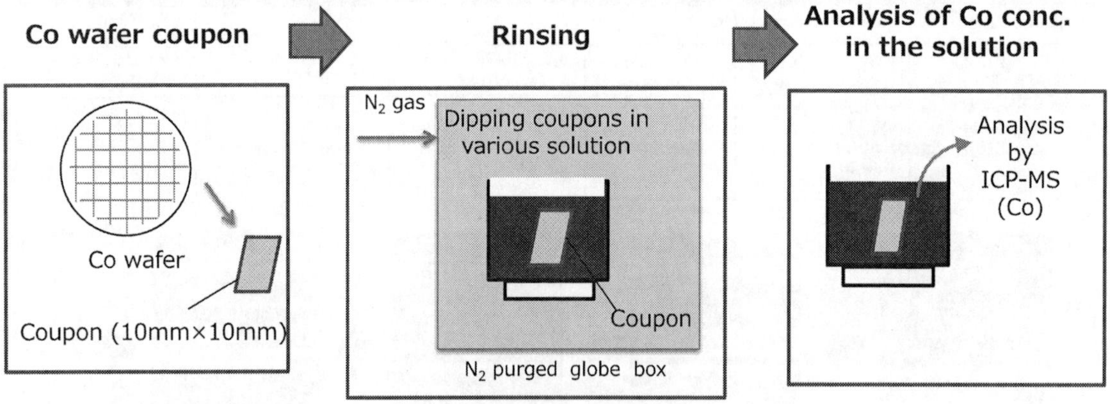

Condition	Range
pH	2-10 pH2:DHF pH5:CO_2 water PH8-10: NH_4OH water

<Co wafer>
Method :PVD
Co thickness: 28nm

Coupon test① (pH) Result

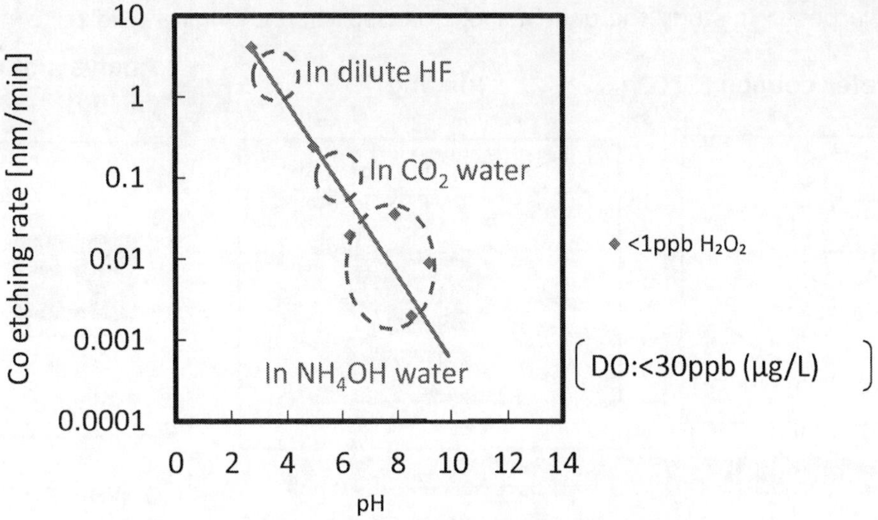

The influence of pH on the etching rate of Co

The Co etching rate significantly decreased as pH increased.

7

Coupon test① (pH)

Why can Co dissolution be avoided by increasing pH?

Condition	Co dissolution mechanism
In acid	As standard oxidation-reduction potential of Co is -0.277V, Co is etched easily. 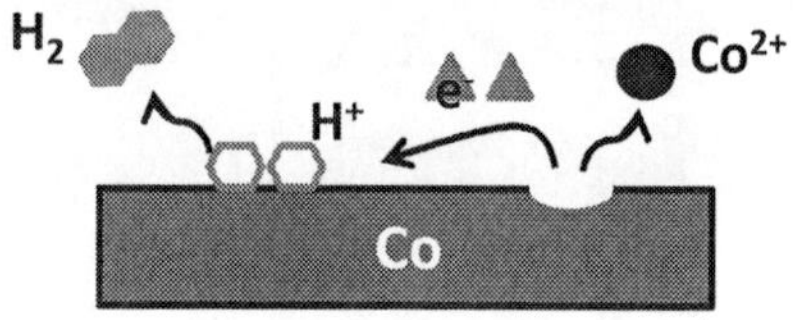
In alkali	The concentration of OH^- is higher in higher pH. →Co reacts with OH^- and forms a passivating $Co(OH)_2$ layer. →The Co will not be dissolved. 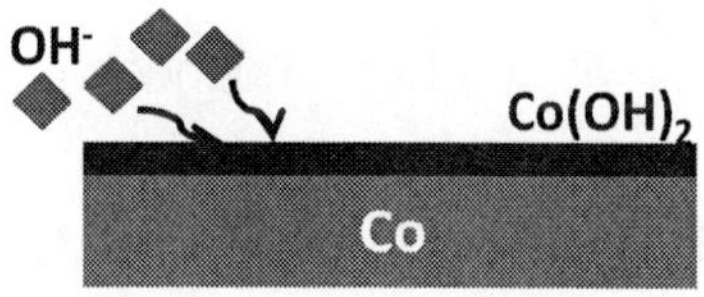

8

Coupon test② (Oxidizer) Experimental Procedure

KURITA

<Purpose>To study the behavior of Co dissolution changing oxidizer concentration

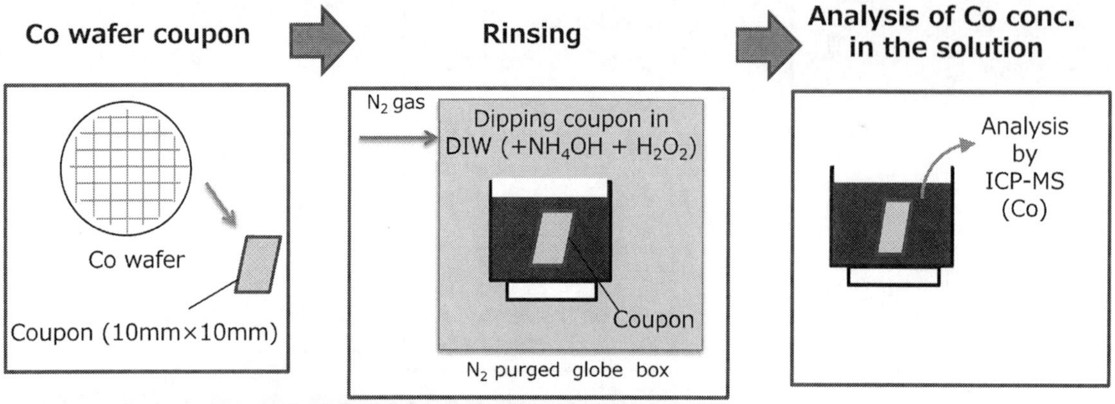

Condition	Range
pH	6-10 The pH was adjusted by adding NH_4OH
H_2O_2 conc,	**0.01-1000ppm (mg/L)**

<Co wafer>
Method :PVD
Co thickness: 28nm

Coupon test② (Oxidizer) Results

KURITA

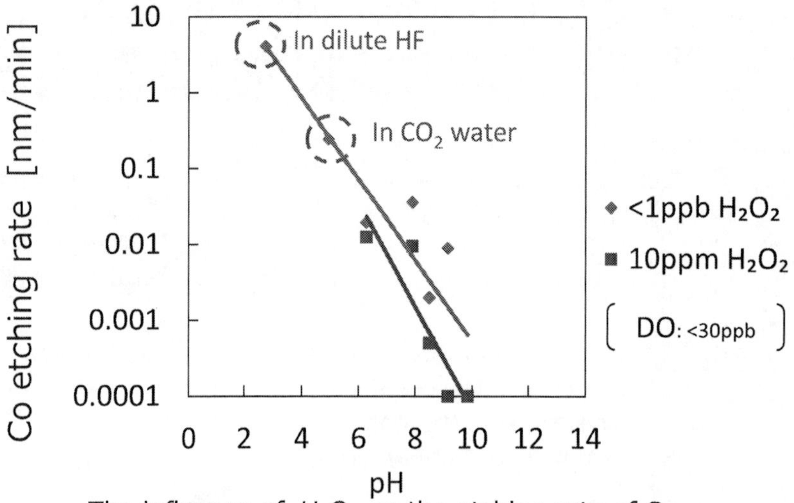

The influence of H_2O_2 on the etching rate of Co

By adding 10 ppm H_2O_2, the etching rate of Co gets lower.

Coupon test ② (Oxidizer) Results

KURITA

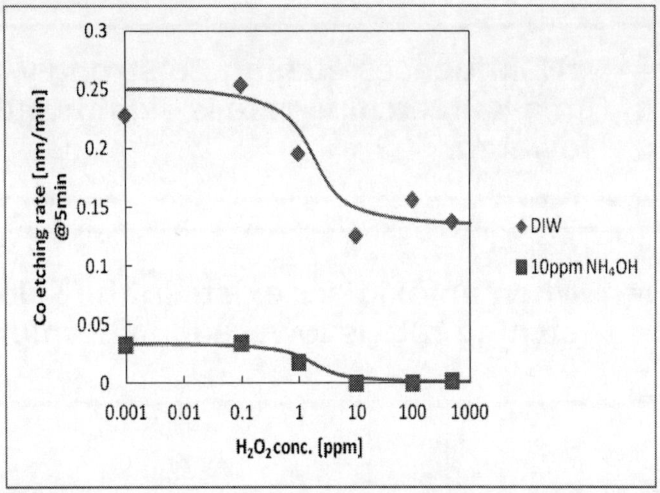

10ppm NH_4OH
:pH=9

The correlation between the concentration of H_2O_2 and the etching rate of Co.

About both solutions

· 0 → 10ppm : The etching rates of Co decreases as H_2O_2 concentration increases.

· 10ppm → :The etching rates is stable at lower level.

Agenda

KURITA

1. Introduction
2. Coupon test
3. Surface analysis
4. Mechanism
5. 300mm wafer test
6. Summary

Surface analysis

Coupon test ① ⇒ pH influences etching Co strongly. In alkali, etching rate is significantly lowered.

Coupon test ② ⇒ When an oxidizer exists in the solution, etching rate is lowered furthermore.

To clarify the mechanism of Co dissolution, we analyzed etched Co surface by XPS (X-ray photoelectron spectroscopy).

Surface analysis

<Preparation of XPS> XPS spectrum for data analysis

The binding energy concerning Co compounds are close each other,

We obtained the binding energy of $Co(OH)_2$, CoO and Co_3O_4 originally. Each sample was prepared as follow.

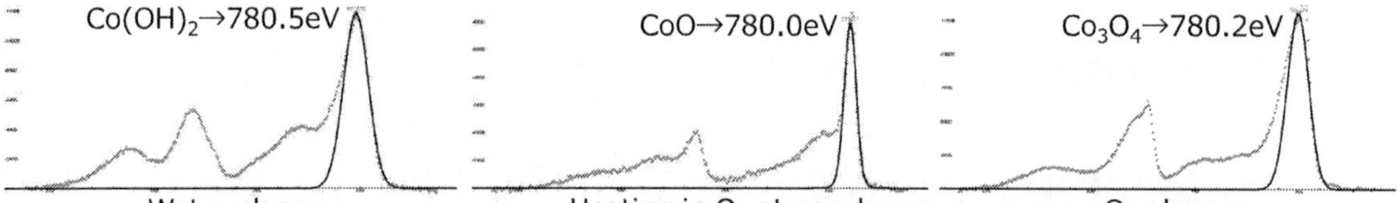

Water plasma | Heating in O_2 atmosphere | O_2 plasma

XPS result of Co surface

Compounds	Deposition method	Binding energy (eV)
$Co(OH)_2$	Water plasma	780.5
CoO	Heating in O_2 atmosphere	780.0
Co_3O_4	O_2 plasma	780.2

Surface analysis Experimental Procedure

KURITA

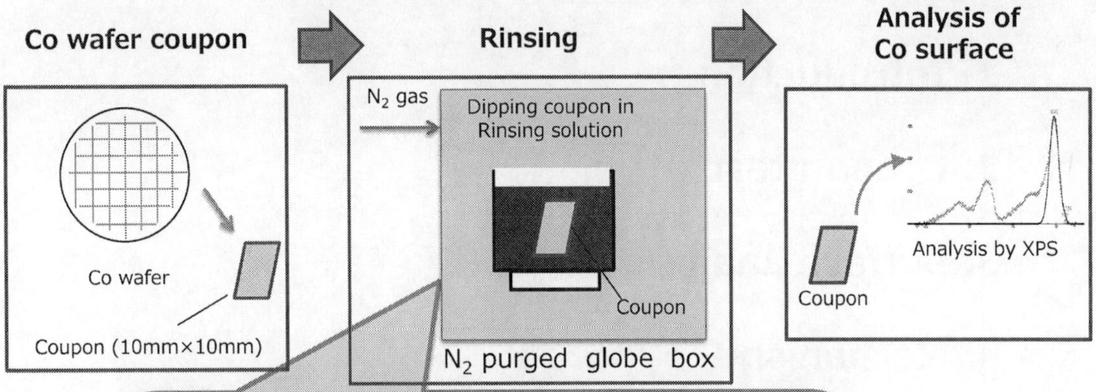

No.	Condition
1	No rinsing
2	0.05% DHF
3	0.05% DHF → 10ppm NH_4OH
4	0.05% DHF → 10ppm H_2O_2
5	0.05% DHF → 10ppm NH_4OH+10ppm H_2O_2

15

Surface analysis Results

KURITA

Rinsing condition and Co compound ratio on the surface after rinsing

No.	Condition	Co compound ratio [%]		
		Co	CoO	$Co(OH)_2$
1	No rinsing	N.D.	0.7	99.2
2	0.05% DHF	24.2	N.D.	75.8
3	0.05% DHF → 10ppm NH_4OH	3.2	N.D.	96.8
4	0.05% DHF → 10ppm H_2O_2	1.1	2.9	96.1
5	0.05% DHF → 10ppm NH_4OH+10ppm H_2O_2	0.6	3.0	96.5

Based on these results, we assumed reactions by rinsing.

16

Agenda

KURITA

1. Introduction
2. Coupon test
3. Surface analysis
4. Mechanism
5. 300mm wafer test
6. Summary

Mechanism (DHF)

KURITA

No.	Condition	Co compound ratio [%]		
		Co	CoO	$Co(OH)_2$
1	No rinsing	N.D.	0.7	99.2
2	0.05% DHF	24.2	N.D.	75.8

Hypothetical mechanism of Co surface rinsed with DHF

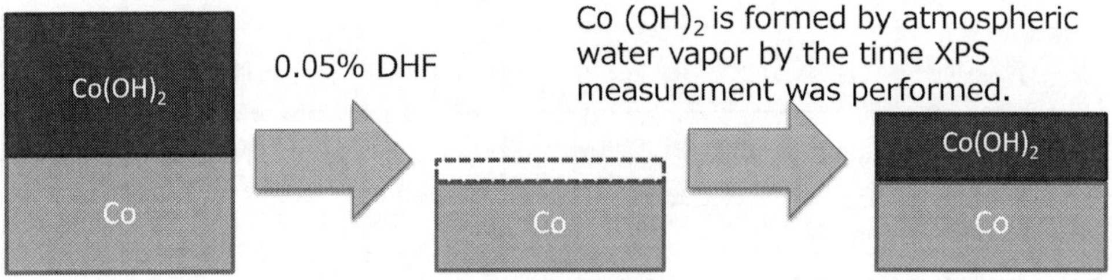

$Co (OH)_2$ is formed by atmospheric water vapor by the time XPS measurement was performed.

$Co(OH)_2 + 2H^+ \rightarrow Co^{2+} + 2H_2O$
$Co + 2H^+ \rightarrow Co^{2+} + H_2$

$Co + 2H_2O \rightarrow Co(OH)_2 + H_2$

Mechanism (NH_4OH)

KURITA

No.	Condition	Co compound ratio [%]		
		Co	CoO	$Co(OH)_2$
1	No rinsing	N.D.	0.7	99.2
2	0.05% DHF	24.2	N.D.	75.8
3	0.05% DHF → 10ppm NH_4OH	**3.2**	**N.D.**	**96.8**

Hypothetical mechanism of Co surface rinsed with NH_4OH water.

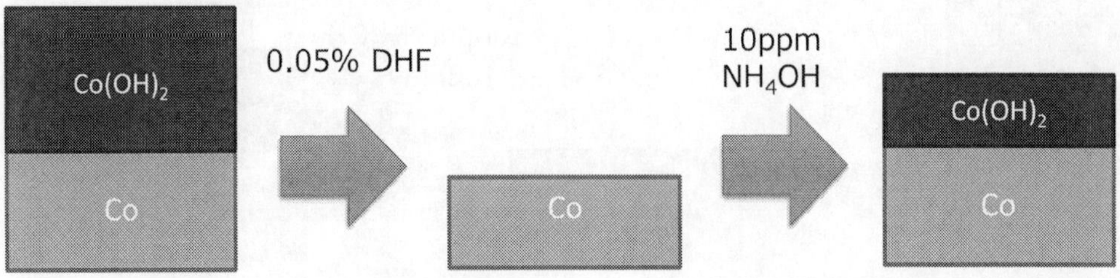

$Co(OH)_2 + 2H^+ \rightarrow Co^{2+} + 2H_2O$
$Co + 2H^+ \rightarrow Co^{2+} + H_2$

$Co + 2H_2O \rightarrow Co(OH)_2 + H_2$ ○
$Co + 2H^+ \rightarrow Co^{2+} + H_2$ ×

○:Fast , ×:No react

Mechanism (H_2O_2)

KURITA

No.	Condition	Co compound ratio [%]		
		Co	CoO	$Co(OH)_2$
1	No rinsing	N.D.	0.7	99.2
2	0.05% DHF	24.2	N.D.	75.8
4	0.05% DHF → 10ppm H_2O_2	**1.1**	**2.9**	**96.1**

Hypothetical mechanism of Co surface rinsed with H_2O_2 water.

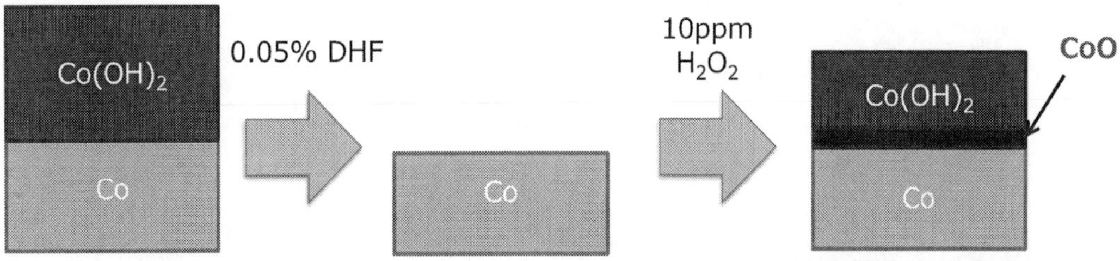

$Co(OH)_2 + 2H^+ \rightarrow Co^{2+} + 2H_2O$
$Co + 2H^+ \rightarrow Co^{2+} + H_2$

$Co + H_2O_2 \rightarrow CoO + H_2O$ ○
$CoO + H_2O \rightarrow Co(OH)_2$ ○
$Co + 2H^+ \rightarrow Co^{2+} + H_2$ △
$Co(OH)_2 + 2H^+ \rightarrow Co^{2+} + 2H_2O$ △

○:Fast , △:Slow

Mechanism ($NH_4OH+H_2O_2$)

KURITA

No.	Condition	Co compound ratio [%]		
		Co	CoO	$Co(OH)_2$
1	No rinsing	N.D.	0.7	99.2
2	0.05% DHF	24.2	N.D.	75.8
5	0.05% DHF → 10ppm NH_4OH+10ppm H_2O_2	**0.6**	**3.0**	**96.5**

Hypothetical mechanism of Co surface rinsed with NH_4OH and H_2O_2 water.

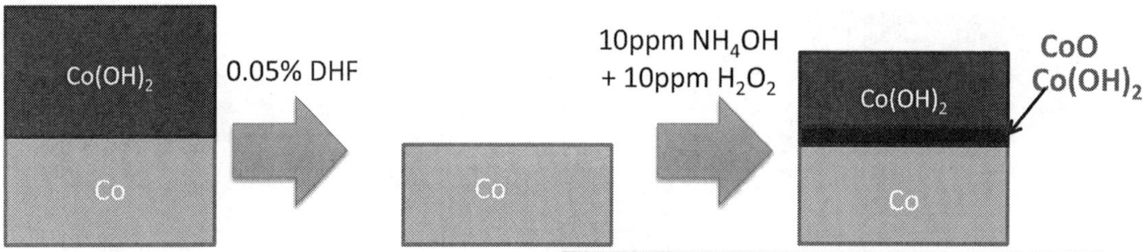

$Co(OH)_2 + 2H^+ \rightarrow Co^{2+} + 2H_2O$
$Co + 2H^+ \rightarrow Co^{2+} + H_2$

$Co + 2H_2O \rightarrow Co(OH)_2 + H_2$	○
$Co + H_2O_2 \rightarrow CoO + H_2O$	○
$CoO + H_2O \rightarrow Co(OH)_2$	○
$Co + 2H^+ \rightarrow Co^{2+} + H_2$	×
$Co(OH)_2 + 2H^+ \rightarrow Co^{2+} + 2H_2O$	×

○:Fast , ×:No react

Agenda

KURITA

1. **Introduction**
2. **Coupon test**
3. **Surface analysis**
4. **Mechanism**
5. **300mm wafer test**
6. **Summary**

300mm Wafer test

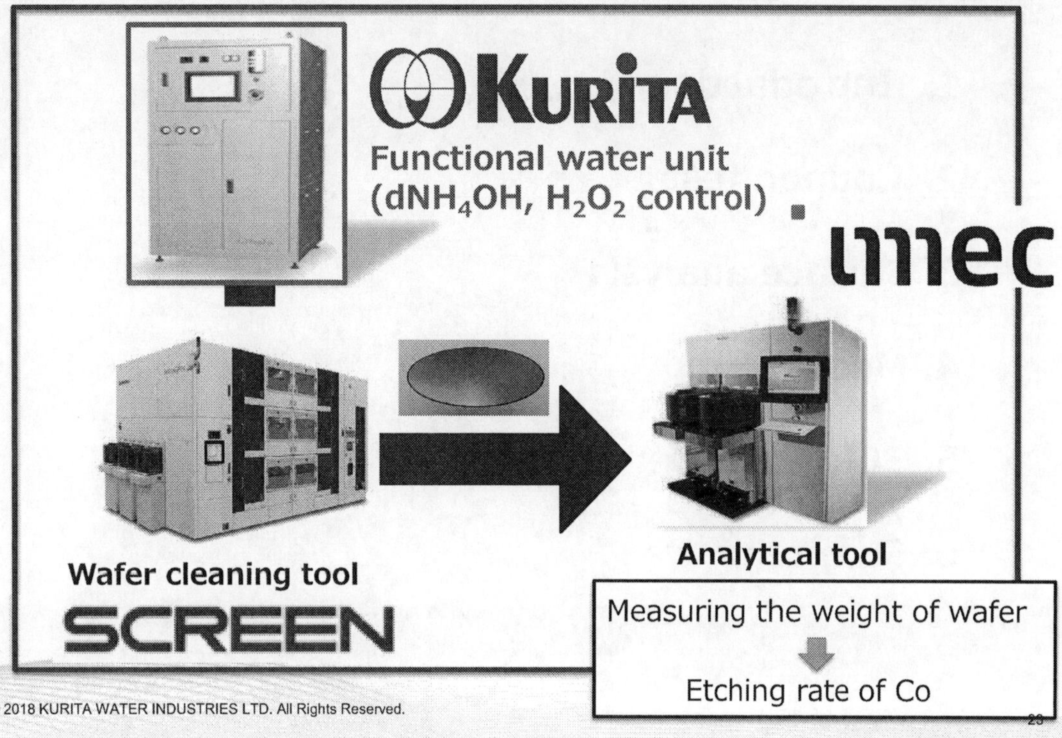

300mm Wafer test

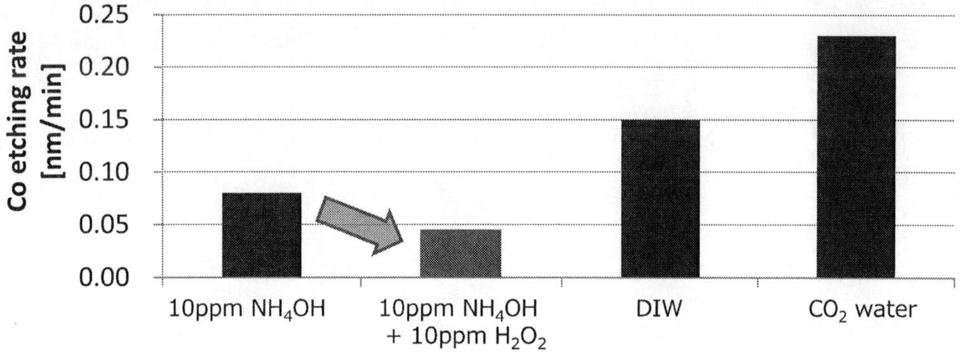

Result of 300mm wafer test

Condition	Conductivity
10ppm NH_4OH +10ppm H_2O_2	20-30 μS/cm
CO_2 water	10 μS/cm

Co dissolution is avoided by rinsing with NH_4OH and H_2O_2 water in the wafer test.

300mm wafer test reproduced the result of coupon test.

Agenda

KURITA

1. Introduction
2. Coupon test
3. Surface analysis
4. Mechanism
5. 300mm wafer test
6. Summary

25

Summary

KURITA

- pH and/or H_2O_2 existence influence Co dissolution strongly.
- CoO, which is formed by oxidizer, functions as the passivation film.
- 300mm wafer test reproduced the result of coupon test.

26

KURITA

Thank you for your attention.

Contact:
Yuichi Ogawa (Kurita)
yuuichi.ogawa@kurita.co.jp
+81-280-54-2697

Developing Wet Cleans for a Cobalt Contact Integration Scheme

Akshey Sehgal, Elango Balu and Scott Beasor
GLOBALFOUNDRIES Inc.
400 Stonebreak Road Extension, Malta, NY 12020 USA
Akshey.Sehgal@GlobalFoundries.com

Abstract – Due to ever shrinking dimensions at advanced technology node and the use of FinFET devices, the contact resistance has gone up significantly. There are various resistance contributions that impact device performance, such as contact resistance, silicide resistance, schottky barrier height, etc. (Figure 1[1]). Efforts are underway to reduce these resistance(s). In this paper, we will detail our wet clean process development for one of the potential approaches to reduce contact resistance (by changing the contact fill metal from W to lower resistance metal, Co).

Apart from forming silicides, Co has not been widely used in advanced technology nodes. This requires development of new wet cleans for Co contact integration. In the Cobalt integration scheme, both W and Co are simultaneously exposed during the post etch residue removal step. The ideal clean for this step needs to remove all the post-etch residue and not attack the exposed W and Co metals, while keeping the contact dimensions unchanged.

This proved to be a problem because all of the existing middle of line (MOL) wet cleans were unable to meet all of the ideal clean conditions simultaneously. Review of the Pourbaix diagrams for W and Co revealed a common passivation area in highly reducing conditions with no oxidizer present (Figure 2). These highly reducing, no oxidizer conditions are difficult to create. Even if these conditions were created and maintained, complete removal of the mixed metal-organic post-etch residue would be slow and/or incomplete. Fig. 2 shows that in strongly acidic solutions, W is passivated and Co dissolves. W and Co both dissolve in weakly acidic and neutral solutions. W dissolves in alkaline pHs while Co is passivated in alkaline pHs. Therefore, one cannot perform W post etch residue removal wet cleans without attacking Co.

To summarize, the problem is that thermodynamically and kinetically there is no wet clean that can be used when W and Co are simultaneously exposed at the bottom of the contact. Without developing a new wet clean for this step, one will have high contact resistance and defeat the motivation of integrating a Co contact.

This paper will detail how this wet clean was developed.

REFERENCES:

1) Kelin Kuhn IEDM 2008

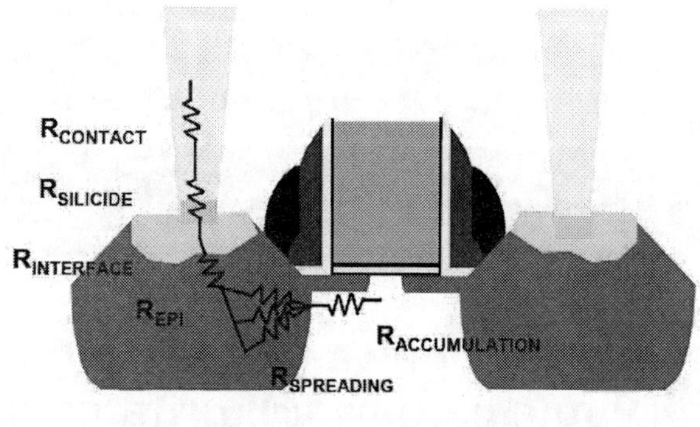

Figure 1: Resistances in a planar device[1]

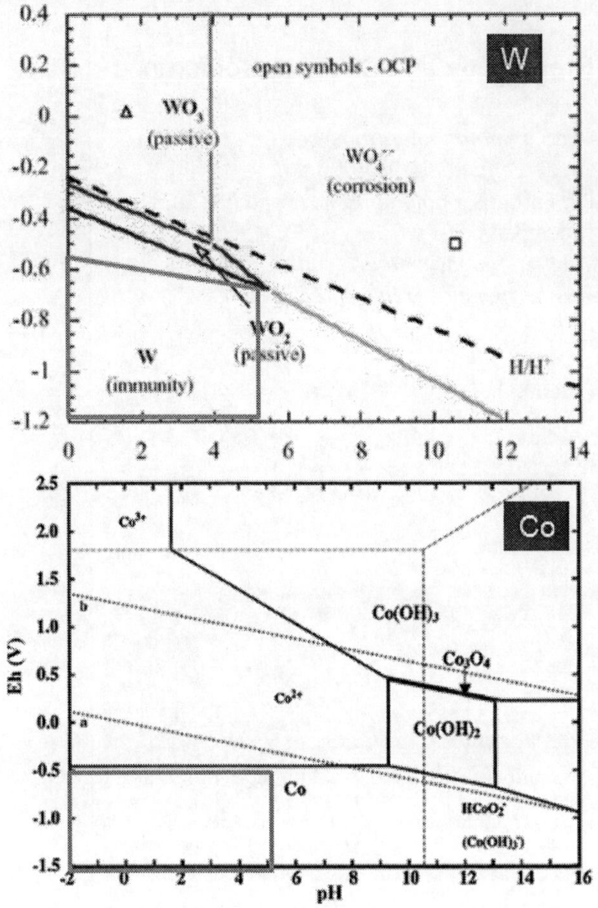

Figure 2: Pourbaix diagrams of W-H_2O and Co-H_2O at 25 °C. Green boxes show approximate area where both W and Co are passivated.

2018 SPCC – Surface Preparation and Cleaning Conference

Minh-Phuong TRAN, CEA-Leti, *minh-phuong.tran@cea.fr*
Hervé FONTAINE, CEA-Leti, *herve.fontaine@cea.fr*
Paola GONZÁLEZ-AGUIRRE, Entegris, *paola.gonzalez@entegris.com*
Carlos BEITIA, CEA-Leti, *carlos.beitia@cea.fr*
Sung-In MOON, Entegris, *sungin.moon@entegris.com*
Jorgen LUNDGREN, Entegris, *jorgen.lundgren@entegris.com*

Deposition of volatile chlorohydric acid on Copper wafer depending on humidity and HCl airborne concentration

April 10-11th, 2018 in Cambridge, MA, USA

CEA–Leti, University Grenoble Aples, 17 rue des Martyrs, 38054 Grenoble, France
Entegris, SAS, Parc Centr'Alp Ouest, 196 rue du Rocher de Lorzier, 38430 Moirans, France

OUTLINE

1. **Introduction**
2. **Experimental setup**
3. **Result - HCl deposition onto Cu 200 mm wafer**
   - Relative Humidity (RH<1%, RH 40% and RH 70%)
   - Airborne concentration (range of 20 – 400 ppbv)
4. **XPS characterization of contaminated Cu surfaces**
5. **HCl - Cu mathematical model**
6. **Conclusion**

INTRODUCTION

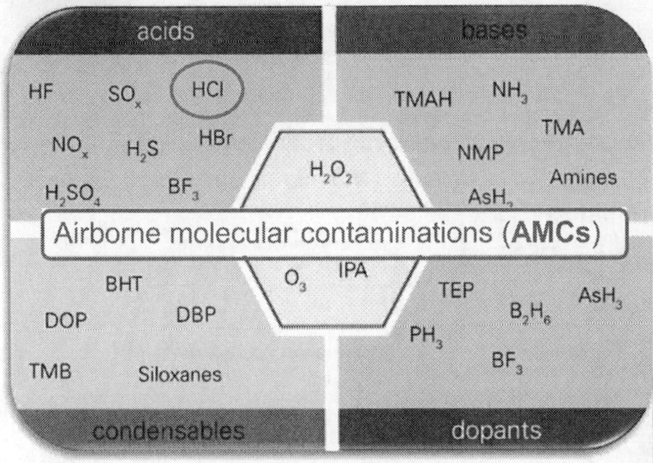

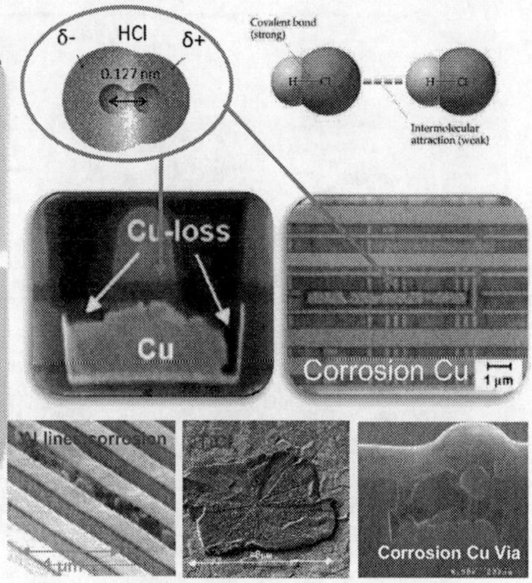

- Cleaning: IPA, acetone, ethanol, terpenes HCl, HNO_3, NH_4OH, etc.
- Plasma etching gases/etched materials: (Cl_2/CCl_4 + Ar), HCl gas (36%) for Al
- Silicon vapor-phase epitaxy – HCl gas

- **Yield losses**
- **Limit AMCs effect**

INTRODUCTION

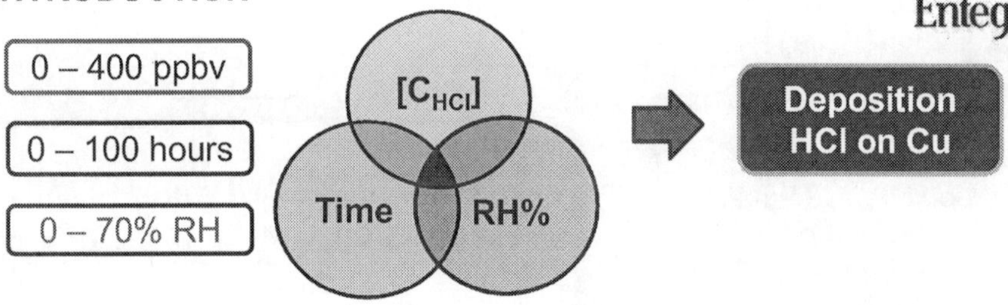

- **Fundamental research deposition HCl on Cu surface (200mm wafer)**

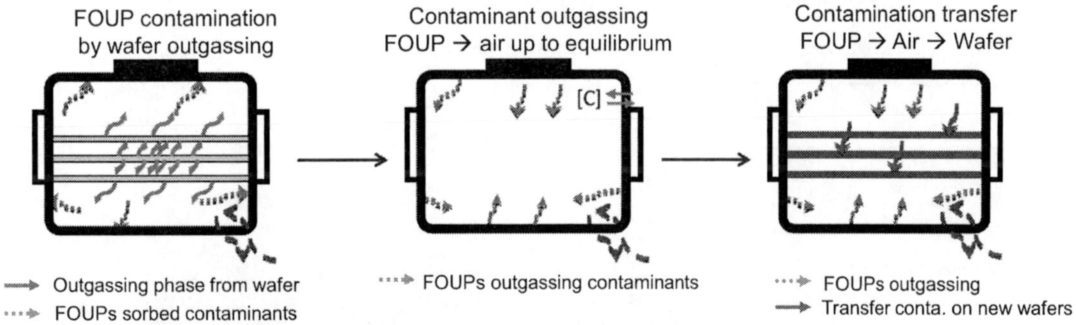

- **Knowledge of [C_{HCl}] in airborne → minimizing defect**
- **Predicting storage time**

EXPERIMENTAL SETUP

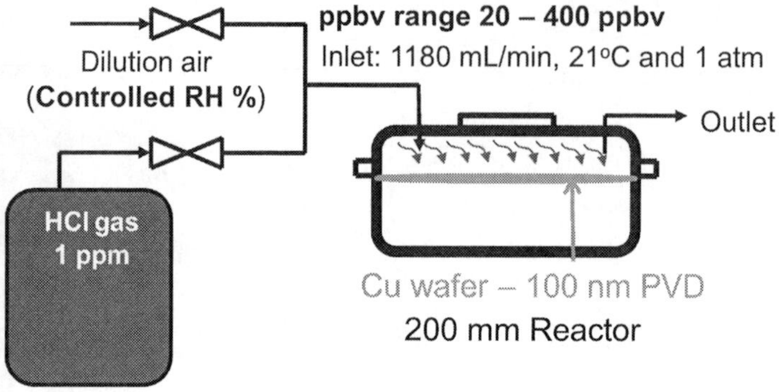

- Copper wafers: Physical Vapor Deposition (PVD) - 100nm copper
- Stored Cu wafer in polypropylene box (PVD process - HCl deposition)
- Diluted HCl gas by clean air

1st

EXPERIMENTAL SETUP

Liquid Phase Extraction (LPE process) Collect DI water on Cu wafer

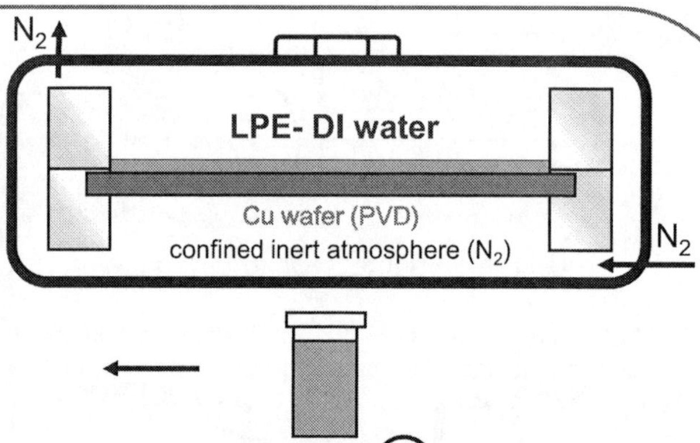

Quantify ion Cl^- by Ionic Chromatography (IC)

Cl^- ion deposition on Cu wafer

$HCl + H_2O \rightarrow Cl^- + H_3O^+$

Efficiency of LPE-IC

- Collection by LPE is > 90% at first extraction(LPE1)
- LLD for Chloride: **< 1.0 x 10^{11} ion/cm^2 (0.05 ppbw)**

2nd

RESULT – HCL DEPOSITION ON CU WAFER

HCl Deposition Kinetic on Cu at 20 ppbv, **varying RH**

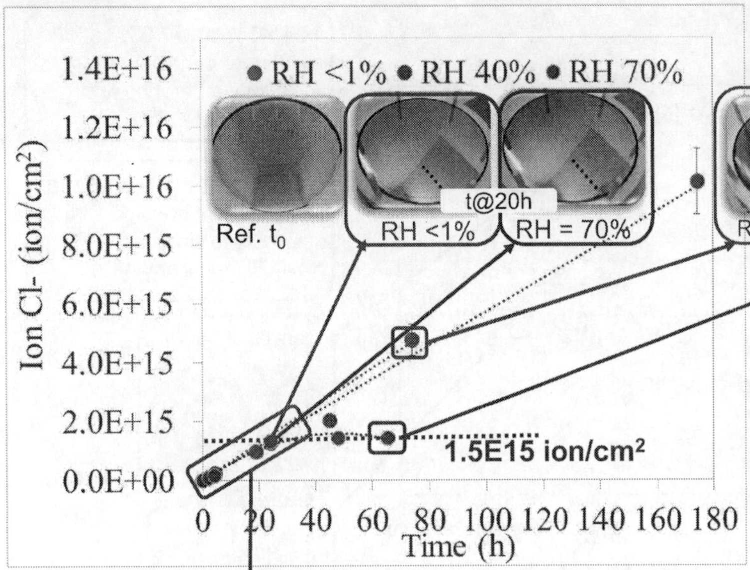

RH <1% and 40%

- Linear kinetic
- Same slope

RH **70%** (<25h)

- Linear kinetic up to saturation threshold
- Same slope < E15 ion/cm²

$[Cl^-]_{HCl@20ppbv}$ = 5.3E13**[Time]* + *[HCl]*$_{to}$

- ❑ Deposition HCl independent on RH (<40%)
- ❑ Threshold 1.5E15 ion/cm² (RH 70%) (monolayer)

RESULT - XPS ANALYSIS - EFFICIENCY OF LPE-IC

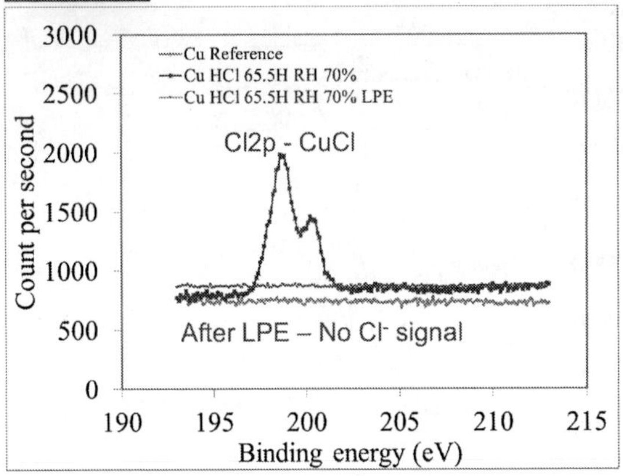

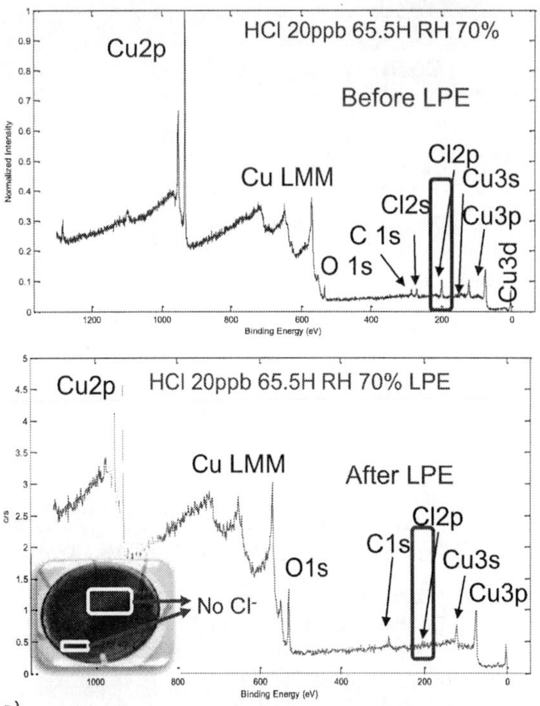

- ❑ Collection by LPE is > 95%
- ❑ Confirm by XPS (Before LPE – After LPE)
  - Before LPE: Cl2p peaks – double peak
  - After LPE: No Cl2p signal
- ❑ After LPE: No Cl2p detected signal - different spots of Cu wafer

Saturation level measured by IC is accurate
(no artifact measurement due to an insoluble Cl⁻ species)

RESULT – HCL DEPOSITION ON CU WAFER

HCl Deposition on Cu for 2h vs. **HCl Airborne Concentration**

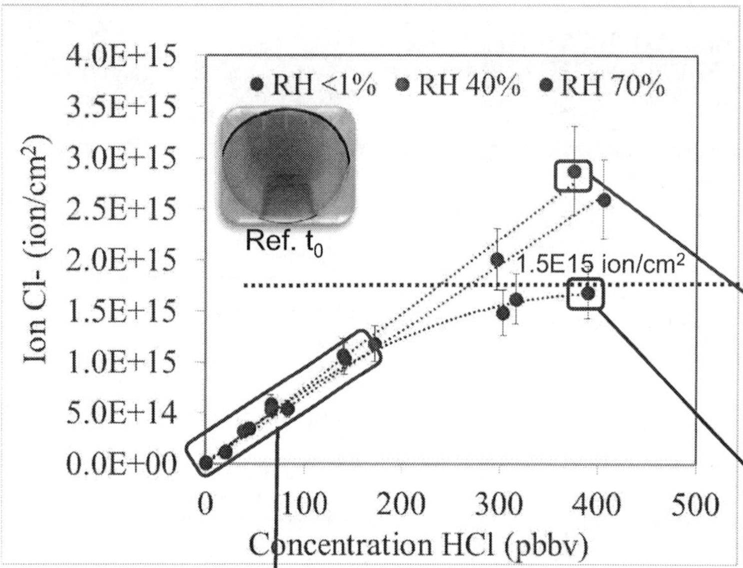

RH <1% and **40%**
- Linear behavior
- Slope difference not signifiant > 250 ppbv

RH = **70%** (< threshold)
- Linear behavior
- Same slope as RH <1% and **40%**

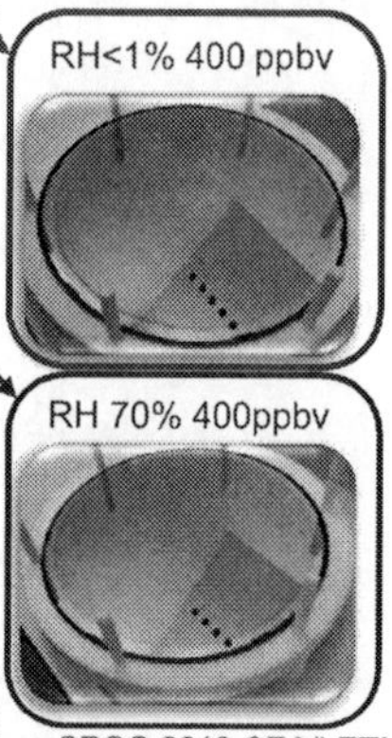

$[Cl^-]_{2H,\ time} = 6.9E12*[C_{HCl}] + [HCl]_{to}$

HCl threshold concentration showed at 1.5E15 ion/cm² (monolayer)

RESULT - XPS ANALYSIS

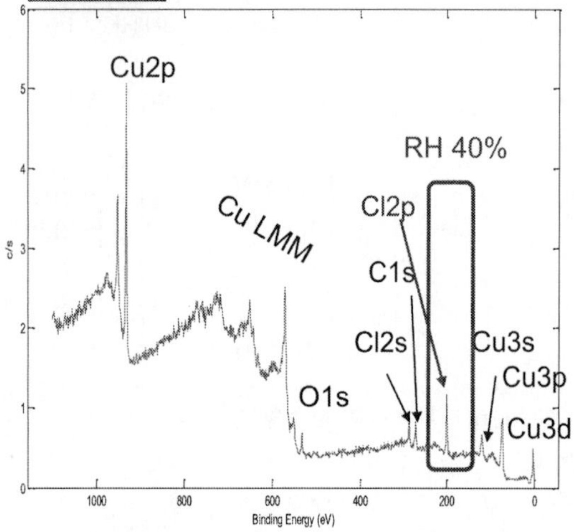

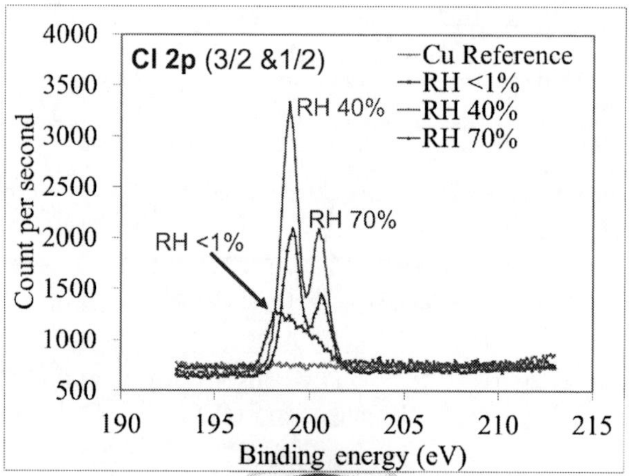

Cu contaminated at 20 ppbv, 24h, varying RH

→ IC measurement showed 1.5E15 Cl⁻/cm² whatever RH%

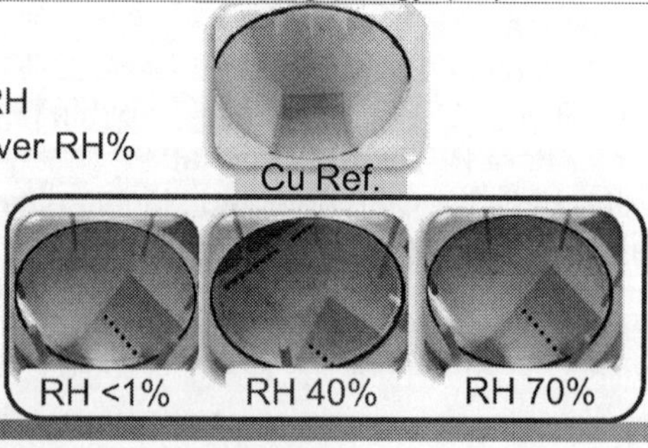

- XPS before LPE
  - → Bonding nature of Cl⁻ on Cu surface
- RH <1%: HCl_{gas} adsorption and Cl⁻ as CuCl
- RH 40%: Cl⁻ as CuCl } → New Cu compounds (Color change)
- RH 70%: Cl⁻ as CuCl }

RESULT - XPS ANALYSIS

XPS results Cu-HCl – Extended exposure time: 24h vs. 66h

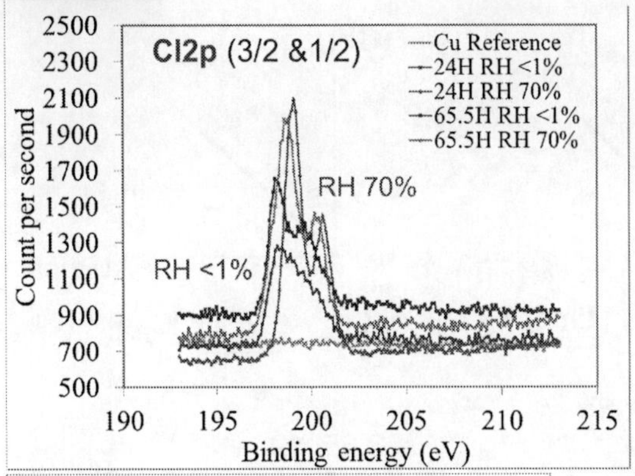

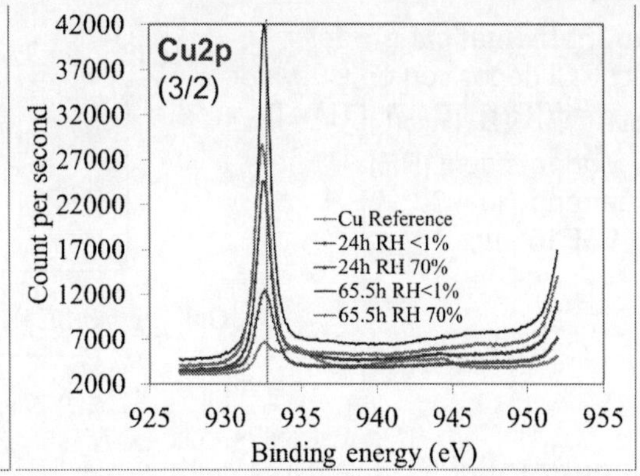

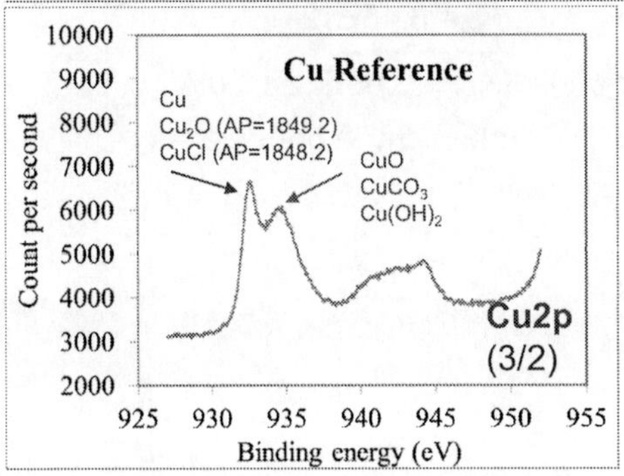

- ❑ RH 70%: Cl^- ~1.5E15 ion/cm² (by IC measurement)
- ❑ Extended exposure time
  - RH >1%: Cl^- as CuCl (65.5h), possibly HCl_{gas}
  - RH 70%: Cl^- as CuCl (65.5h)
- Color
  - RH 70%→ Strong color change (multiple colors) → new Cu compound ($CuCl_x$, Cu_xO, Cu_xCO_3, $Cu(OH)_x$, etc.
  - RH >1% → slight color change (mostly CuCl)
- Oxidation state
  - XPS showed only Cu, Cu^+ peak (RH<1% and RH 70%) while Cu Ref. has Cu, Cu^+ and Cu^{+2}

XPS ANALYSIS

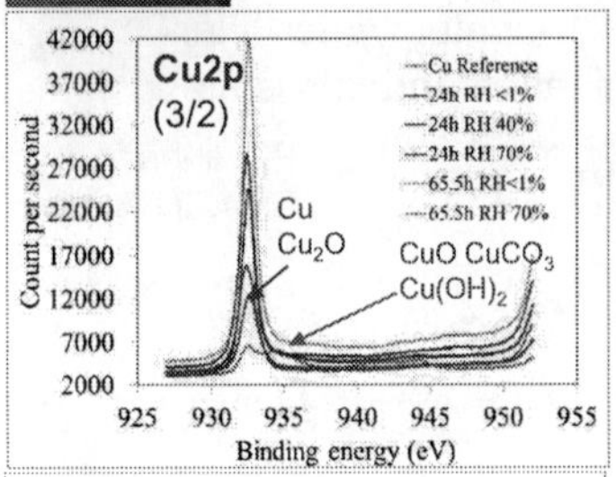

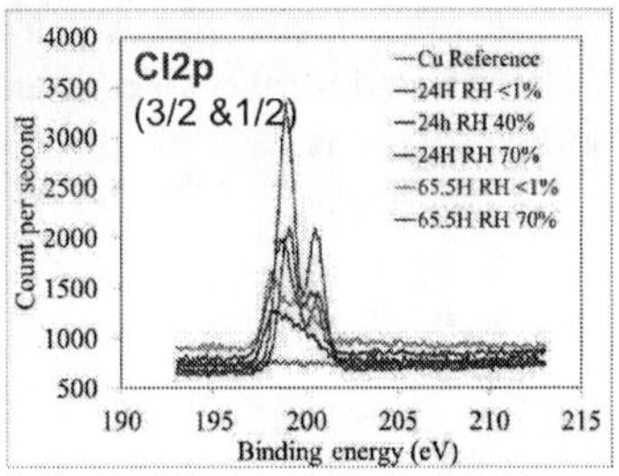

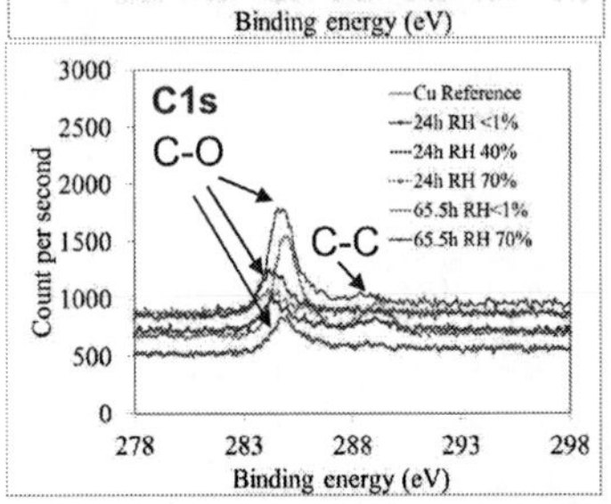

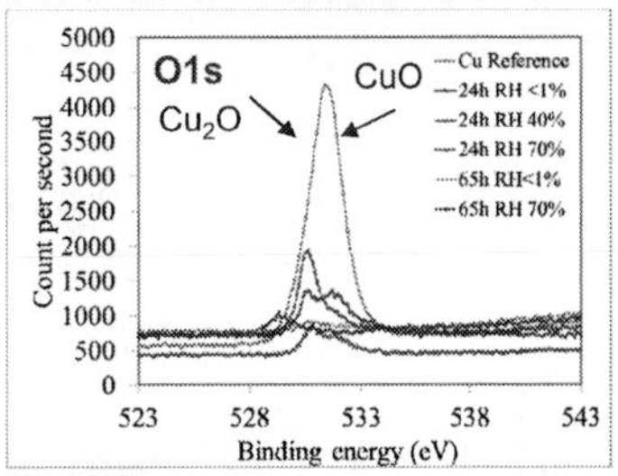

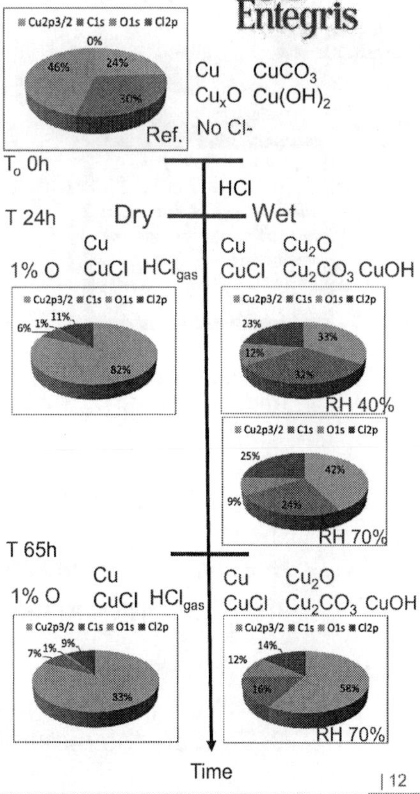

Dry condition: RH<1% Cu|Cu_xO → Cu|CuCl

- HCl, Cl^- replaced Oxygen on Cu surface (Cu, CuCl, HCl, C-C organic); only Cu (I)

Wet condition: RH ≥ 40% Cu|Cu_xO → Cu|CuCl|Cu_xO|Cu_xCO_3, etc.

- HCl, Cl^- reacts CuO & Cu_2O; mainly Cu (I) (Cu, Cu_2O, CuCl, CuOH, Cu_2CO_3, possibly $CuCl_2^-$ (complex))

RESULT – HCL DEPOSITION ON CU WAFER

Linear mathematical model

Predict HCl deposition on Cu wafer

$\mathbf{[Cl^-]_{surf} = f([RH], [C_{HCl}], [Time])}$

$[Cl^-]_{surf}$ independent [RH]

On the range [0 – 70%] RH

up to 1.5E15 at/cm²

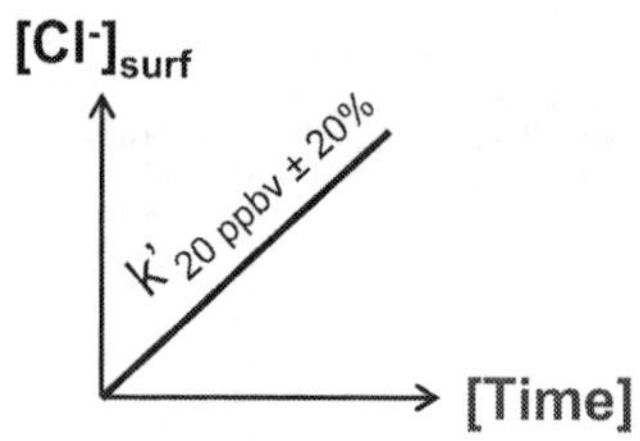

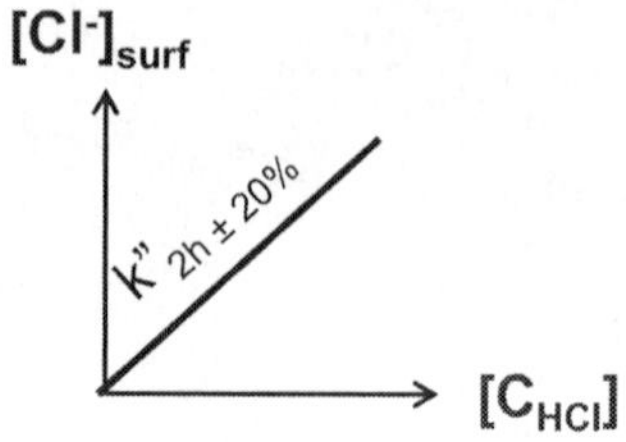

$[Cl^-]_{20ppbv} = 5.3E13*[Time] + [Cl^-]_o$ $\quad k'_{20\,ppbv} = k_c*20_{ppbv}$

$[Cl^-]_{2h} = 6.9E12*[C_{HCl}] + [Cl^-]_o$ $\quad k''_{2h} = k_h*2h$

$K_c = 2.65E12$

$K_h = 3.45E12$

$K_{c,h} = 3.05E12 \pm 20\%$ atom/cm².h.ppbv

Linear model: $C_{surf} = K.C_g t$, with K = ± 20%

$[Cl^-]_{surf} = K_{C,h}[C_{HCl}][Time] + [Cl^-]_o$

With $K_{c,h} = 3.05E12 \pm 20\%$

- ❑ For RH [0 – 70%] Linear model up to 1.5E15 Cl⁻/cm²
- ❑ RH < 40% : valid up to few E16 at/cm²

APPLICATION LINEAR MODEL HCL – CU

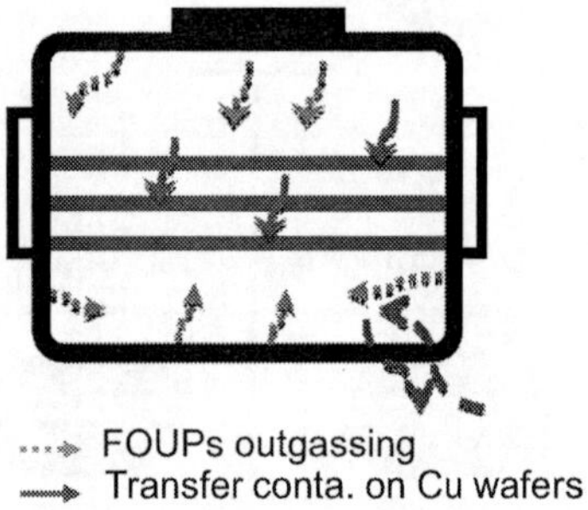

HCl transfer kinetic to Cu-coated wafer after FOUPs intentional contamination and purge step [*]

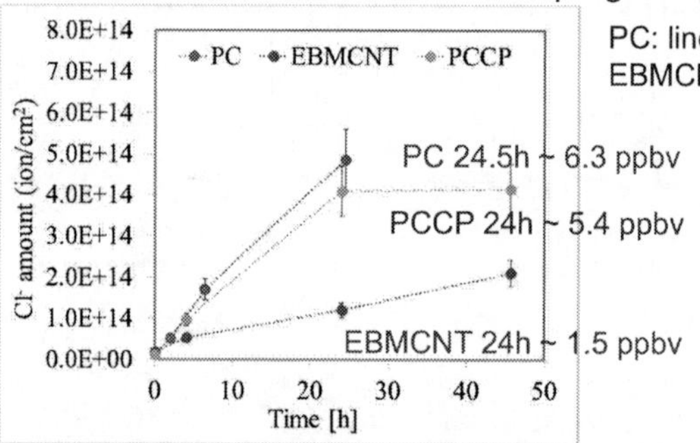

PC: linear
EBMCNT: linear

[*] F. Herran *et al. Micro electronic Engineering 169 (2017) 34 - 38*

- ❑ Example 1: Storage Cu wafer in different FOUPs (PC, PEI, PEEK, EBMCNT, etc.)
  - Known the [Cl⁻] amount transferred on Cu wafer at given time
    → Estimate the **average HCl airborne concentration** in FOUPs

    $[Cl^-]_{surf} = k*[C_{HCl}]*[Time] + [Cl^-_o]$
- ❑ Example 2: ITRS 2016 – limit Cl⁻ on Cu surface E13 Cl⁻/cm²
  - Known the $[C_{HCl}]$ outgassing in certain FOUPs
    → Estimate the **maximum storage time**

CONCLUSIONS

Entegris

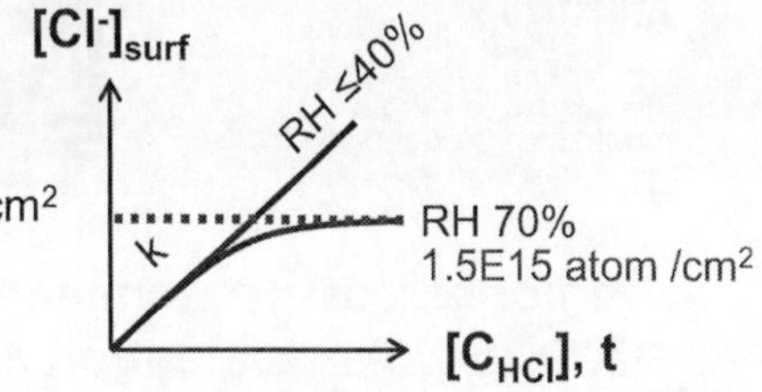

- ❑ HCl deposition on Cu surface
  - Linear law versus $[HCl]_{air}$, Time
  - No dependent on RH% [0-70%] up to 1.5E15 atom /cm²

  $[Cl^-]_{surf} = k^*\mathbf{[C_{HCl}]^*[Time]} + [Cl^-_o]$
  With k = 3.05E12 (Cl^-/cm^2.h.ppbv) ± 20%

- ❑ Oxidation state of Cu after HCl exposition is mostly Cu^+

  Dry condition: RH <1%
  - HCl, Cl^- replaced Oxygen on Cu surface (Cu, CuCl, HCl, C-C organic); only Cu (I)

  Wet condition: RH ≥ 40%
  - HCl, Cl^- reacts CuO & Cu_2O; mainly Cu (I) (Cu, Cu_2O, CuCl, CuOH, Cu_2CO_3, possibly $CuCl_2^-$ (complex))

- ❑ Not recommend to store Cu wafer in high humidity (RH 70%)
- ❑ Model can be used as industrial tool to predict and to limit the yield losses caused by HCl in cleanroom conditions

SPCC 2018 CEA/LETI-ENTEGRIS | 15

Thanks for your attention

Leti, technology research institute
Commissariat à l'énergie atomique et aux énergies alternatives
Minatec Campus | 17 rue des Martyrs | 38054 Grenoble Cedex | France
www.leti.fr

INSTITUT CARNOT CEA LETI

Entegris
50 YEARS OF PURE ADVANTAGE

SPCC 2018
Surface Preparation and Cleaning Conference

leti
CEA tech

CAMBRIDGE 10 & 11th APRIL 2018

Ammonia cross contamination: FOUP to wafer evaluation and its volatile acids comparison

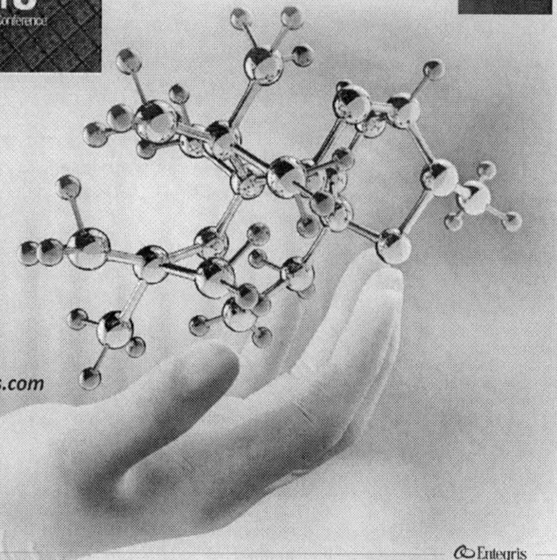

Surface Preparation and Cleaning Conference

Paola González-Aguirre, Entegris, *paola.gonzalez@entegris.com*
Hervé Fontaine, CEA-Leti, *herve.fontaine@cea.fr*
Sung In Moon, Entegris, *sungin.moon@entegris.com*
Minh Phuong Tran, CEA-Leti, *minh-phuong.tran@cea.fr*
Jorgen Lundgren, Entegris, *jorgen.lundgren@entegris.com*
Carlos Beitia, CEA-Leti, *carlos.beitia@cea.fr*

Entegris

OUTLINE

- **Introduction**
- **Issues/objectives**
- **Experimental protocol**
- **Results**
  - Contaminant sorption by the FOUPs
  - Contaminant outgassing from the FOUPs
  - Contaminant transfer to wafers
- **Conclusions**

Entegris leti

FOUP/WAFER MOLECULAR CONTAMINATION

- Demonstration of a cross-contamination chain between FOUP & wafers

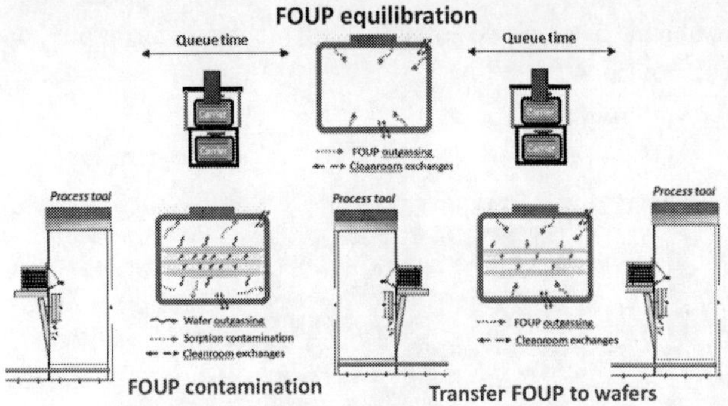

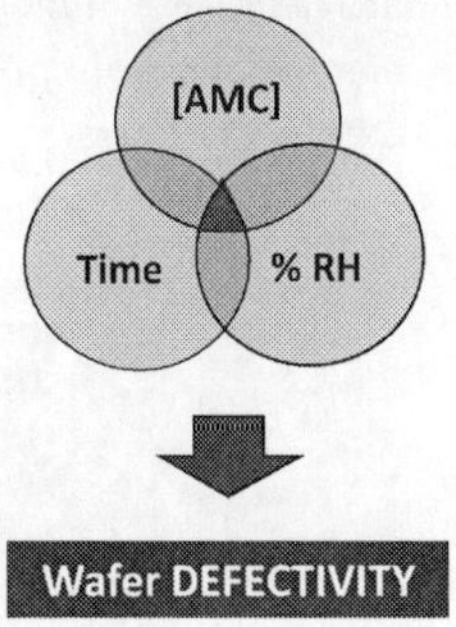

- FOUPs must protect the wafer environment from AMCs
  - Cross-molecular contamination: **AMCs → FOUP → Wafer**

Sorption of AMCs from CR air, released from stored wafers, equipment connection…

Sensitive subsequent outgassing, transfer to wafers (promoting defectivity)

INTRODUCTION: AMC

- Airborne molecular contamination (AMC) may be responsible for severe yield losses
- Among AMCs, **HX** are well known as **root cause of defectiveness**

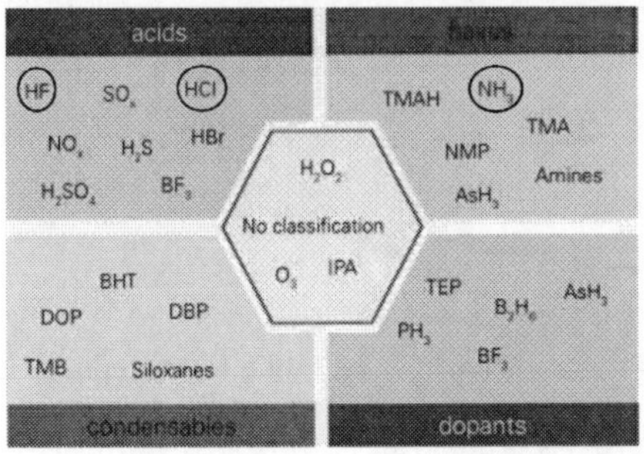

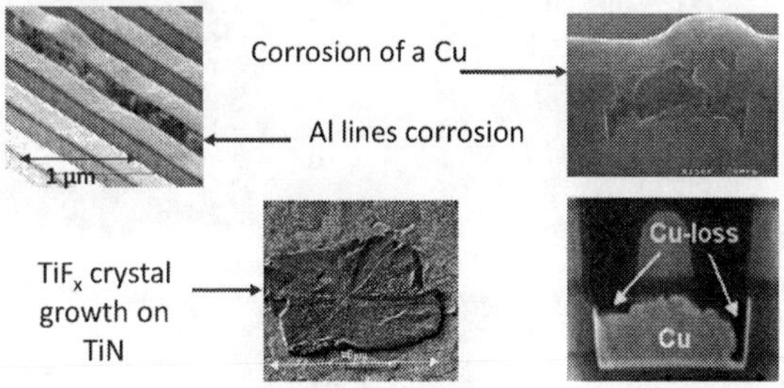

- Fabs feedback concerning **ammonia**
  - Photolithography (T-topping)
  - Crystal growth un photomask
  - Time dependent Haze
  - Post CMP

Many sources of **ammonia** in fabs:

- Staff members
- Surface functionalization from priming age
- Cleaners
- Polishing slurries materials

GOALS

- Compare critical NH_3 molecular contamination behavior in two Entegris FOUP models, in terms of:
  - Contaminant, sorption and subsequent release
  - Contaminant cross-contamination on stored wafers

☑ SPECTRA PC

Polycarbonate

☑A300 EBM/CNT

EBM/C-nanotubes

EXPERIMENTAL PROTOCOL

0) FOUP CONDITIONING
Clean room equilibration
(21°C 40% RH) 7days

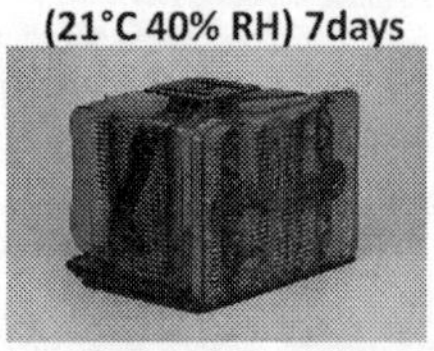

1) INTENTIONAL CONTAMINATION
10µL- droplet 24h

2) PURGE
(N2 gun) 5 min

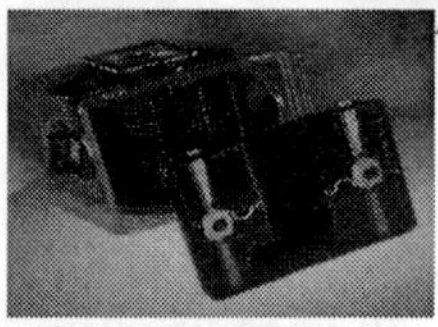

3) STORAGE of Cu-WAFERS, LPE-IC

3) OUTGASSING MONITORING
IMPINGER-IC
(once per day) 8 days

HF 2.0% 11.2ppm
HCl 3.7% 11.6ppm
NH3 2,9% 14.7ppm
& 0.29% 1.47ppm

IC = Ion Chromatography, LPE = Liquid Phase Extraction
Test carried by twice

SORPTION/OUTGASSING BY PC FOUPs

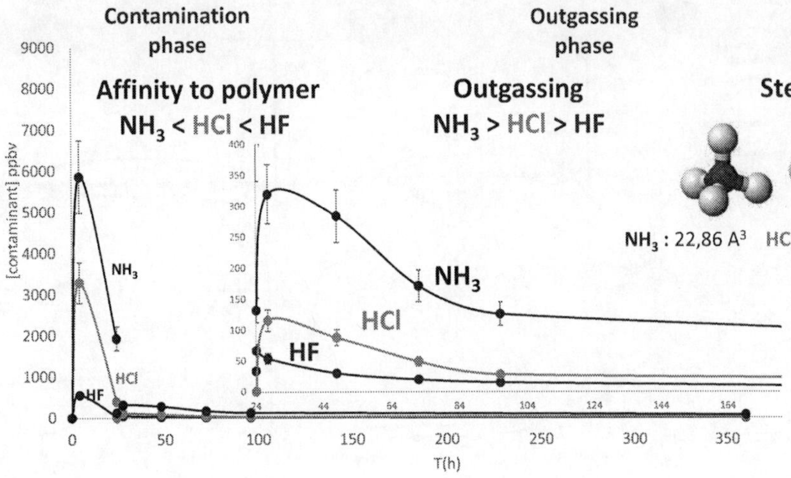

Steric effect

NH_3 : 22,86 A³ HCl : 22,45 A³ **HF** : 15,77A³

- **Immediate Strong outgassing** after purge, followed of a **concentration equilibrium**
- **Solubility** is the outgassing driving force

NH_3 < HCl < HF

SORPTION/OUTGASSING BY EBMCNT FOUPs

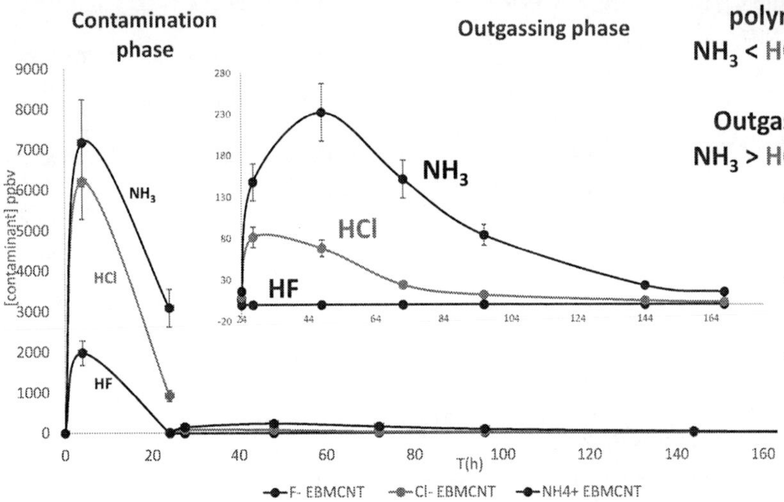

Affinity to polymer

NH_3 < HCl < HF

Outgassing

NH_3 > HCl > HF

- **Strong outgassing** for HCl and NH_3 after purge, followed of a **concentration equilibrium**
- No detectable outgassing in HF case **LLD (low limit detection) for F^-** 0.62ppbv
- In the NH_3 case, in respect to PC material the maximum outgassing is detected some hours later after purge

$D_{EBMCNT} < D_{PC}$

COMPARATIVE SORPTION/OUTGASSING FOUPs ABILITY

Contamination /outgassing

Outgassing zoom

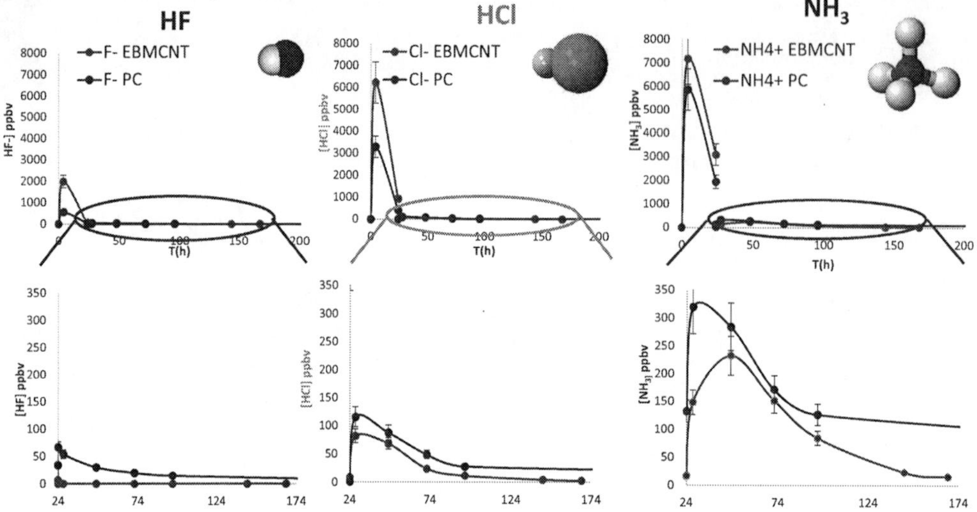

Contaminant affinity to polymer PC > EBMCNT

Outgassing PC > EBMCNT

VEHICLE TEST FOR CONTAMINATION TRANSFER EVALUATION

Need of a proper vehicle test for ammonia cross contamination

- Affinity to contaminant
- Knowledge of a kinetic model

ACIDS (HF, HCl)

- Cu wafers
- AlCu wafers

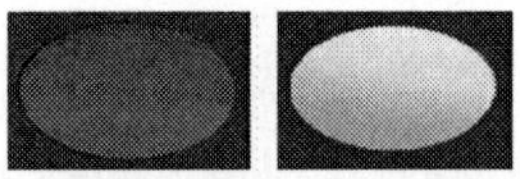

$[F^-] = [F^-_{max}]*(1 - exp(-k't))$ **[1]**

K=f(RH%, [HF]), F^-_{max}=f(RH%)

$[Cl^-] = (k*[C_{HCl}])*[t] + [Cl_0^-]$ **[2]**

K= kinetic constant

BASES (NH_3)

- ~~Cu wafers~~
- ~~AlCu wafers~~
- ~~Si wafers~~
- Cr wafers

- Cu wafers not suitable: NH_4^+ analysis not available
- Surface AlCu is not very sensitive to NH_3.
- Si surfaces no sensitive to NH_3 contamination

Langmuir deposition model – Cr-NH_3

$$C = C_{max} - (C_{max} - C_o)e^{-k_{ads}C_g t}\ [3]$$

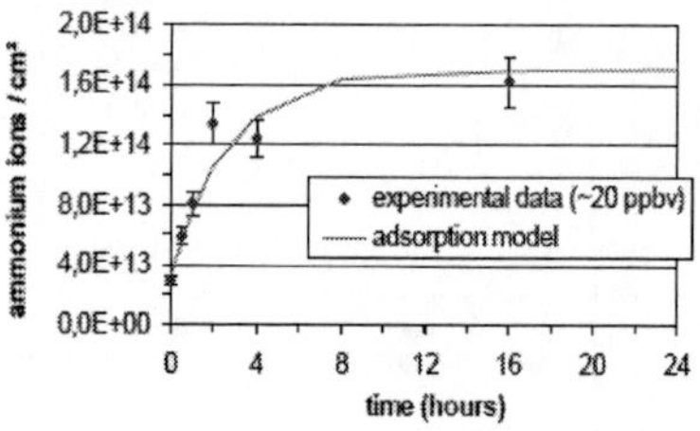

C_o = 1.5E11 (ion/cm²)
C_{max} = 1.9E14 (ion/cm²)
RH = 40%

[1] F. Herran, H. Fontaine, P. Gonzalez-Aguirre, C. Beitia, J. Ohlsen, J. Lundgren. A mathematical model forecasting HF adsorption onto Cu-coated wafers as a function of airborne concentration and moisture, Solid State Phenomena (2016), 255-323

[2] M-P Tran, P Gonzalez-Aguirre, C Beitia, J Lundgren, S-I Moon, H Fontaine Deposition of volatile chlorohydric acid on copper wafer depending on humidity and HCl airborne concentration, SPCC2018

[3] H. Fontaine; G. Demenet; V. Enyedi; S. Cetre. Study of the airborne SO2 and NH3 contamination on Cr, MoSi and quartz surfaces of photomasks, Photomasks Technology 2010, Vol 7823, pp28

CONTAMINATION TRANSFER FOUP TO WAFER

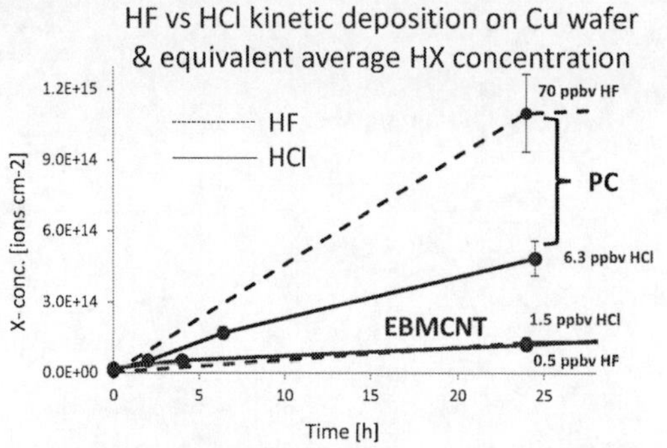

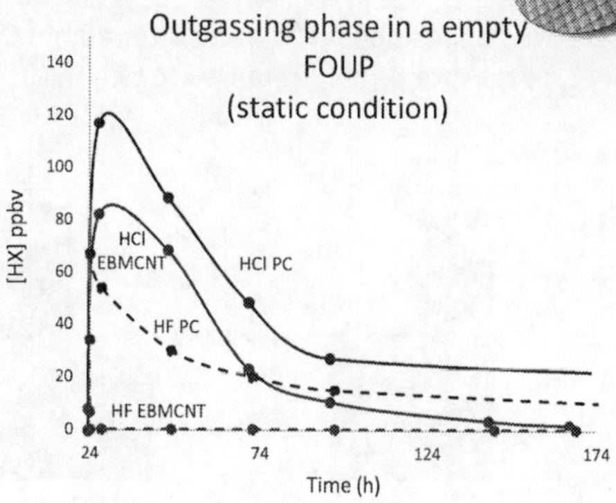

International Roadmap for Devices and Systems (IRDS 2013)

●HF 1E14 ● HCl 1E13

- HX transfer to Cu-coated wafers: **HF > HCl**
- Large FOUP outgassing rates depend on HX reservoir and diffusion kinetics, and then FOUPs exhibits a strong HCl outgassing but low transfer to copper wafers

$$S_{HCl} < S_{HF}$$

$$D_{EBMCNT} < D_{PC}$$

AMMONIA SORPTION/OUTGASSING OF FOUPs

Initial contamination concentration 14.7 ppmv vs 1.47ppmv

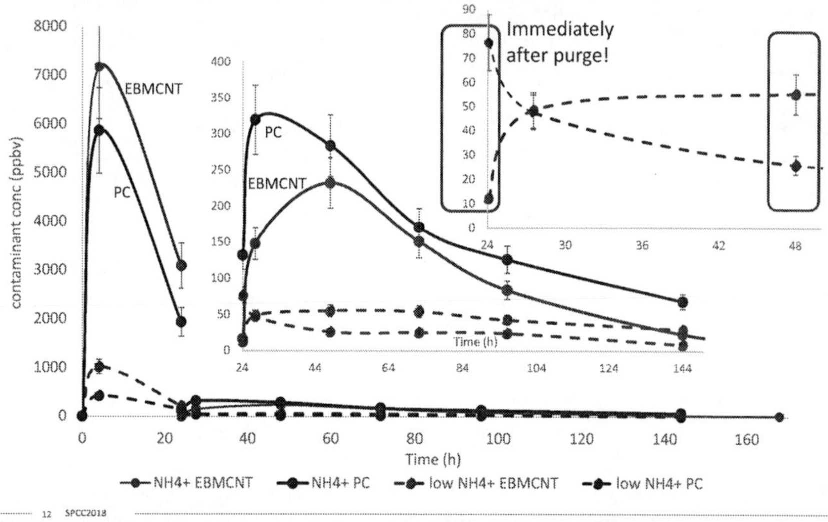

- **Strong outgassing** after purge, followed of a **concentration equilibrium**
- Given the poor NH_3 polymer affinity (comparing with HF, HCl)

  Solubility: NH_3 < HCl < HF
- Contaminant concentration located **mainly in the surface**

  D_{NH3}: $D_{PC} > D_{EBMCNT}$

NH_3 kinetic deposition on Cr wafer

ppb air	Time to 90% saturation
2	26.7h
10	5.4h
24	2.3h
73	42 min
270	12 min

C_{max} = 1.9E14 (NH_4^+/cm²

CONTAMINATION TRANSFER FOUP TO WAFER

NH_3 deposition on Cr wafer after 40min Cr wafer exposure & $[NH_3]$ required to rich this transfer

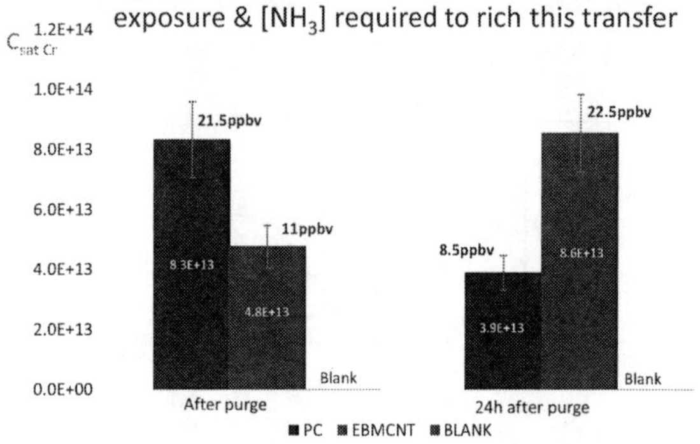

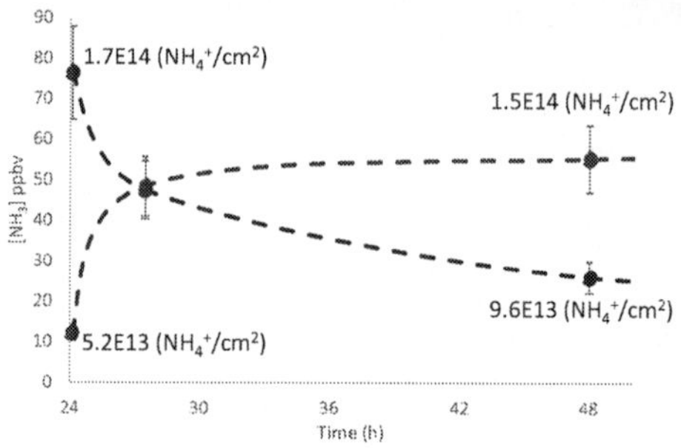

- Consequence of the NH_3 **Immediate PC strong outgassing** respect to the belated NH_3 outgassing in **EBMCNT** the **low transfer to Cr wafers** (as in HCl case) is due to a **low NH_3 solubility**
- This transfer is in agreement to the assumption $\mathbf{D_{EBMCNT} < D_{PC}}$

Entegris leti

PC & NH_3 REACTION

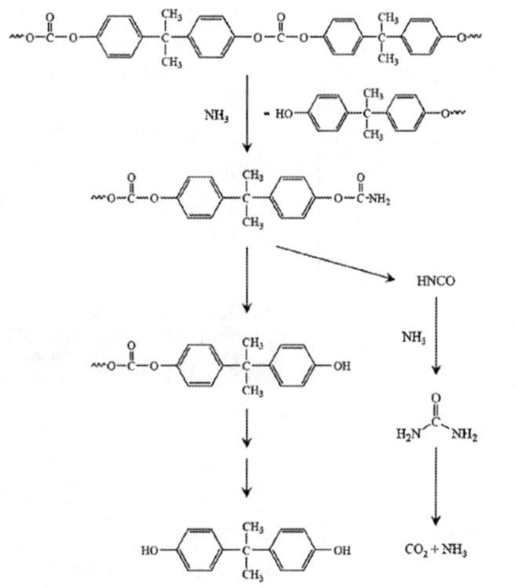

K. Hatakeyama et al. Chemical recycling of polycarbonate in dilute aqueous ammonia solution under hydrothermal conditions, J. Mater. Cyces Waste Manag. 16 (2014) 124–130

PC and **EBMCNT** compatibility with NH_3 solution (**TGA** characterization)

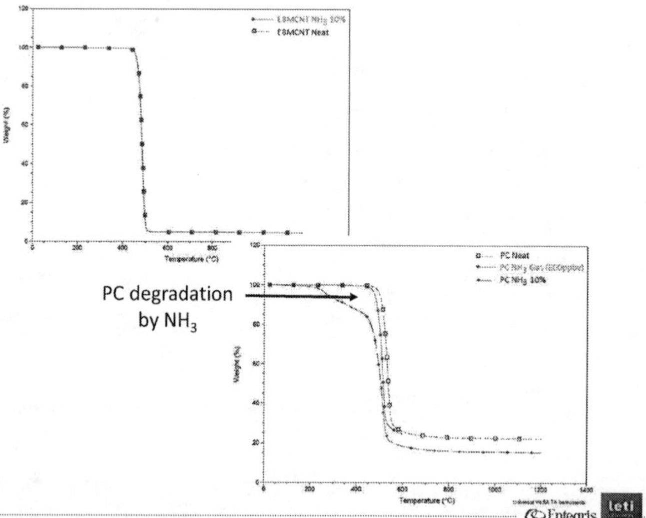

Entegris leti

SUMMARY

→ Contaminant FOUP <u>affinity</u> : **NH_3** < HCl < HF, FOUP contamination <u>f (S,D)</u> & **PC > EBMCNT**

→ Contaminant <u>outgassing</u> after purge: **NH_3** > HCl > HF , mainly <u>f (solubility) &</u> & **PC > EBMCNT**

→Contaminant <u>acid transfer</u> to wafer: HCl < HF, mainly f(diffusivity) & **PC > EBMCNT**

→**Assuming** same vehicle test the expected contaminant <u>transfer</u> to wafers will follow:

NH_3 < HCl < HF & **PC > EBMCNT**

→ **PC** use is not suitable in **NH_3** environments due it's reactivity

Among the tested FOUPs, EBM/CNT is the most efficient to limit NH_3, HF & HCl contamination transfer to wafers → reduced wafer defectiveness is expected

50 YEARS OF
PURE ADVANTAGE

RESULTS ANALYSIS

From literature:

- NH_3 D in PFA are greater than HCl
- HCl D in PC is greater than **HF**
- HCl S in PC is lower than **HF**
- S in PC & EBMCNT follows NH_3 < HCl < **HF**
- D & S are lower in EBMCNT than PC*

We postulate:

Contamination

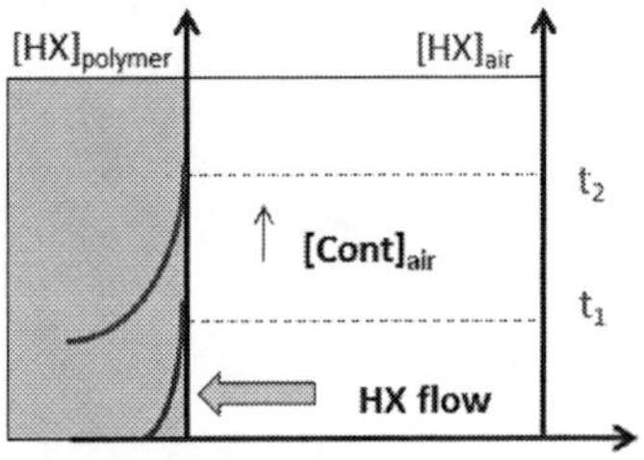

- $[Cont]_{air}$ increases and penetrates the polymer membrane
- $[Cont]_{surf}$ is defined by **Solubility**
- **Penetrant's flow governed by diffusivity**
- FOUP's contamination **f (D, S)**

purge

Outgassing

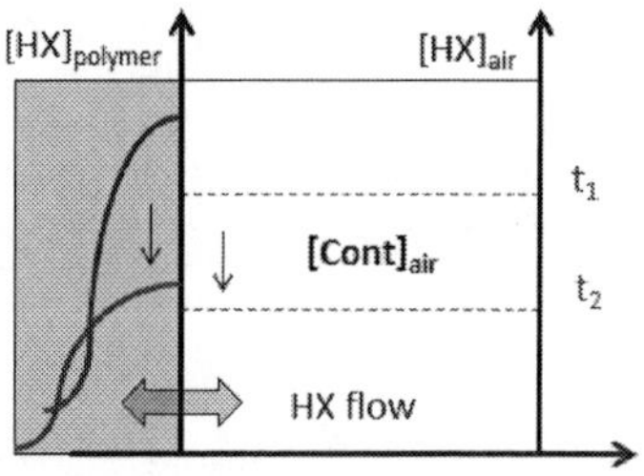

- $[Cont]_{air}$ increases until $[Cont]_{surf}$ reach the equilibrium **solubility**
- Outgoing diffusion continues: $[Cont]_{surf}$ is progressively reduced → $[Cont]_{air}$ is pulled down
- **$[Cont]_{air}$ is mainly solubility-dependent**

Cross contamination to wafer

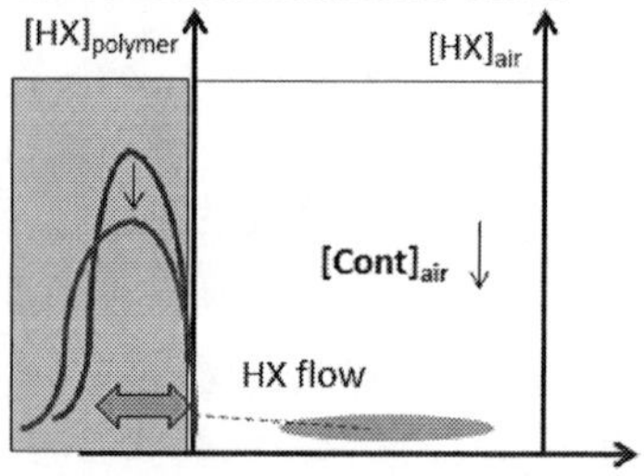

- Any contaminant molecule released will be retained by the wafer
- $[Cont]_{air}$ → 0 (cont affinity to wafer) then $[Cont]_{surf}$ → 0 as well
- **$D_{air} >> D_{polymer}$ → Cont transfer to Cu mainly governed by $D_{polymer}$**

M Cualfield, C.W Extrand, S.I Moon, Estimating Hydrochloric acid and Ammonium Hydroxide Loss, Controlled Environments Magazine, 2011
T.Q Nguyen PhD Thesis (2012)
P. Gonzalez-Aguirre, H. Fontaine, R. Pastorello, C. Beitia, J. Ohlsen, J. Lundgren HF transport coefficients in polymers used for microelectronic applications. Defect and Diffusion Forum (2016), 68-76

* Public soon

SPCC 2018

Hanyang University NEMPL

Acidic Cleaning Solutions for Post InGaAs CMP Cleaning

Jin-Goo Park[1,2] *, Jung-Hwan Lee[a], In-chan Choi[1], Hyun-Tae Kim[1], Lieve Teugels[3], and Tae-Gon Kim[3]

[1] Department of Bio-Nano Technology and [2] Materials Science and Chemical Engineering, Hanyang University, Ansan, 15588, Republic of Korea.
[3] imec, Kapeldreef 75, Leuven, B-3001, Belgium

April 11, 2018

NEMPL Hanyang University

Nano-Bio Electronic Materials and Processing Lab.

CONTENTS

01 **Introduction**

- High mobility semiconductor for Logic Device

02 **Research objective**

03 **Experimental procedure**

04 **Results and discussion**

- Evaluation of etch rate
- Component analysis
- Zeta potential measurement
- PRE (Particle Removal Efficiency) Analysis

05 **Summary**

NEMPL Hanyang University

Nano-Bio Electronic Materials and Processing Lab.

High mobility semiconductor for Logic Device

Technology development trend of semiconductor industry

- ✓ **Demand of high speed**
- ✓ **Decreasing the power consumption**
- ✓ **Shrinking device size (high density of integration)**

Logic Devices Roadmap

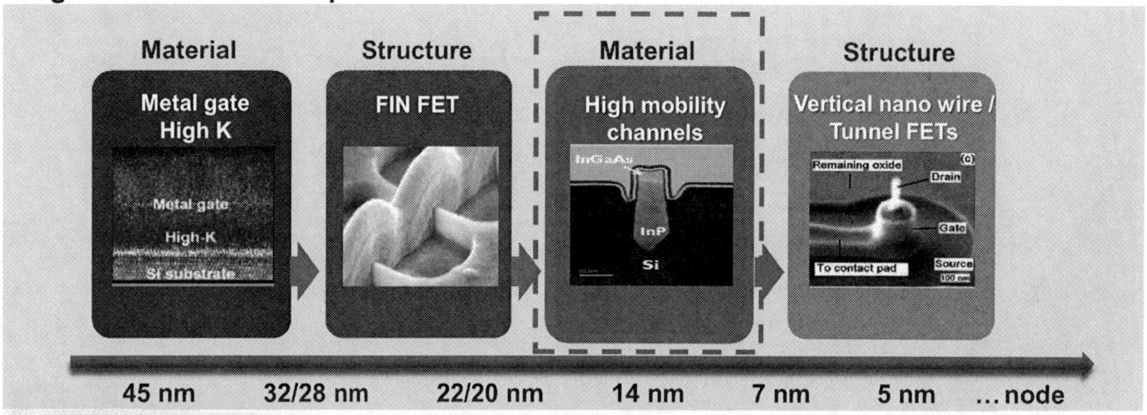

Recently new material is required for high mobility channels

Babu, Suryadevara. *Advances in Chemical Mechanical Planarization (CMP)*. Woodhead Publishing, 2016.

Properties of semiconductor materials

CMOS Structure

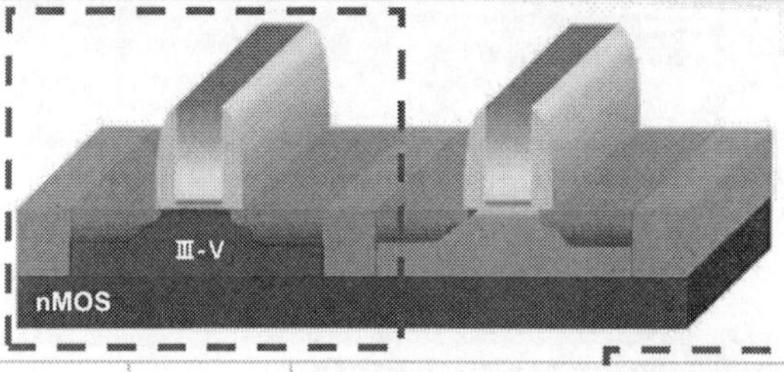

Material	Si	III / V				
		GaAs	InP	InGaAs	InAs	InSb
Electron mobility (cm^2/Vs)	1,600	9200	5400	**11,200**	40,000	77000
Hole mobility (cm^2/Vs)	430	400	200	**300**	500	850
Lattice constant (Å)	5.43	5.65	5.86	**5.87**	6.06	6.48
Band gap (eV)	1.11	1.42	1.34	**0.74**	0.36	0.17

➢ **Combined elements from group III and V** shows higher electron mobility and it is considered as an **alternative to Si as a channel material**, specifically for nMOS.

➢ Considering the lattice mismatches and band gap, **InGaAs** may consider as a promising material for nMOS.

http://www.yokoyama-gnc.jp/english/research/cmos.html

CMP process for III-V Fins

❖ STI first

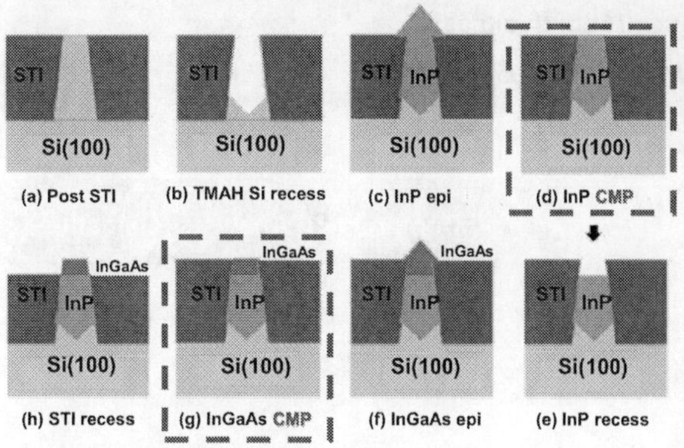

✓ ART (Aspect-Ratio Trapping) : Trapped defects in the III-V layer

✓ **CMP is needed to reduce the overburden post epitaxial growth**

❖ STI last

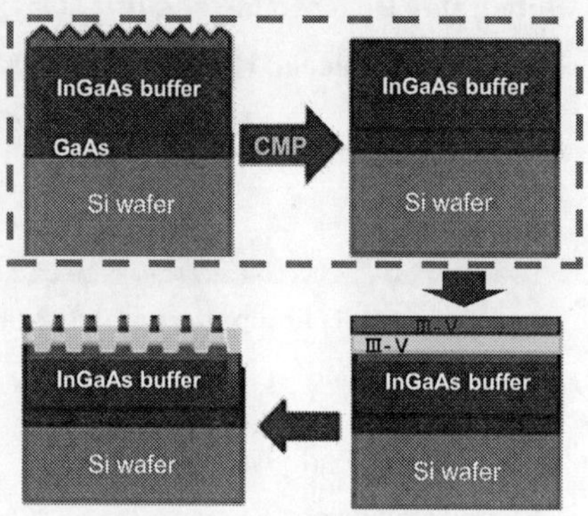

✓ SRB (Strain Relaxed Buffer): Reduced lattice mismatch with Si

✓ **CMP is needed before epi growth of new layer**

In both approaches, InGaAs requires at least 1 CMP step using silica slurry

Lieve Teugels et al., Semicon Korea 2015, 4-6 February 2015

Motivation

- After CMP process, **post InGaAs CMP cleaning** process was performed using megasonic, brush and **diluted ammonia solutions** to remove slurry contaminants.

❖ Issues of post InGaAs CMP cleaning

- The **conventional post InGaAs CMP cleaning process is not effective** for removing nano silica slurry particles from InGaAs surface.

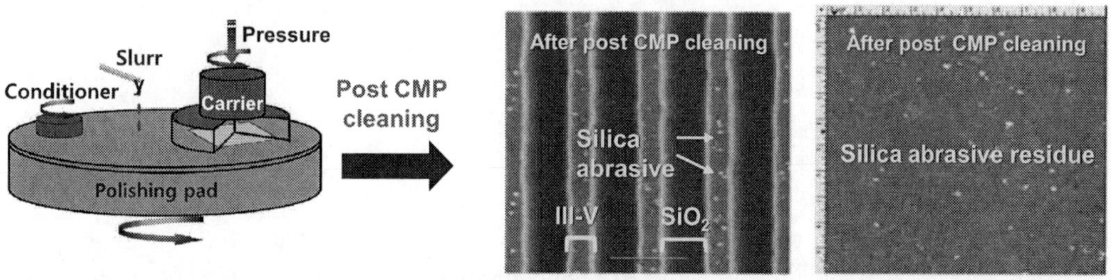

➢ **Development of new cleaning chemical is required for post InGaAs CMP process**

✓ **High cleaning efficiency**

✓ **Minimal material loss with less oxidized surface**

Lieve Teugels et al., Semicon Korea 2015, 4-6 February 2015

Experiments for etch rate and oxidation

❖ Materials & Experimental condition

- **Substrate : Polished InGaAs ($In_{0.53}Ga_{0.47}As$; ~75nm thickness)**
- **Cleaning Chemicals: HCl** (0.001 M, 0.005 M, 0.01 M, 0.05 M)
  H_2O_2 (0.001 M, 0.005 M, 0.01 M)

Etch Rate (Material Loss)

- **Photoresist (PR) line patterns of 110 μm**
  - Selective Etching
- **AFM (**XE 100, Park System, Korea**)**
  - Etched step height

Composition Analysis (Oxidation)

- **XPS (**Sigma Probe, Thermo, UK**)**
  - Oxide percentage

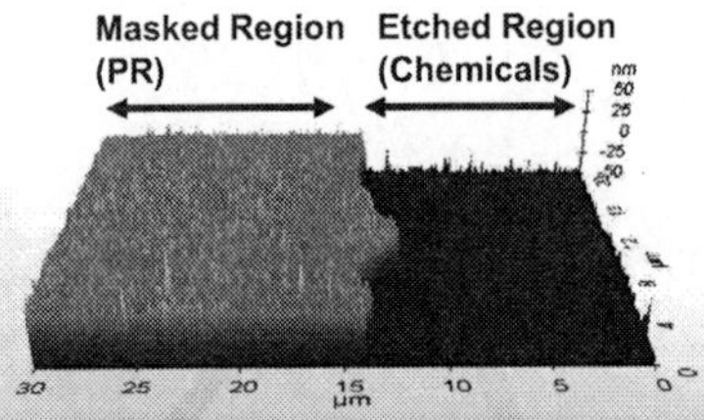

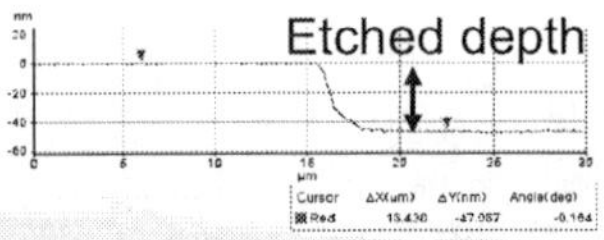

$$\text{Etch rate } \left(\frac{\text{Å}}{\text{min}}\right) = \frac{\text{Etched step height}}{\text{Etching time}}$$

Etch rate analysis

❖ Etch rate vs. various concentration of chemical

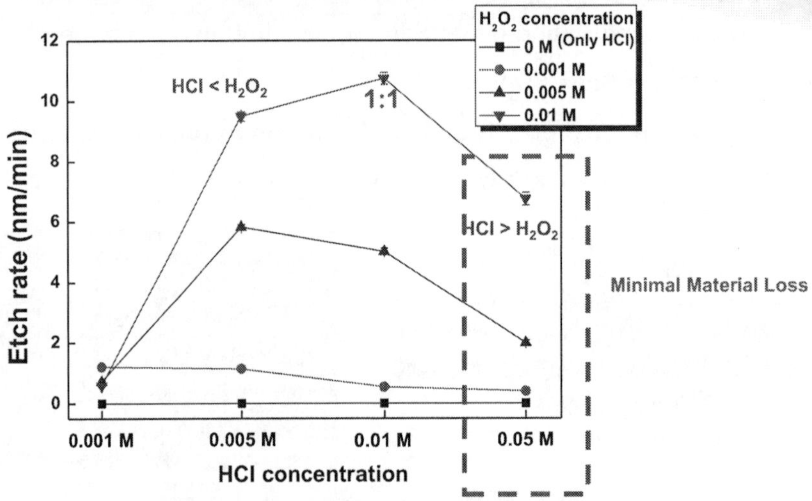

- ➢ **Only HCl doesn't etch the InGaAs surface without H_2O_2**
- ➢ **Highest etch rate** was achieved at **HCl : H_2O_2 = 1 : 1**
- ➢ **Lower concentration of HCl than H_2O_2 (ER ↑)**
- ➢ **Higher concentration of HCl than H_2O_2 (ER ↓)**

XPS analysis for oxide content

❖ **XPS analysis was carried out for analyzing the oxide percentage of InGaAs**

✓ Ref. (As-received wafer)

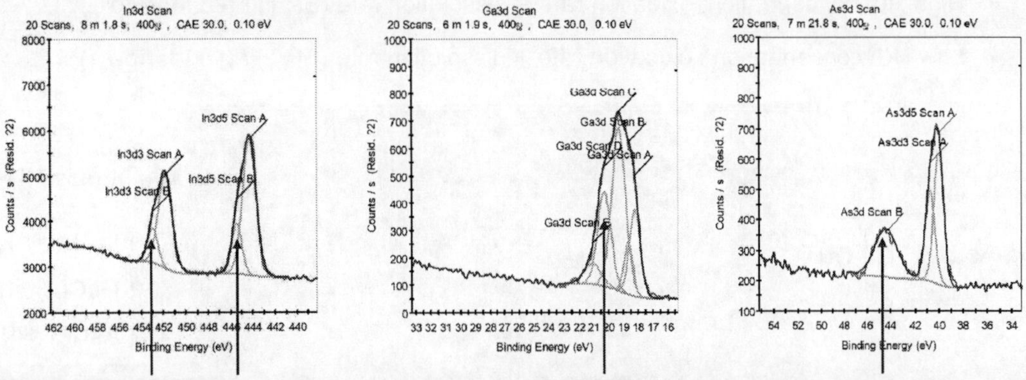

Chemical state	In_2O_3	Ga_2O_3	As_2O_3
O Atomic %	21.28 %	23.43 %	33.8 %

➢ **XPS was used to evaluate the cleaning chemical effect on InGaAs oxidation**

XPS analysis for oxide content

❖ **Oxide content vs. HCl Conc.**

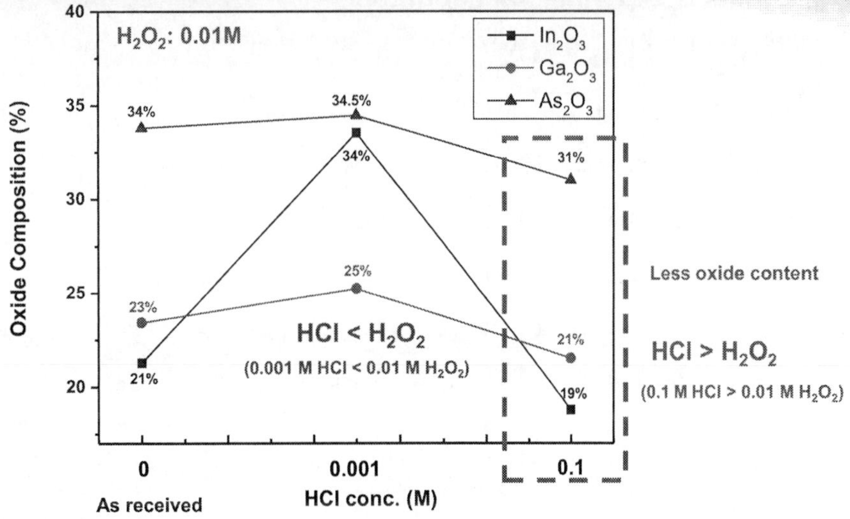

➢ **Lower conc. of HCl increases the oxide composition**

➢ **Higher conc. of HCl decreases the oxide composition**

➢ **High conc. of HCl than H_2O_2 can minimize the oxide layer formation**

Etching mechanism

❖ Etching mechanism according to the HCl concentration

✓ High HCl concentration: oxidation rate **<** dissolution rate (**ER** ↓) / (**Oxidation** ↓)

✓ Low HCl concentration : oxidation rate **>** dissolution rate (**ER** ↑) / (**Oxidation** ↑)

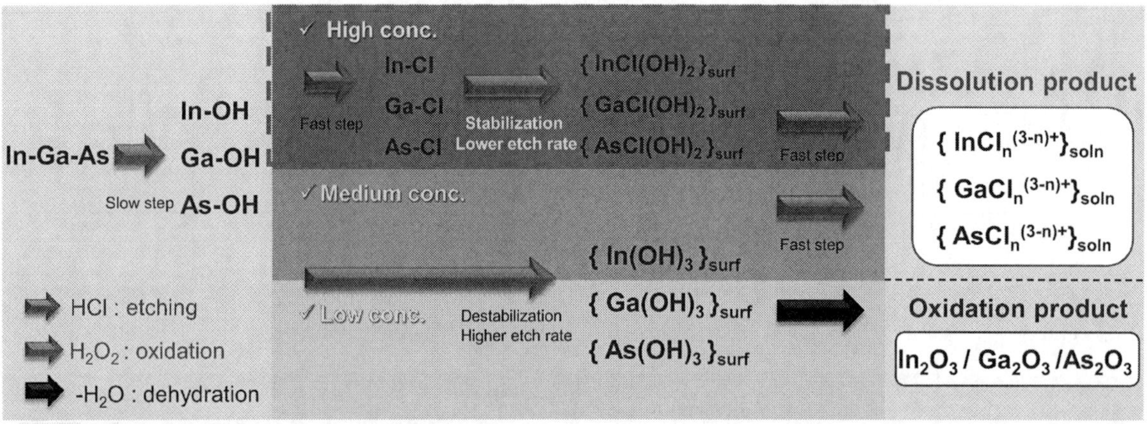

➢ **Based on the material loss and surface oxidation results, higher conc. of HCl (0.05 M) and lower conc. of H_2O_2 (0.001M) is considered as a optimized cleaning solution**

Experiments for the evaluation of PRE

❖ Materials & Experimental condition

- Substrate : InGaAs ($In_{0.47}Ga_{0.53}As$)
- Particle : Silica (130 nm, 289 nm)
- Optimized Conc. of cleaning solution: **HCl (0.05M) + H_2O_2 (0.001M)**
- **Surfactant : ADS** (Ammonium Dodecyl Sulfate) (0 – 0.5mM)
- **Megasonic: batch type, 1MHz, 600W**

Particle contamination	Cleaning process	PRE Analysis
• **Particle contamination** - Wet deposition - 2 x 10^{-5} wt% Silica particle solution	• **Chemical cleaning** - Only Cleaning Chemical - Cleaning Chemical + ADS • **Physical cleaning** - Cleaning Chemical + MS - Cleaning Chemical + ADS + MS	• **Optical microscope (LV100D, Nikon, Japan)** - Particle counting (289 nm) • **FE-SEM (MIRA3, TESCAN, Czech)** - Particle counting (130 nm)

PRE vs. Etch amount

❖ **PRE and Etch amount vs. Cleaning Time**

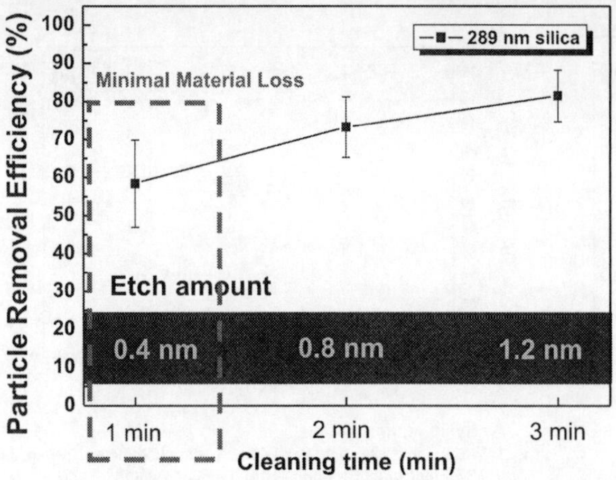

❖ **Cleaning condition**

- Cleaning solution :
  HCl : 0.05M, H_2O_2 : 0.001M
  pH : 1.36
- Dipping process

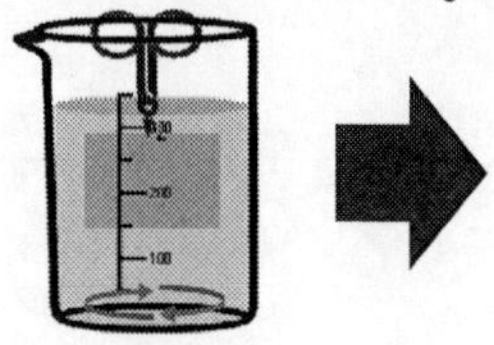

- Dipping time
  - 1 min
  - 2 min
  - 3 min

- Cleaning time ∝ Etch amount
- **PRE is increased with increase in the etch amount**
- **Modification of zeta-potential is required to increase the PRE with minimal material loss.**
  - ✓ [Zeta-potential] particle : (-) , InGaAs : (+) ➡ particle : (-) , InGaAs : (-)

Mechanism of surfactant cleaning

❖ **Modification of electrostatic force using surfactant**

- Introduction of strong electrostatic repulsion between particle and substrate can enhance the PRE during the post CMP cleaning process, which was achieved by using surfactant

Surfactant

✓ [Zeta-potential] particle : (-) InGaAs : (+) ➡ particle : (-) InGaAs : (-)

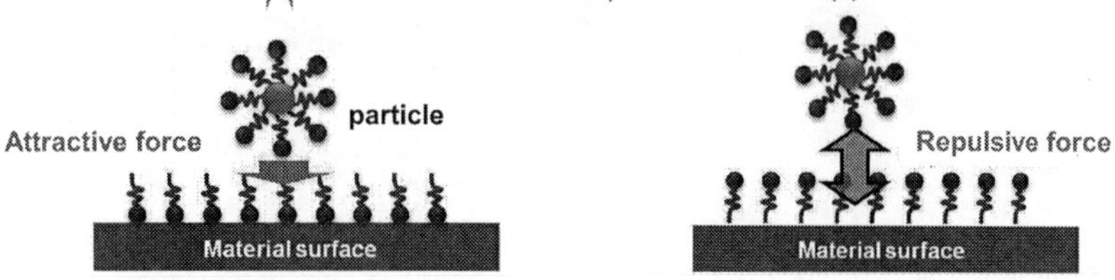

❖ **Ammonium Dodecyl Sulfate (ADS)**

- ADS is one of an anionic surfactant widely used for cleaning process

< ADS Properties >

Chemical Formula : $C_{12}H_{29}NO_4$

CMC value: 6.6 mM

R. Vos, M. Lux et al., Journal of The Electrochemical Society, 148 (12) G683-G691 (2001)

Zeta potential Measurement

❖ Particle zeta-potential (silica)[1]

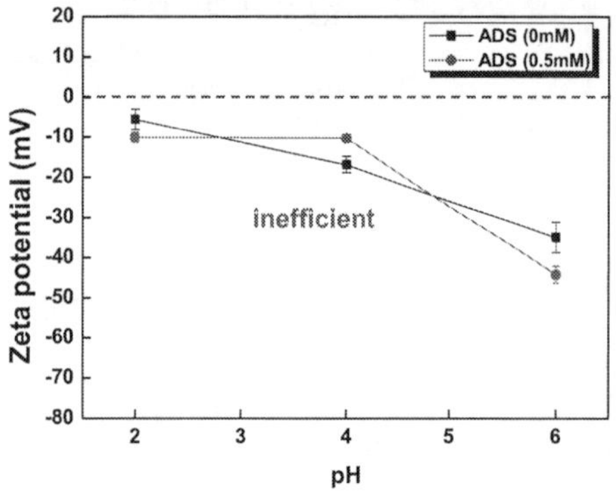

❖ Surface zeta-potential (InGaAs)[2]

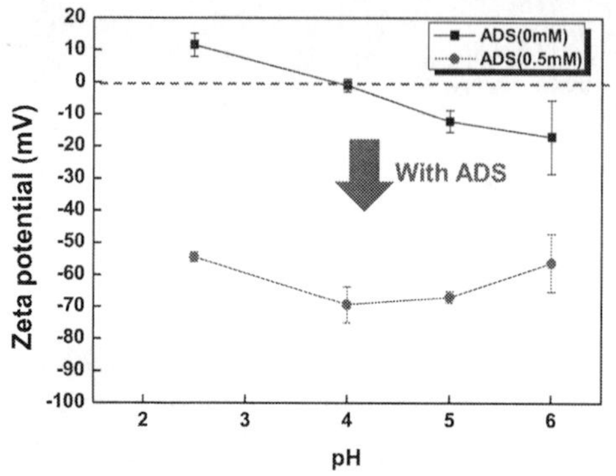

- **Analysis equipment**
  ELS-Z (Electrophoretic Light Scattering Spectrophotometer) Otsuka, Japan[1]
  SurPASS Electrokinetic Analyzer Anton Paar, Austria[2]

➢ Zeta potential of InGaAs surface changed significantly after addition of ADS

✓ (+) Z.P. —ADS→ (-) Z.P. in acidic region

➢ **Addition of ADS introduces strong repulsive force between silica particle and InGaAs substrate**

✓ **Silica particles can be easily removed from the InGaAs surface**

PRE vs. ADS conc.

❖ ($HCl + H_2O_2$) vs. ADS

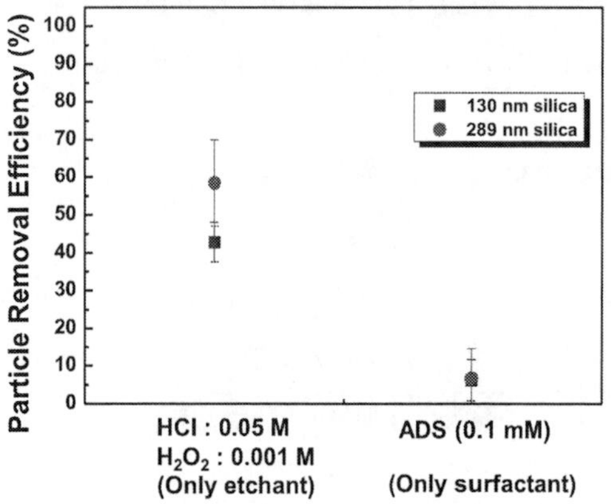

❖ ($HCl + H_2O_2$) + ADS

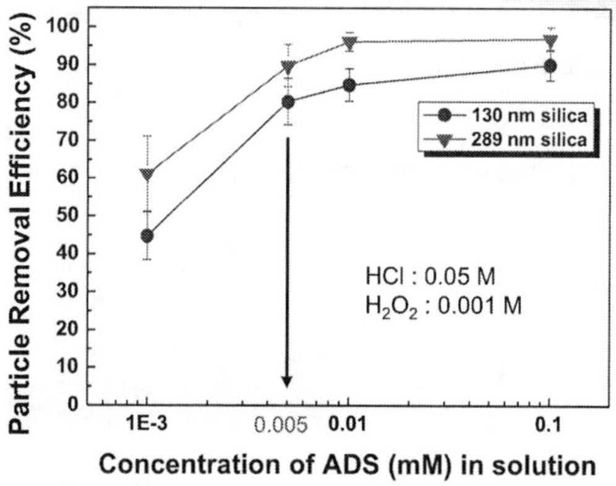

➢ Using ADS alone is ineffective to remove particles without etchant

✓ Etching process is necessary to remove the particles

➢ **PRE increases dramatically with the addition of ADS**

✓ Combined effect of surface etching and electrostatic repulsive force increases the PRE

✓ PRE was increased drastically at the addition of 0.005 mM conc. of ADS

Etch rate vs. ADS conc.

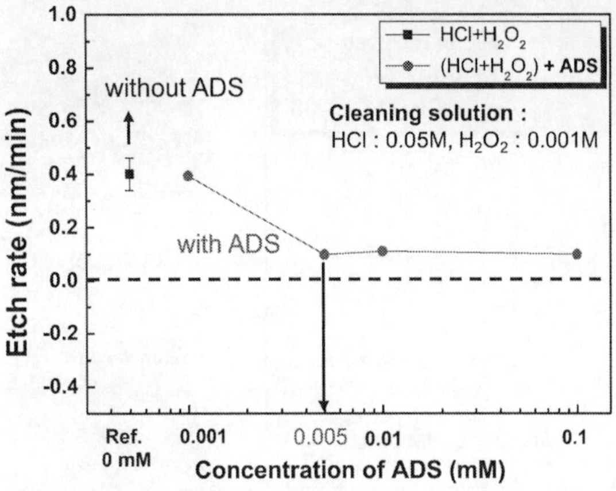

- **Addition of ADS decreases the etch rate of InGaAs**
  - ✓ Etch rate was decreased drastically at 0.005 mM and above conc. of ADS
  - ✓ Introduction of ADS not only increasing the PRE but also decreasing the material loss

Nano-Bio Electronic Materials and Processing Lab. 17

PRE vs. Megasonic

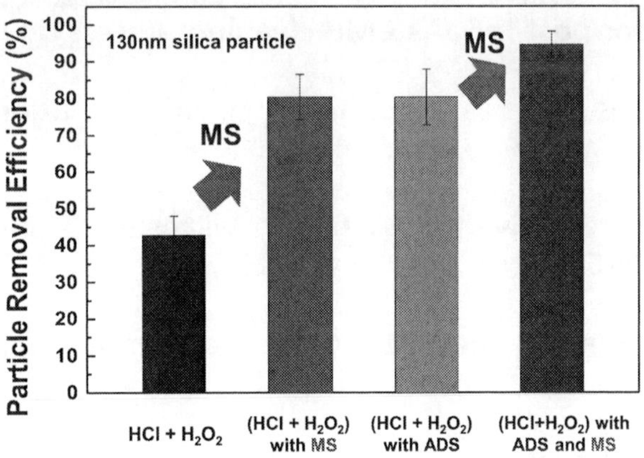

❖ Cleaning condition

- ✓ Cleaning solution :
  HCl : 0.05M, H_2O_2 : 0.001M
  ADS (0.005 mM)
- ✓ Megasonic (1 MHz, 600 W)

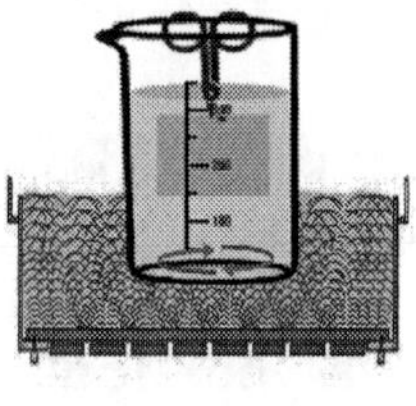

- Megasonic increases the PRE of various cleaning solutions
- **HCl and H_2O_2 cleaning solution with ADS and MS achieved above 95% PRE**
  - ✓ Mechanism of megasonic cleaning : acoustic streaming, bubble collapse
  - ✓ Combined effects of physical force, surface etching and electrostatic repulsive force

Nano-Bio Electronic Materials and Processing Lab. 18

Surface roughness

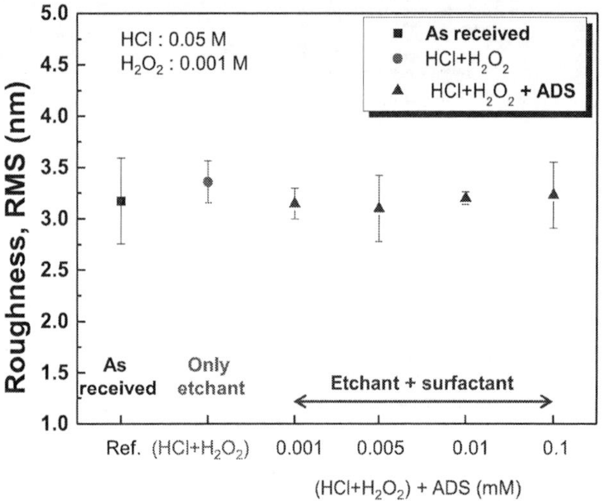

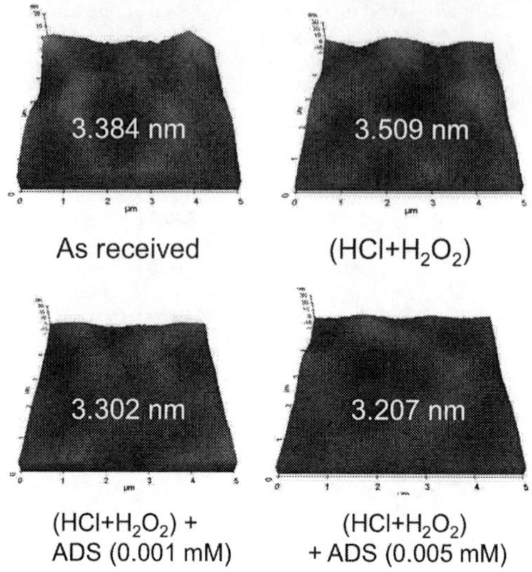

- After cleaning process, there is no significant change in roughness compared to the initial state
  - ✓ Etch amount was negligible to affect the roughness (minimized surface damage)

Summary

- **Acid cleaning solution** was used for **post InGaAs CMP cleaning** process.
- **Higher HCl conc. than H_2O_2** is considered as a **optimized cleaning solution** based on the results of material loss and surface oxidation.
- **ADS** was introduced for **increasing repulsive force** between silica particle and InGaAs substrate to **enhance the PRE**.
- Introduction of **ADS** not only **increasing the PRE** but also **decreasing the material loss**.
- There was **no noticeable roughness change** was observed after cleaning process compared to the initial state.
- **Acid cleaning with surfactant and megasonic achieved above 95% PRE.**

imec
SPCC 2018
Thank You for
Your Attention !!
NEMPL
NEMPL
Hanyang University
Nano-Bio Electronic Materials and Processing Lab.

20TH SURFACE PREPARATION AND CLEANING CONFERENCE (SPCC) – 2018

Fundamentals of Post-CMP Cleaning of Dielectric Surface Contaminated with Ceria (Nano-to-Micro) Particles

Atanu Das, Daniela White,
Wonlae Kim, Michael White

R&D Surface Preparation and Integration
Danbury, CT
T +1 203 207 9397
Email: atanu.das@entegris.com

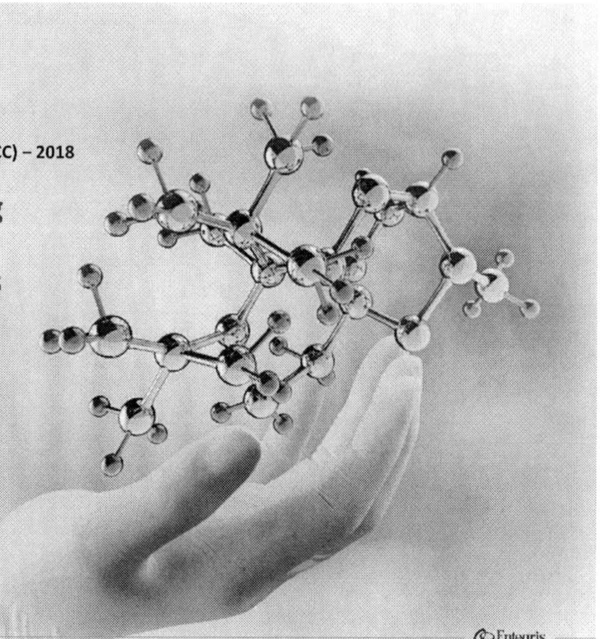

AGENDA

01	Fundamentals of Ceria-Silica CMP Process
02	Reactivity Between Ceria and SiO_2 and Si_3N_4
03	Commodity Cleaner vs. Formulated Cleaners
04	Formulation Design Concept and Components
05	Ceria Cleaning Mechanism – Spectroscopic Evidences
06	Positive vs. Negatively Modified Ceria Cleaners
07	Entegris Recommendations
08	Conclusions

POLISHING OF SiO_2 FILMS BY CeO_2 PARTICLES – PROPOSED MECHANISM

Ceria – CMP process

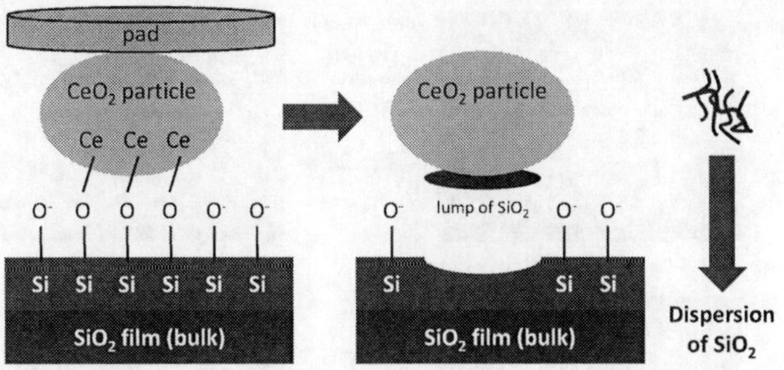

Overall reaction

$$R\text{—}\mathbf{Si}\text{—}OH + Base \longrightarrow R\text{—}\mathbf{Si}\text{—}O^- + BaseH^+$$

$$R\text{—}\mathbf{Si}\text{—}O^- + HO\text{—}\mathbf{Ce}\text{—}R' \longrightarrow R\text{—}\mathbf{Si}\text{—}O\text{—}\mathbf{Ce}\text{—}R' + OH^-$$

- Key steps and factors determine silica removal:
  - Si–O–Ce bond formation
  - Particle size distribution
  - Hydrodynamic forces acting at the interface
  - Pad properties
  - Fluid Rheology
- Challenges:
  - CMP doesn't always remove Si – O – Ce from surface
  - An effective post-CMP formulation needed to clean the surface

Ref: T. Hoshino et al. / Journal of Non-Crystalline Solids 283 (2001) 129-136

Entegris

REACTIVITY BETWEEN CERIA AND SILICON NITRIDE (WHY MODIFY CeO_2?)

CeO_2–Si_3N_4 surface interactions (3 key steps)

Step 1: Silicon nitride hydrolysis

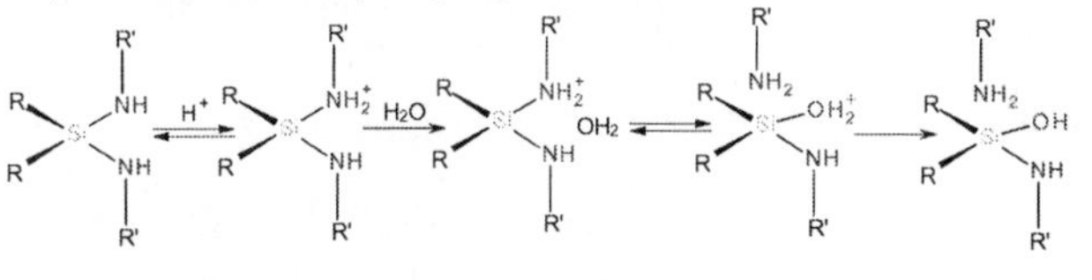

R = subsurface neighboring atom
R' = surface atom covalently bonded to nitrogen (silicon or hydrogen)

Step 2: Reaction of the surface silanols/oxoanions
Step 3: Removal of the surface oxide layer with ceria particle

Appropriate organic base can influence silicon nitride hydrolysis

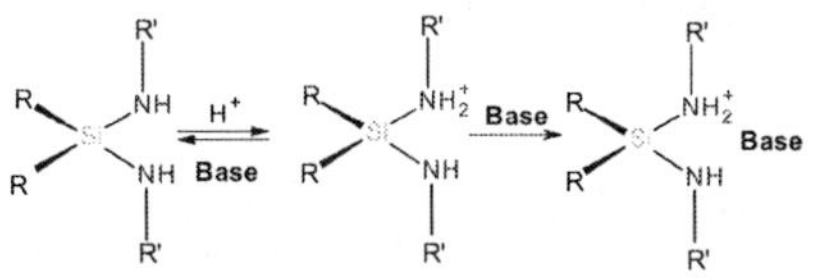

- Organic base can create a new acid-base equilibrium
- Deprotonate the surface amide
- Site-block water from silicon metal

CeO_2 is modified with appropriate organic base to tune the Si_3N_4 removal

T. P. Johns et al. Electrochemical and Solid-State Letters, 8 q8r G218-G221 (2005)

Entegris

WHY FORMULATED CERIA CLEANERS VS. COMMODITY?

1. EHS/Safety concerns with traditional cleans (hot SPM, dHF, SC-1, TMAH + dHF)
2. One-step clean process requirement for throughput improvement
3. Need for improved particle removal
4. Need for improved metal removal
5. No damage to dielectric substrates

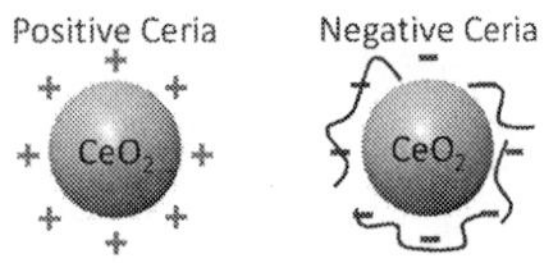

1. Many different ceria particles and chemistries are on the market
2. Cleaning challenges vary greatly depending on slurry type → no universal cleaner so far
3. **Best approach is for slurry supplier to collaborate directly with cleaning supplier and customer to develop the best BKM**

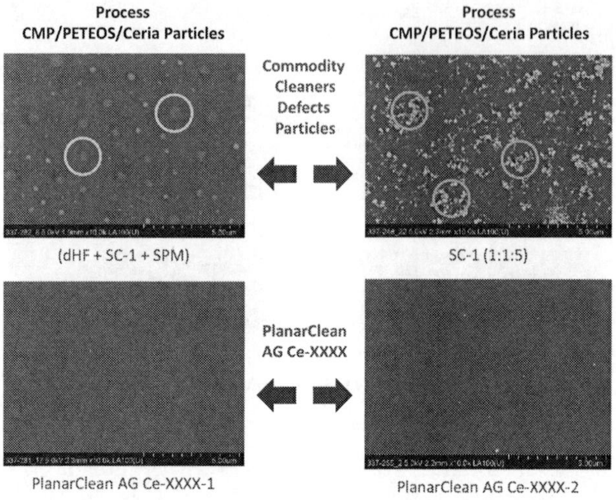

INTERACTIONS OF ORGANIC MOLECULES WITH CERIA SURFACE AND SURFACE CHARGE VS. pH

Surface-modified ceria in CMP slurries

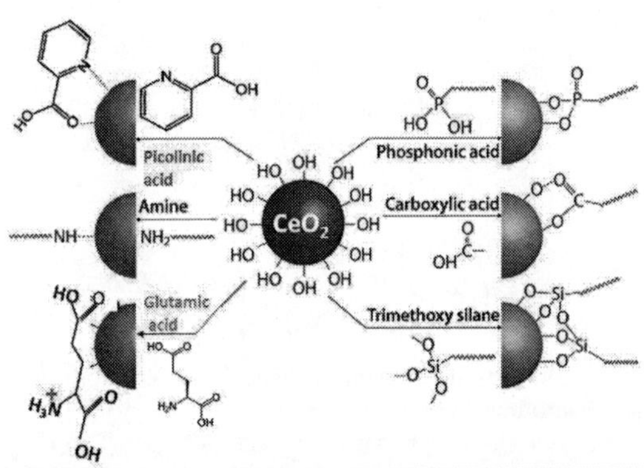

Various CeO_2 Particles Tested

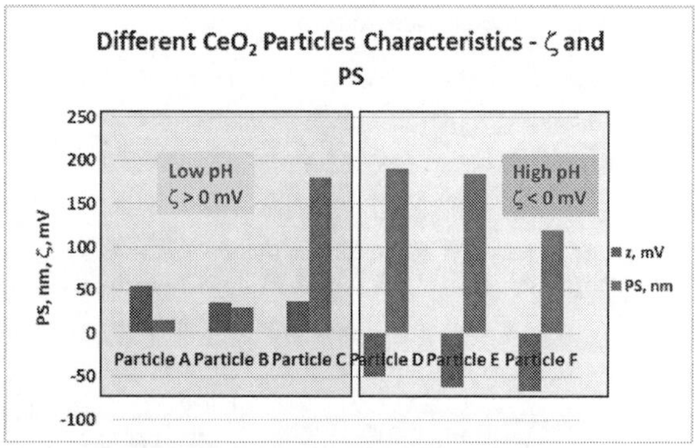

- **Surface modified CeO_2 particles in the CMP slurries: positive or negative surface charge;**
- **Particle size: 15 – 200 nm;**
- **Need to understand CeO_2 surface chemistry and types of interactions with the dielectric surfaces;**
- **Six different types of CeO_2 particles tested – can we design an universal cleaner?**

DIFFERENT CLEANING FORMULATIONS FOR DIFFERENT CERIA SURFACE CHEMISTRIES?

Particles E, F

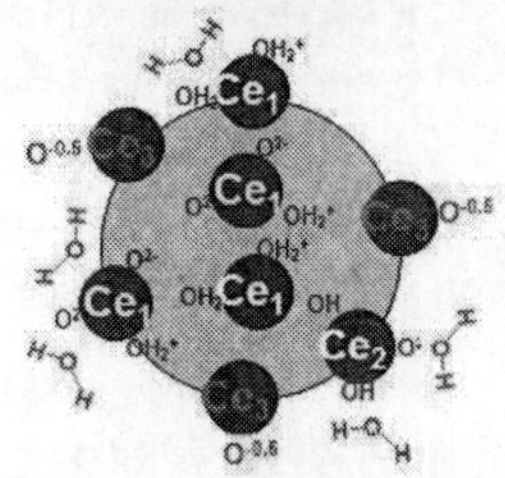

Particles B, C, D

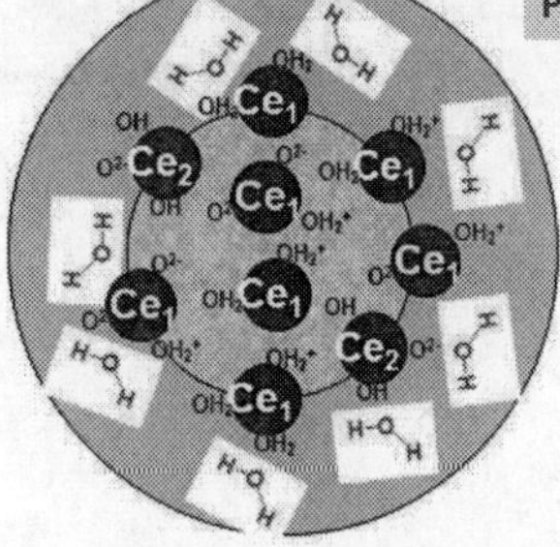

- More acidic surface, partially hydroxylated
- Small amount of surface water H-bonded
- Surface exposed –OH for -Si-O⁻ condensation
- Stronger Ce-O-Si bonds, difficult to break/clean

- More basic surface, more hydroxylated
- Outer-sphere shell of H-bonded water
- Reduced surface reactivity
- Weaker Ce-O-Si bonds, easier to break

Based on the FTIR-ATR, UV-VIS and titration experiments data

Entegris

HOW DOES THE COMMODITY CLEANER WORK? AN EXAMPLE

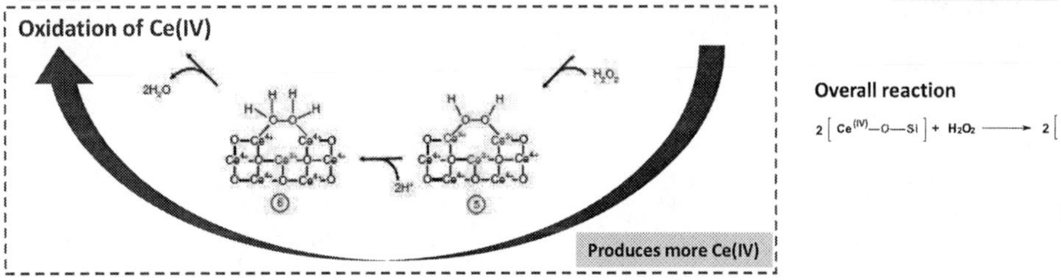

Commodity cleaners as controls:
SC-1 – $H_2O:H_2O_2:NH_3$ (1:1:5)
SPM – $H_2SO_4:H_2O_2$ (1:4), T > 100°C

Drawback:
- H_2O_2 can attack to the dielectric
- Basic – H_2O_2 dissolve more oxide
- SPM – additional step needed for leftover sulfur

Overall reaction

$$2\left[Ce^{(IV)}\text{—O—Si}\right] + H_2O_2 \longrightarrow 2\left[Ce^{(III)}\text{—O—Si}\right] + 2H^+ + O_2$$

H^+ | Complexant (C)

$$\left[Ce^{(III)}\text{— C}\right] + HO\text{—Si}$$

soluble

Wei-Tsu T. et al. Journal of Solid state Science and Technology, 6 (10) P7118

Entegris

SURFACE INTERACTIONS AND KEY BOND-BREAKING STEPS

CeO_2-SiO_2 Surface Interactions

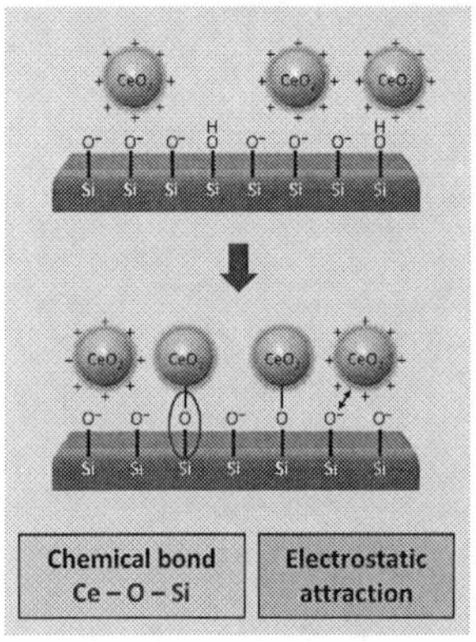

Ce-O-Si Bond-Breaking Mechanisms

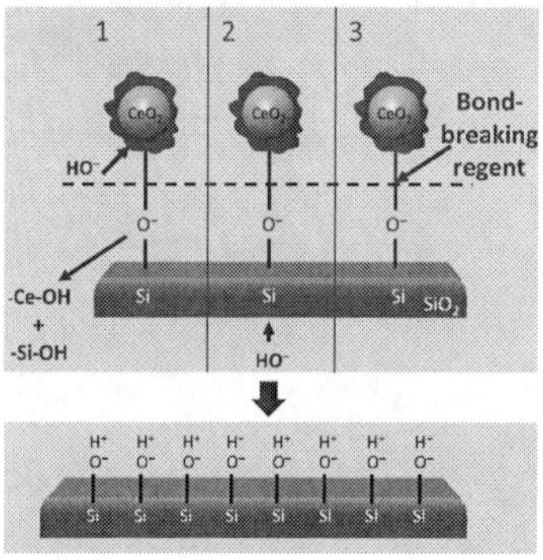

Cleaning formulation (low or high pH) can be designed based on types of ceria and its interaction to the surface

Three possible options:

1. Nucleophilic attack of OH- to Ce-center
2. Partial etch of the surface
3. Breaking of (CeO_2)–SiO_2 bond

Formulation design concepts:

1. High pH hydrolysis of -Ce-O-Si- bonds by HO- nucleophilic attack to Ce-center
2. Additives needed to stabilize –Ce-OH species and prevent redeposition
3. High pH partial etch/dissolution of the surface –Si-O-Ce- groups + redisposition prevention.
4. Low pH: Bond-breaking additives, followed by CeO2 complexation, and particles stabilization and dispersion

INTERACTIONS OF CE(IV) AND CE(III) SALTS WITH CEXXXX-01: MONITORED BY CV EXPERIMENT

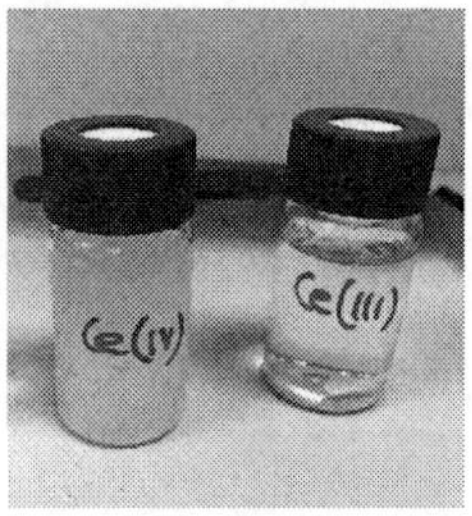

- Ce(IV)/Ce(III) redox couple can't be detected in the solvent potential range
- CeXXXX-01 redox wave remains unchanged after the treatment with Ce(III) or Ce(IV) salts
- Redox property of CeXXXX-01 component remains active for Si – O – Ce bond breaking

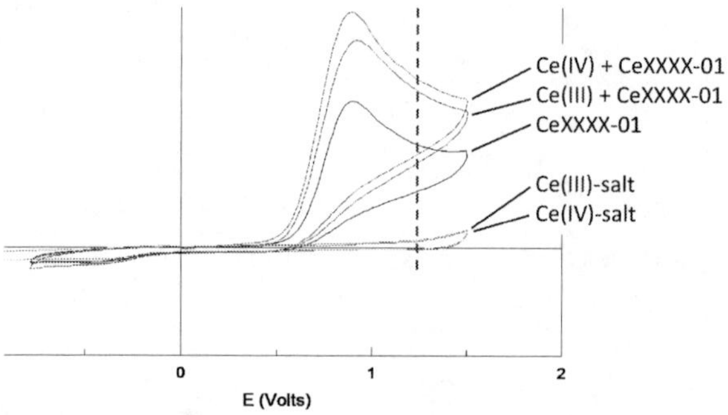

SPECTROSCOPIC EVIDENCE OF ACTION OF BOND-BREAKING COMPONENT

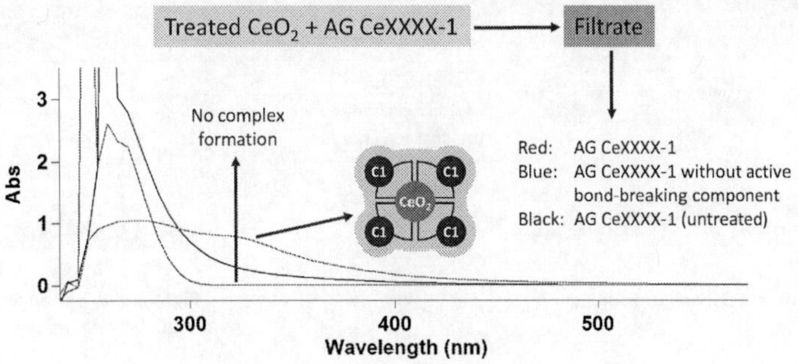

Samples were collected after the PCMP process

Without the active bond-breaking component, no complex formation was observed, indicating component helps to keep the ceria soluble

LOW PH FORMULATION DEVELOPMENT

CeO_2 (powder) + Formulation

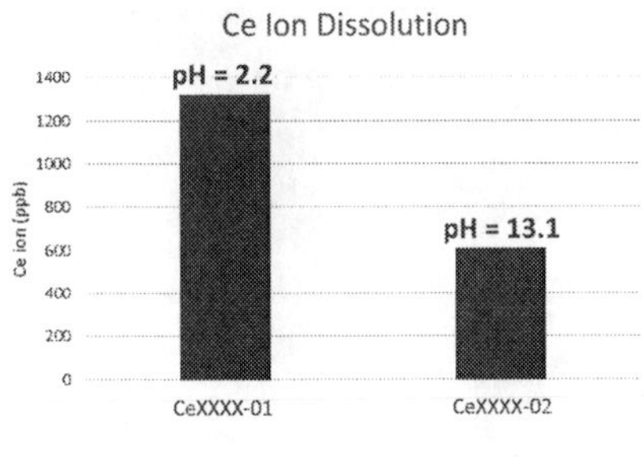

- Purpose
  - Ceria particle dissolution comparison
- Test conditions
  - CeO_2 powder (from Aldrich)
  - Chemistry: CeXXXX-01, CeXXXX-02
  - Chemistry dilution (100:1)
  - Procedure
    - Add CeO_2 in chemistry
    - Stir for 15 min
    - Filter the powder
    - ICP MS analysis for Ce ion

CeXXX-01 and CeXXX-02 can dissolve certain amount of Ceria powder

CeXXX-01 shows better Ceria powder dissolving than CeXXX-02

LOW PH FORMULATION FOR NEGATIVE CERIA

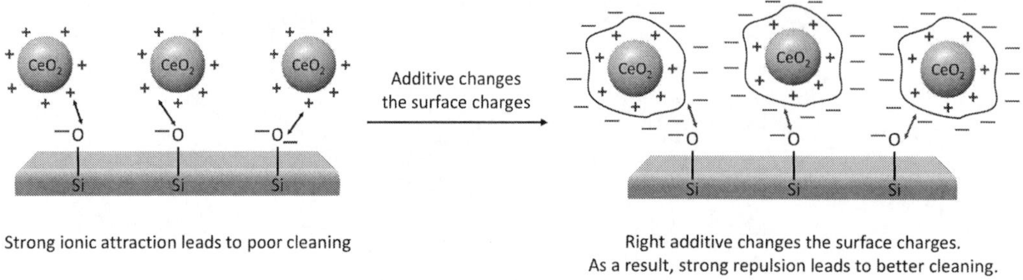

CLEANING PERFORMANCES WITH AND WITHOUT ADDITIVES

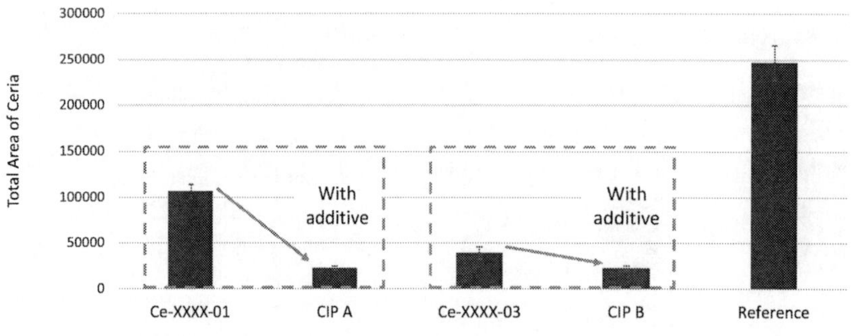

Cleaning additive has strong effect in changing the surface property of CeO_2 particles

Choosing the right additive is one of the key steps to design efficient formulation

Entegris

CLEANING OF POSITIVE-CeO_2 ON PETEOS

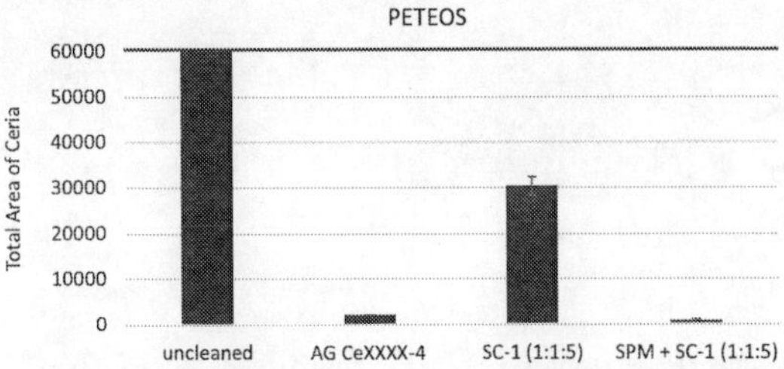

- No H_2O_2 or HF was used in the formulation
- Compare to commodity cleaner (H_2O_2 + NH_4OH), CeXXXX-4 sows 150× improvement
- No CeO_2 agglomeration indicates soluble Ceria-formation or better dispersion after PCMP process

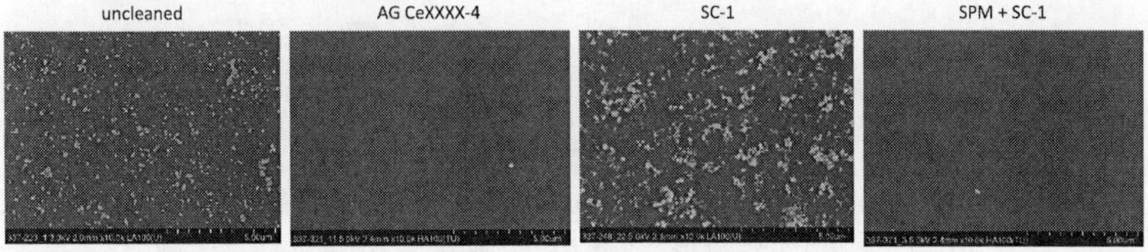

Entegris

TREATMENT OF POSITIVE-CeO_2 WITH CLEANER SOLUTION

Treated-CeO_2 + Cleaner —Stir 5 min→ Measure particle size

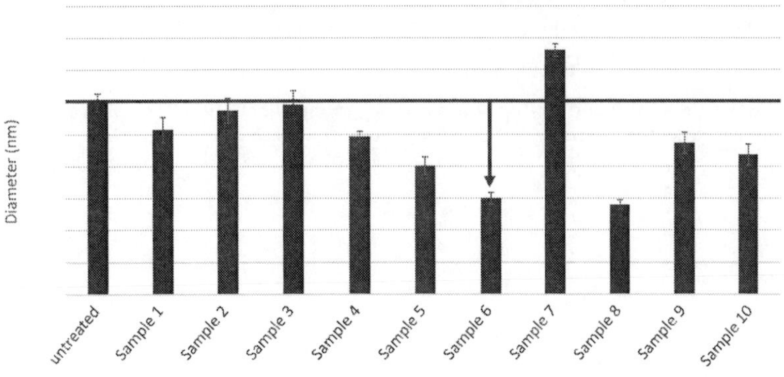

No particle agglomeration was observed

Sample 6 show highest reduction of size; indicates higher degree of surface interaction with the cleaner

Entegris

CONTACT ANGLE EXPERIMENT TO MONITOR ORGANIC RESIDUE REMOVAL EFFICIENCY

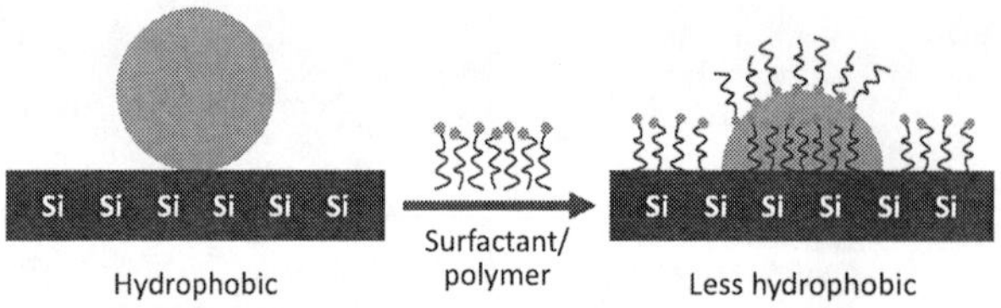

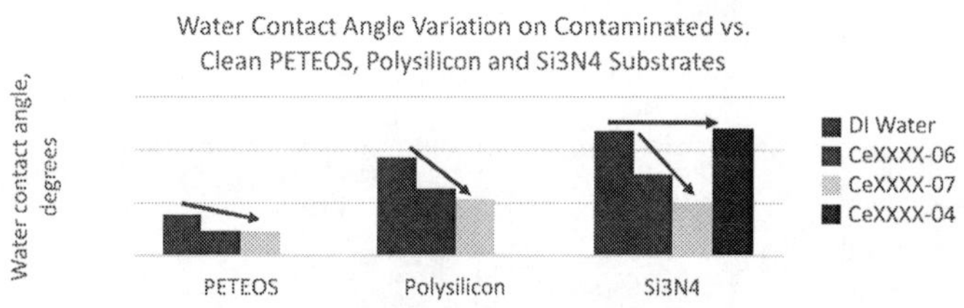

Contact angles are reduced after cleaning with CeXXXX-06 and 07

Surfaces become more hydrophilic, indicates less organic residue leftover after the cleaning

Entegris

CONCLUSIONS

- Understanding the nature of Ceria surface and interaction with the Silica surface are the key to designing the proper cleaner
- Spectroscopic evidence suggests that the right bond-breaking component is essential to remove Ceria from surface
- Different bond-breaking reagent are necessary for each type of Ceria at different pHs
- Several low-pH and high-pH high-performance Ceria cleaning formulation, AG-CeXXXX were developed at Entegris, based on in-depth mechanistic understanding on Silica-Ceria surface interactions
- So far there is no universal ceria cleaner

Entegris

PURE ADVANTAGE

Entegris

19th Surface Preparation and Cleaning Conference (SPCC)

Characterization of incoming PVA brush for 10nm below post CMP cleaning process

April 11, 2018

Jung-Hwan Lee[1], Murhukrishnan Purushothaman[1], Kwang-Min Han[1], Shohei Shima[3], Satomi Hamada[3], Hirokuni Hiyama[3], and Jin-Goo Park[1,2,*]

Department of [1]Bionanotechnology and
[2]Materials Science and Chemical Engineering, Hanyang University,
[3]EBARA Corporation
*jgpark@hanyang.ac.kr

Hanyang Univ., Ansan, 426-791, KOREA

Post CMP Cleaning

❖ **Configurations of post CMP Cleaning Module**

(*FOUP = Front Opening Unified Pod)

PVA Brushes!!

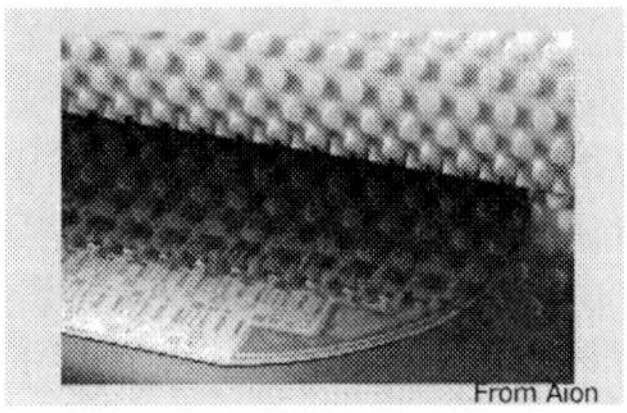

< "E" CMP & P-CMP Tool >

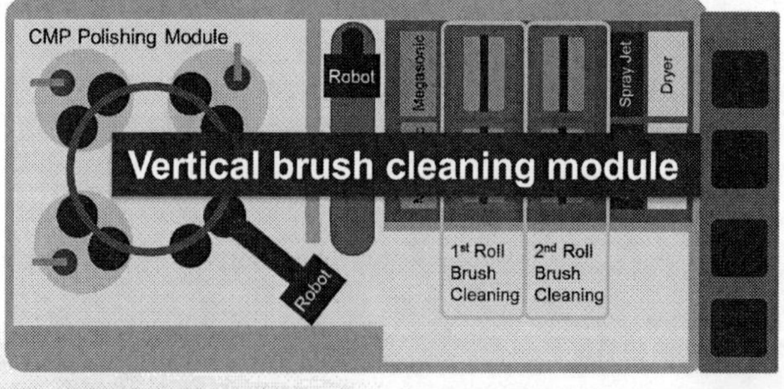

< "A" CMP & P-CMP Tool>

Cassette Line

PVA Brush Cleaning Process

❖ Brush Cleaning Module

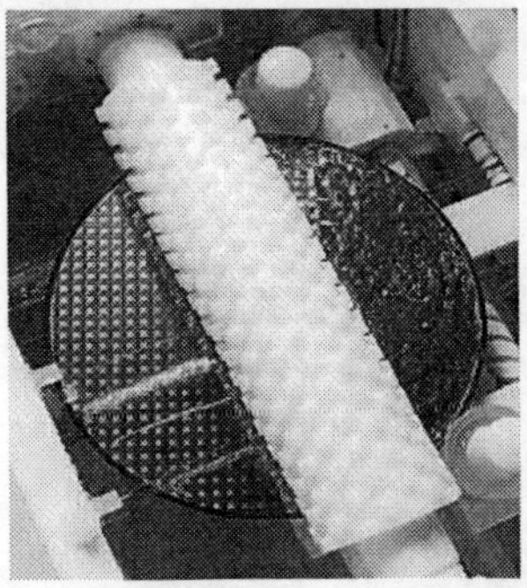

❖ Schematic diagram of Brush Cleaning Module

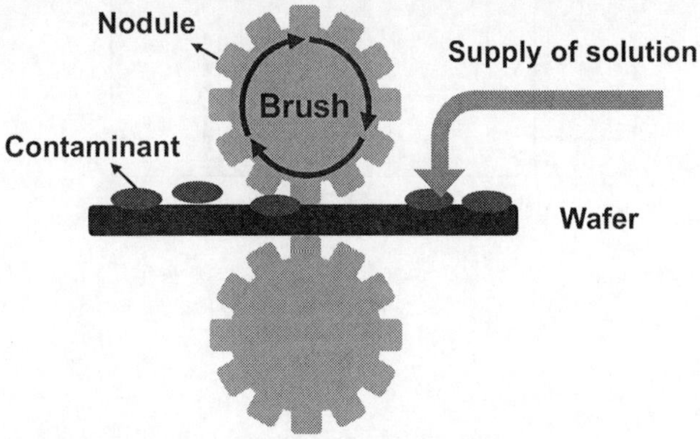

❖ **Advantages of Brush Cleaning**

- **High cleaning efficiency** due to its physical force by direct contact between brush and wafer surface.
- **Effective low cost of ownership (COO)**
- **Process flexibility with various solutions**

Defects from PVA Brush as a function of process time

❖ Defect Count Vs. Wafer Run No.

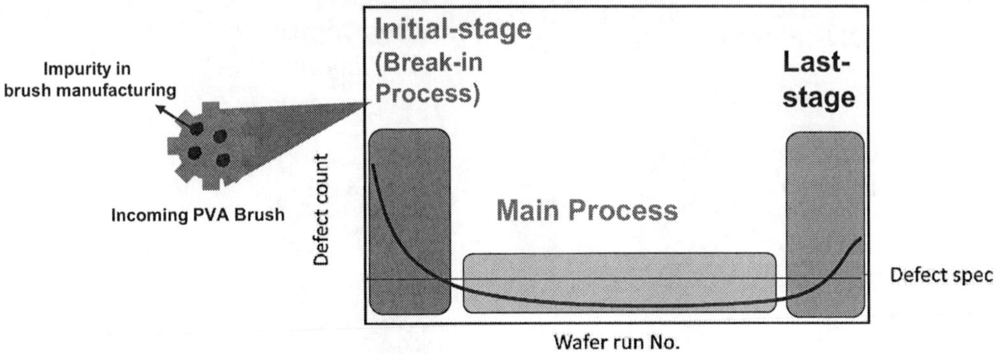

- **Defect Level: Initial-stage > Last-stage >> Main Process**
- **An incoming brush shows higher defect level at initial-stage due to the presence of residual impurities inside the brush.**
- **Pre-treatment process (break-in process) for the removal of impurities from incoming brush is necessary before using.**

Break-in Process of Incoming Brush

❖ **Effect of Brush Break-in Process**

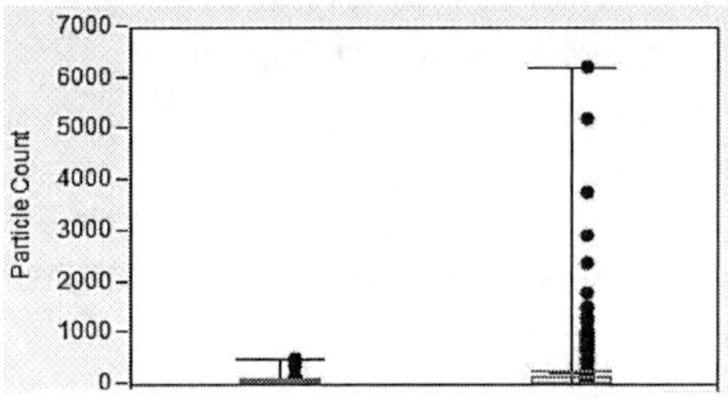

(a) With Break-in, (b) Without Break-in

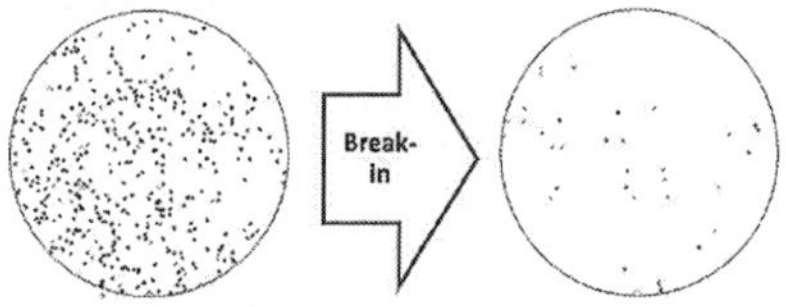

(a) Defect map before and after Break-in

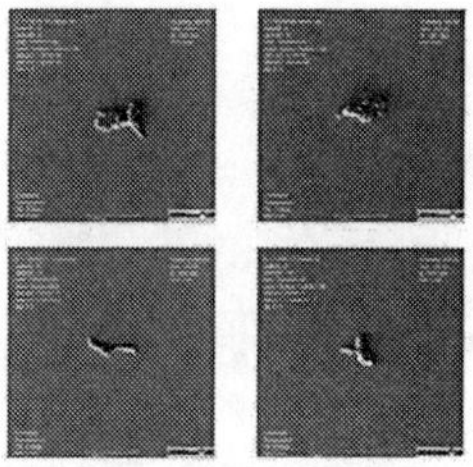

(b) SEM images of organic residue defect

- Conventional break-in process applies **DIW flow through** method and **scrubbing** on dummy wafer to **remove impurities from incoming brush**.
- Break-in process significantly reduces impurities of new brush and increases efficiency of post CMP cleaning.

***Ref: Hong Jin Kim, Korea Cleaning UGM 2016**

Conventional Analysis Methods of Impurities

❖ **Analysis of brush Impurities after DIW Flow Through Break-in**

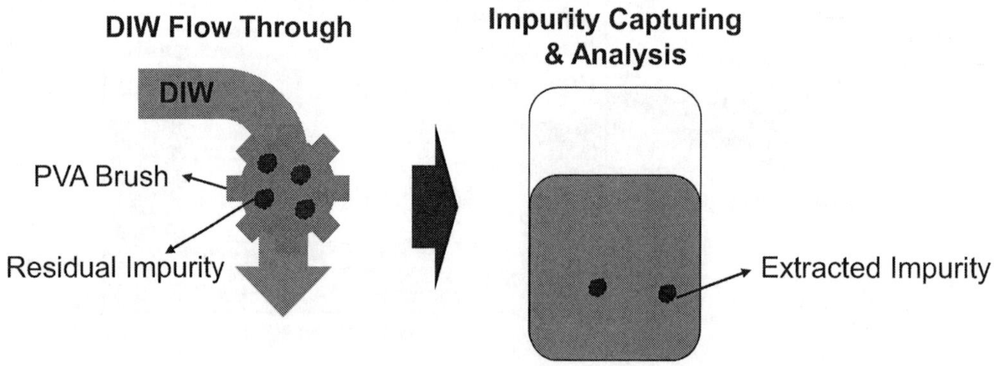

Extremely Low Conc. Impurities

- **Composition of impurities** from incoming brush was not studied due to **extremely low concentration of impurities in conventional break-in method**.

Developed Analysis Methods of Impurities

❖ **Analysis of brush Impurities after Ultrasonication process**

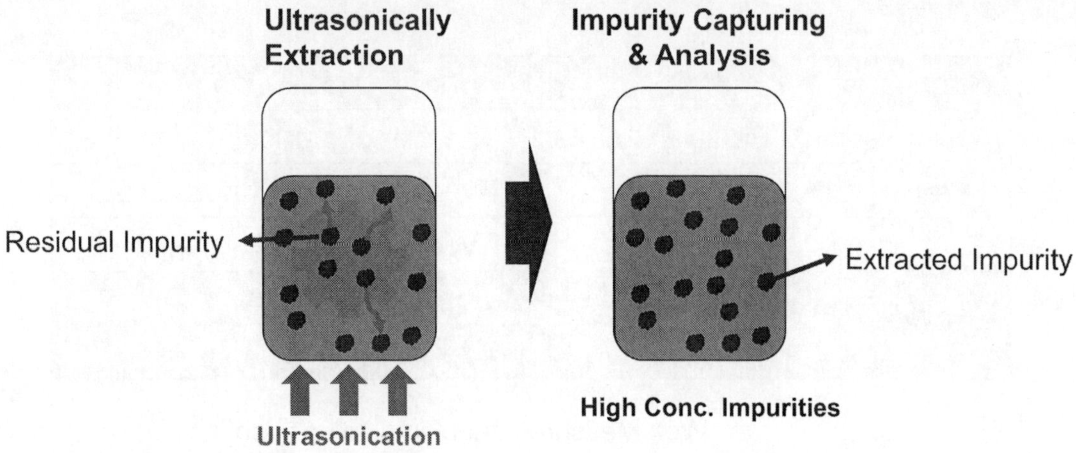

➢ Developed characterization method using **ultrasonication can extract impurities at higher concentration and make it possible to analyze the impurities.**

Characterization Procedure and Equipment

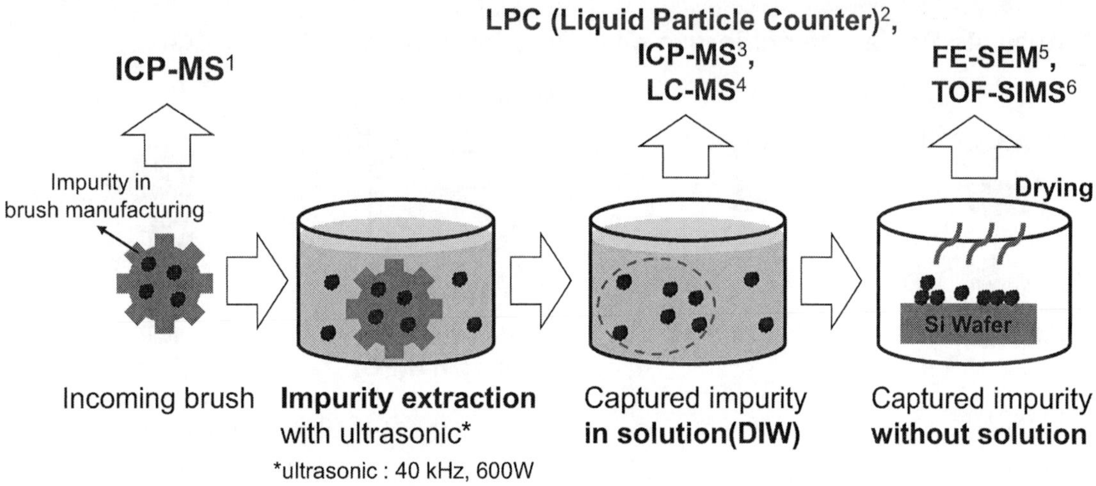

Characterization Procedure and Equipment

Status	Equipment	Analysis
Wet	Liquid Particle Counter[2]	Number of particulate impurity
	ICP-MS[1, 3]	Type of Impurity (Inorganic)
	LC-MS[4]	Type of Impurity (organic)
Dry	FE-SEM[5]	Shape, Size
	TOF-SIMS[6]	Type of Impurity (organic)

Wet: Measurement of impurities in solution
Dry: Measurement of impurities after drying process

1. ICP-MS Analysis of Incoming Brush without Extraction

❖ **Analysis Procedure of Inorganic Impurity from Brush** (Without Extraction)

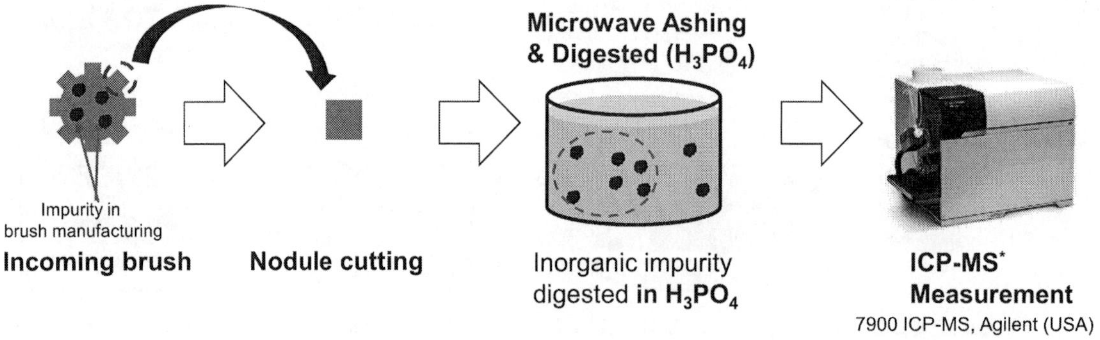

* ICP-MS: Inductively coupled plasma mass spectrometry

➢ **Inorganic impurities in brush w/o extraction process was analyzed by using ICP-MS analysis.**

1. ICP-MS Analysis of Incoming Brush without Extraction

❖ Analyzed Inorganic Impurities from Brush

Element	Concentration (ug/g, ppm)	SD (Standard Deviation)	Relative SD (%)	Composition (%)	Total Amount (ug/g, ppm)
Si	**4278.596**	**157.878**	**3.690**	**88.650**	4,826
Ti	523.721	25.080	4.789	10.851	
W	0.036	0.002	4.162	0.001	
Cu	14.118	0.672	4.764	0.293	
Fe	9.916	0.751	7.575	0.205	

- **ICP-MS analysis shows the presence of Si residues in an incoming brush.**
- **An incoming brush contains high level of Si based impurity.**

2. Liquid Particle Analysis of Extracted Solution

❖ # of Particle Vs. Ultrasonication Time

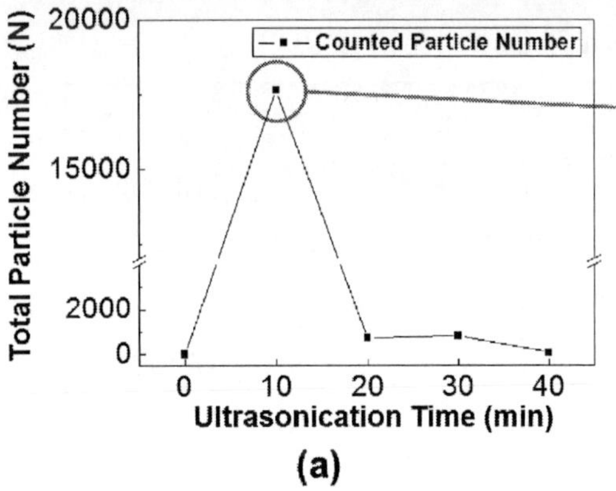

(a)

❖ Particle Size Distribution

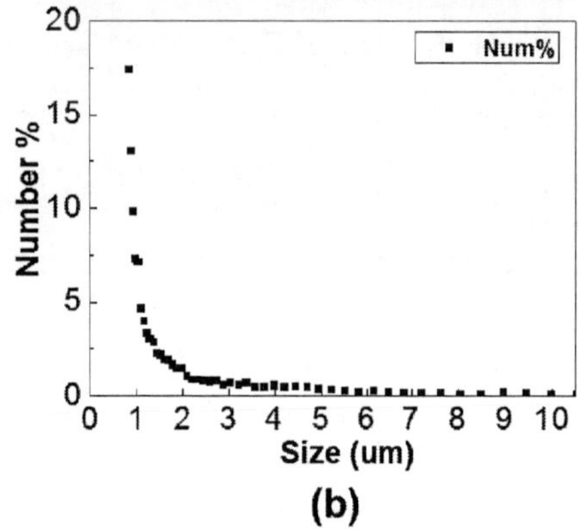

(b)

- **Most of the particulate contaminants were extracted completely within 10 min.**
- Particle size range: X nm ~ 4 um
- **Ultrasonication is very effective and fast process to capture the impurities from the brush.**

3. ICP-MS Analysis of Extracted Solution

❖ **Analysis Procedure of Inorganic Impurities from Extracted Solution**

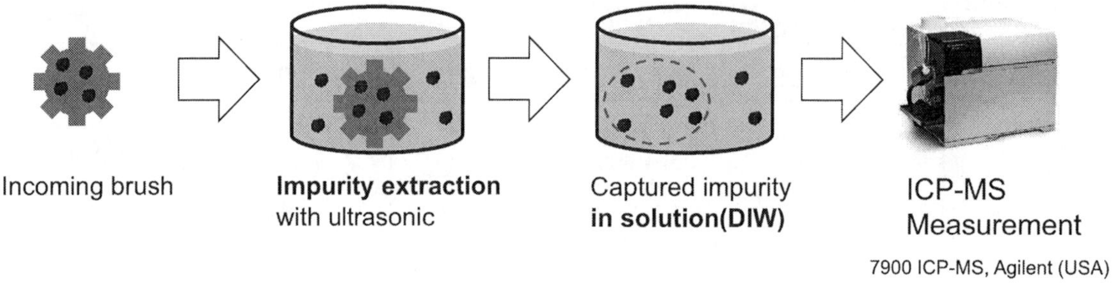

➢ **Inorganic impurities w/ ultrasonically extraction process was analyzed by using ICP-MS analysis.**

3. ICP-MS Analysis of Extracted Solution

❖ **ICP-MS results of Extracted Solution from Brush**

Element	Concentration (ng/mL, ppb)	Relative SD (%)	Composition (%)	Total Amount (ng/mL, ppb)
Si	**35.355**	**2.8**	**74.11**	47.703
Ti	3.422	4.3	7.17	
W	2.403	2.3	3.36	
Cu	4.924	1.8	10.32	
Fe	1.602	2.4	5.04	

➢ **ICP-MS analysis confirms the presence of Si impurities in ultrasonically extracted solution of incoming brush.**

4. LC-MS Analysis of Extracted Solution

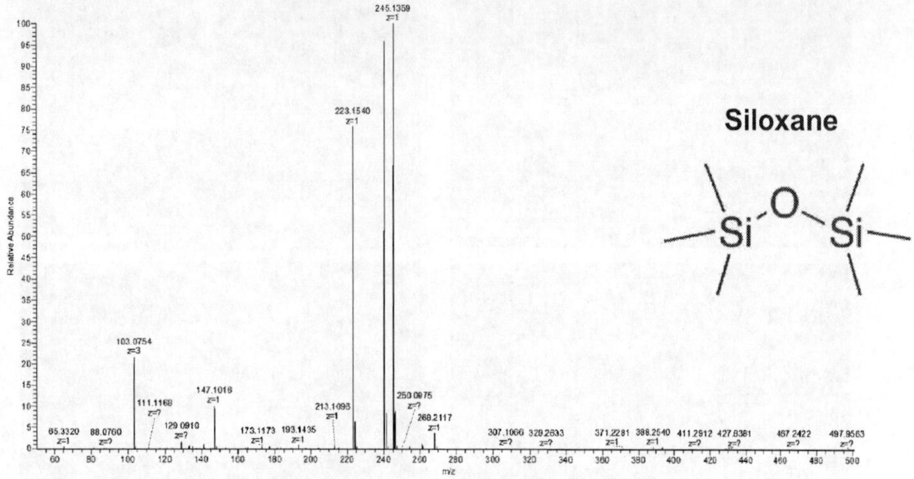

- Siloxane peaks **were observed from extracted solution.**
- This LC-MS result is well matched with ICP-MS results.

5. FE-SEM Measurement of Dried Sample

❖ **FE-SEM Images of Impurities after Drying Process**

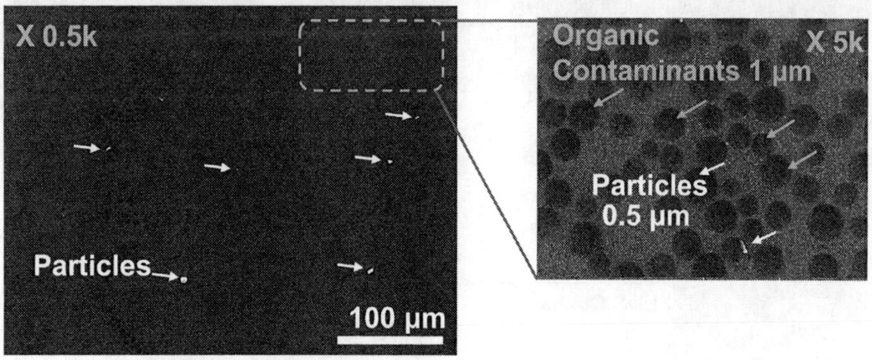

(a) X 0.5K **(b) X 5K**

- **Organic contaminants and particles were observed.**
- **Density : organic contaminants >> particles**

5. FE-SEM Measurement of Particles

❖ **FE-SEM Images of Particulate Impurities**

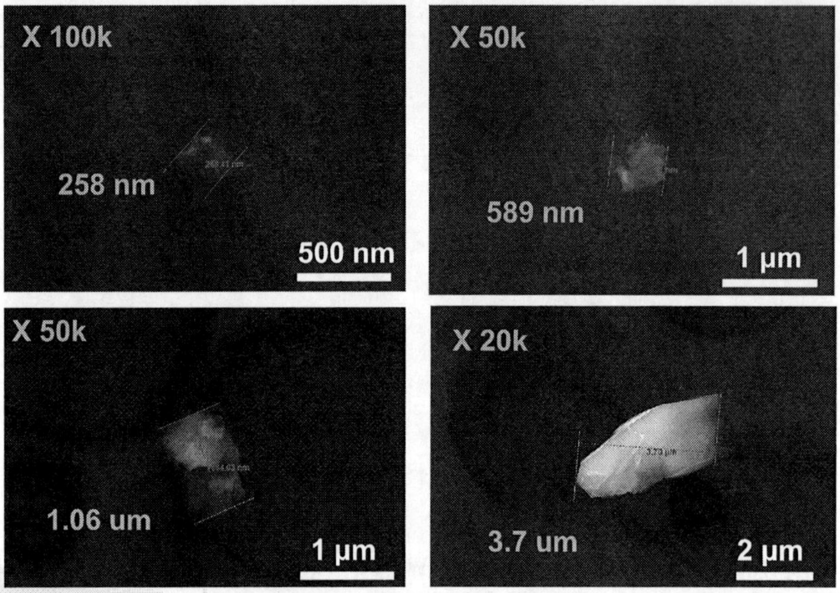

- **Particle size range: 200 nm ~ 4 um**
- These FE-SEM results are well matched with LPC results.

5. EDX Analysis of Contaminants

❖ **Uncontaminated Area**

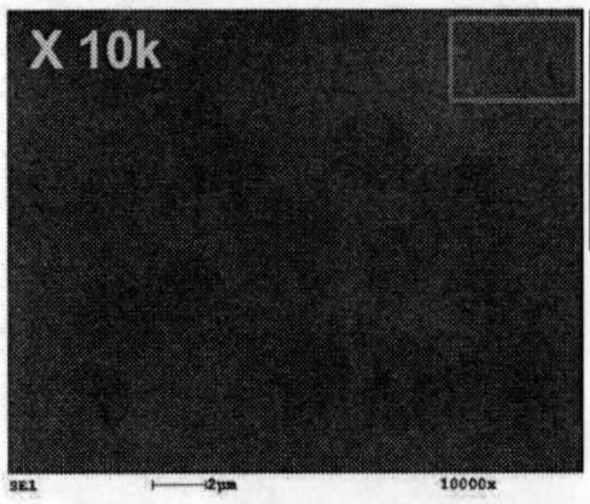

EDS Quantitative Results

Element	Wt%	At%
SiK	100.00	100.00

❖ **Contaminated particle**

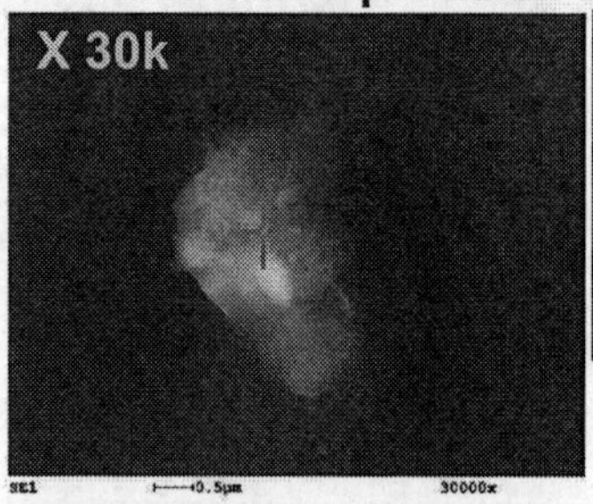

EDS Quantitative Results

Element	Wt%	At%
CK	61.60	73.30
OK	18.62	16.64
SiK	19.78	10.07

❖ **Contaminated Area**

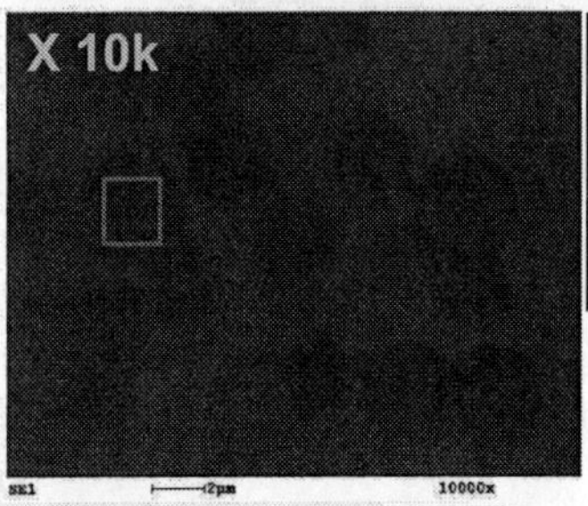

EDS Quantitative Results

Element	Wt%	At%
CK	43.34	64.14
SiK	56.66	35.86

❖ **Contaminated Area**

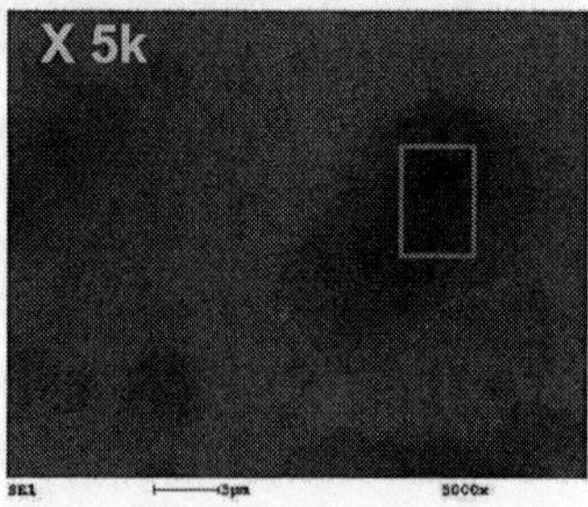

EDS Quantitative Results

Element	Wt%	At%
CK	61.38	74.60
OK	13.57	12.38
SiK	25.06	13.02

- **Organic contaminants and particles show carbon peak.**

6. TOF-SIMS Analysis of Dried Sample

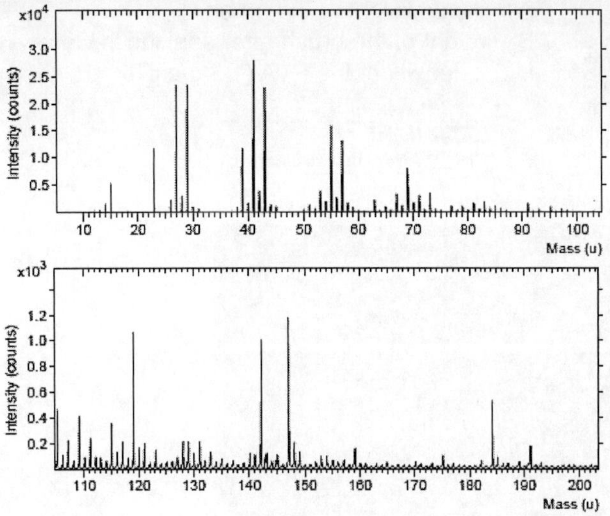

Siloxane

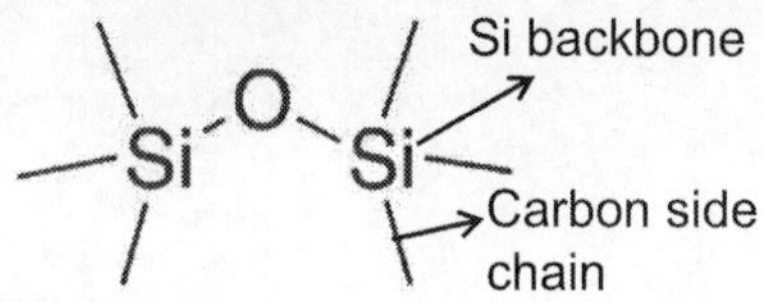

Siloxane: Organosilicon with Si-O-Si linkage

- **Siloxane** peaks were observed from incoming brush after dried on Si wafer substrate.
- This TOF-SIMS result is well matched with ICP-MS and LC-MS results.

FE-SEM Images of Ultrasonicated Brush

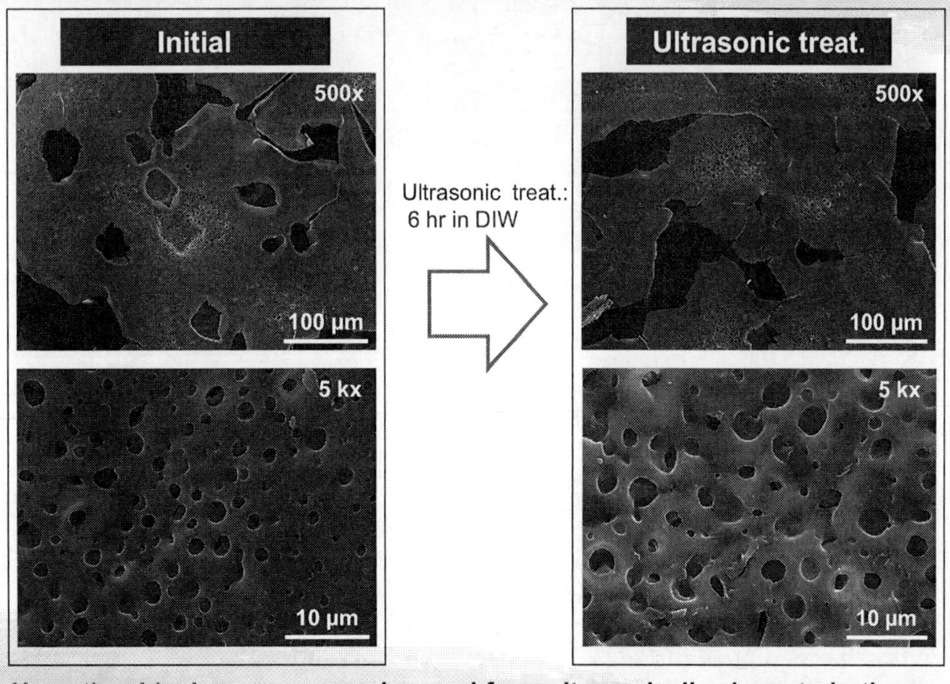

- **No noticeable damages were observed from ultrasonically characterization process**

Porosity Measurement of Ultrasonicated Brush

$$\text{Porosity } \% = \frac{B - A}{(B - A) + \frac{A}{D_{pva}}}$$

A : empty weight of the brush
B : weight of the brush after soaking in water
D_{pva} : density of the PVA (1.3 g/cm^3)

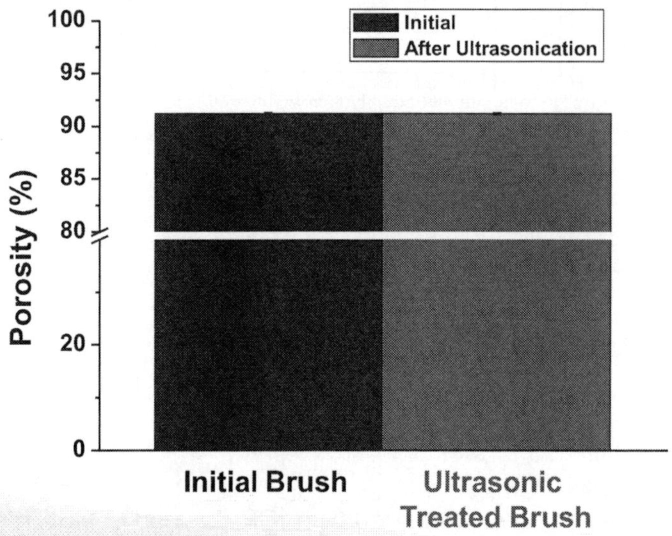

➢ **Porosity of PVA brush was not changed after ultrasonication for 6 hours.**

Analysis Procedure: Effect of Cleaning Chemistry(TMAH) on PVA Brush

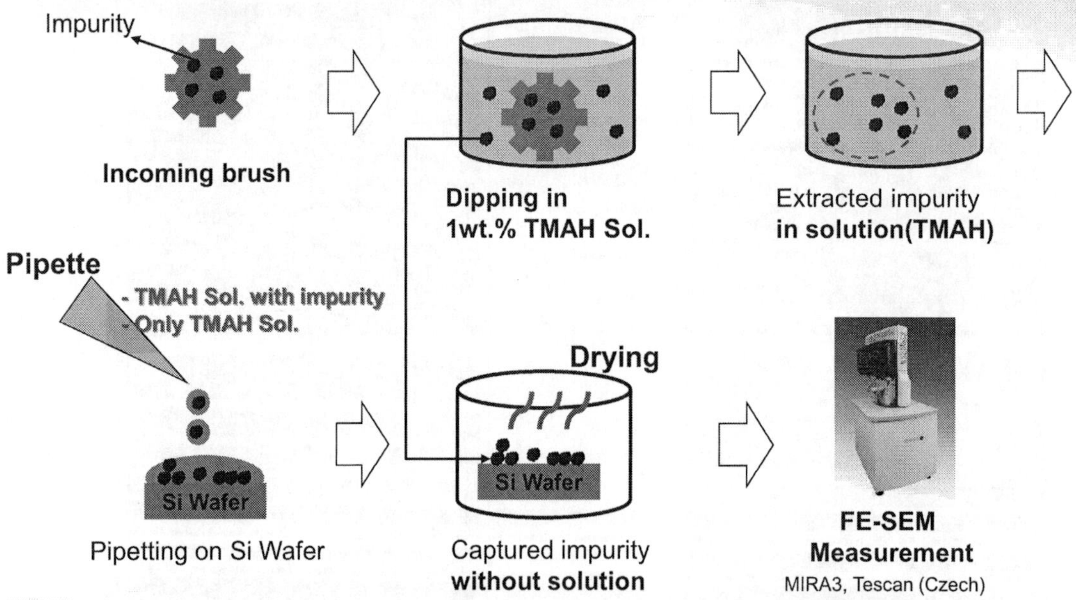

➢ **PVA brush was dipped in 1wt.% TMAH sol. to observe the effect of cleaning chemicals (TMAH) on impurity generation from brush.**

Effect of TMAH chemistry on PVA brush

❖ **FE-SEM Images of Organic Residue from Brush Dipped in 1wt.% TMAH**

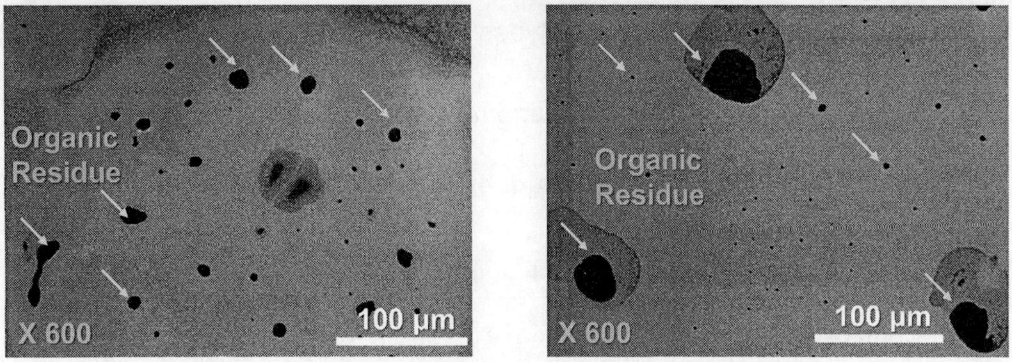

➢ **Brush dipped in 1wt.% TMAH solution shows organic residue contamination.**

Transport Mechanism of Siloxane by Cleaning Chemical

1. In general, organic residues were observed in metal(especially Cu) CMP process.
2. Cleaning chemical of Cu CMP include **TMAH**.
3. **TMAH has high dissolution ability of siloxane.**
4. **Siloxane can be delivered** from inside of brush **to metal surface due to cleaning chemical.**

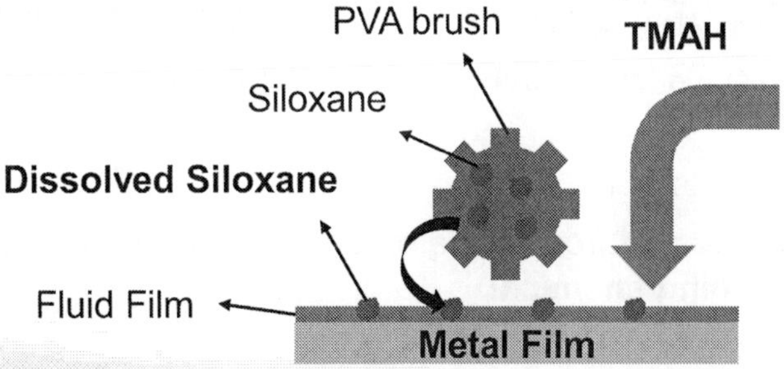

Summary

1. **Ultrasonication** is very effective and fast method to characterize the PVA brush.
2. Incoming PVA brush contains 2 types of impurities (**soluble and insoluble**).
3. Soluble impurities such as **siloxane** may create **organic residues**.
4. Insoluble impurities such as **PVA debris** may create **particle residues**.

	Soluble Impurity	Insoluble Impurity
Composition	**Siloxane**	Weakly bonded **PVA** debris
Shape	Thin and circular organic residue	Particle
Analysis Method	ICP-MS, TOF-SIMS, LC-MS	LPC, FE-SEM, TOF-SIMS

Nano-Bio Electronic Materials and Processing Lab. 25

Summary

5. **Soluble impurities may be a root cause of organic residues** after post CMP cleaning process.

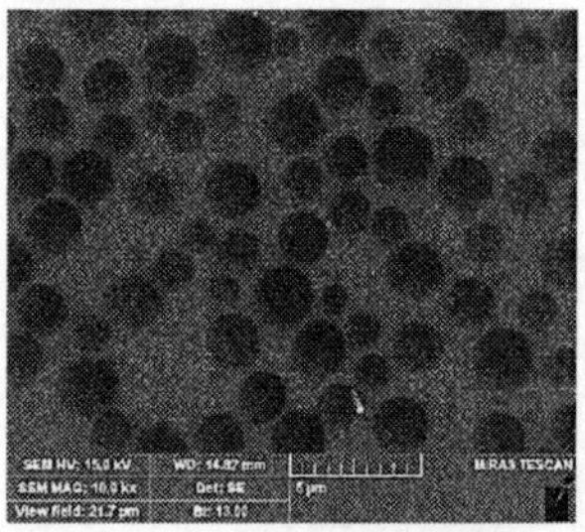

(a) Organic Residue from Brush

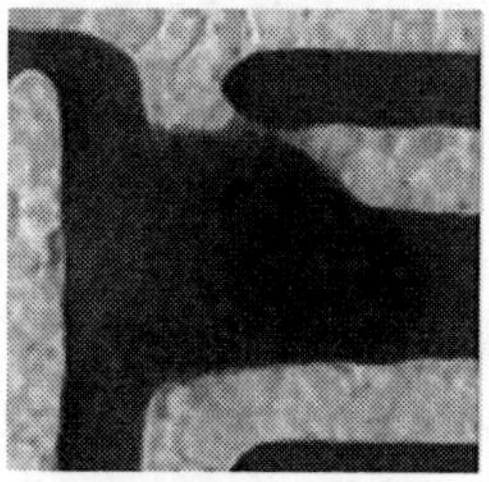

(b) Organic Residue after post Cu CMP Cleaning

*Courtesy from GlobalFoundries

6. **Siloxane can be delivered** from inside of brush **to metal surface due to etching ability of cleaning chemical.**

Nano-Bio Electronic Materials and Processing Lab. 26

HANYANG UNIVERSITY
한양
1939
THANK YOU FOR
YOUR ATTENTION
NEMPL
NEMPL
Hanyang University
Nano-Bio Electronic Materials and Processing Lab.

Post Cleaning Chemical of Tungsten Chemical Mechanical Planarization for Memory Devices

Jun Yong Kim
Cleaning/CMP Technology

SAMSUNG THE NEXT CREATION STARTS HERE

1. CMP Process and Cleaning challenges
2. Problem Statement
3. Results of Cleaning Chemical Evaluation
4. Summary

SAMSUNG THE NEXT CREATION STARTS HERE

Chemical Mechanical Polishing (CMP)

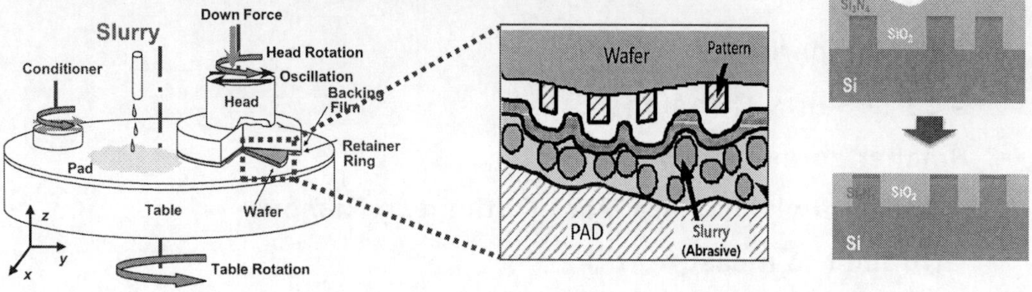

- CMP was first introduced to semiconductor process to planarize ILD.
- With improvements in equipment and consumables, CMP is used in STI CMP for isolation, W, Cu CMP for metallization and Buffing CMP for particle removal.
- The number of W CMP steps is increasing with the introduction of VNAND.

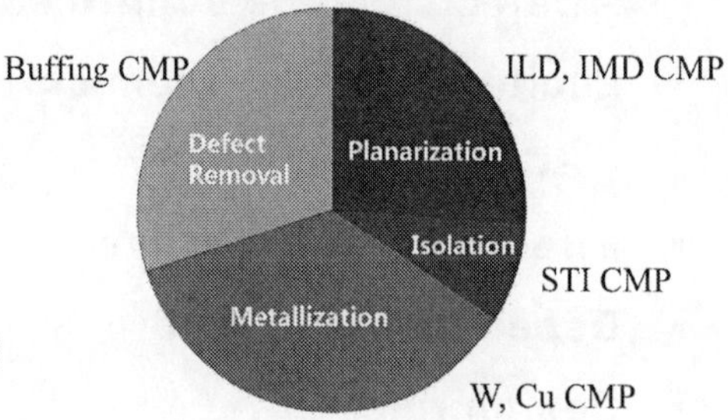

DEFECT control

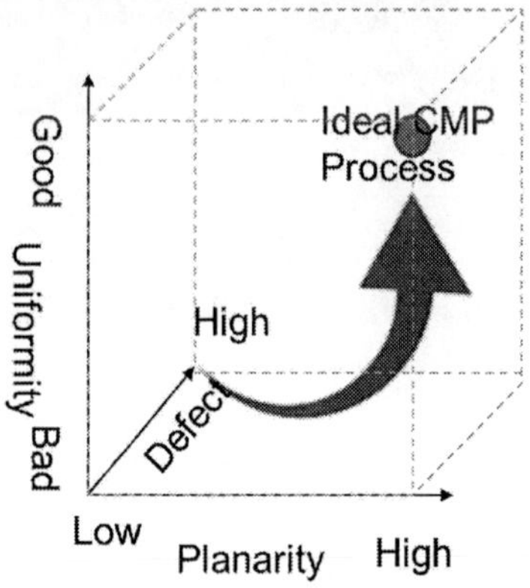

Slurry	Abrasive Size, Shape	Uniform, Round	
Disk	Diamond Height, Arrangement	Even protrusion, Regular	
Pad	Contact, Hardness	Uniform, Soft	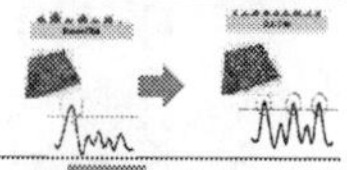
Process /Consumables	Head Pressure, Brush gap, Brush	Low Brushless	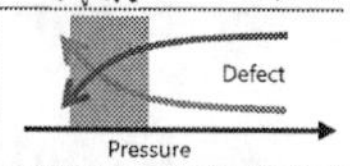
Environment	Foreign Particle, Pattern	Edge Engineering	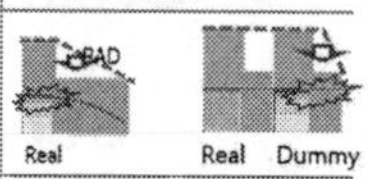
Chemical	Cleaning Time	-	-

- CMP process has been developed through equipment, functional slurries and pad, disk.
- Scratch reduction has been mainly made through pad, slurry changes.
- Particles are reduced through changes in brush and process conditions.
- Post CMP Cleaning Chemicals(SC1,NH4OH, and HF) have been used since CMP process was introduced.

Post CMP cleaning Challenges

- Abrasive particle
  - Ceria Slurry Cleaning
- Smaller abrasive size
  - Smaller Particles are More Difficult to Remove
  - Difficult to measure
- Non-visual defect
  - Defects can be confirmed by Fab-out yield.
- Limited cleaning chemical
  - SC1, NH_4OH, HF
- Effect of brush lifetime
- Dependency on tool
- Design rule

SAMSUNG THE NEXT CREATION STARTS HERE

Results of Cleaning Chemical Evaluation

1. Problem Statement
2. Contamination Source
3. Evaluation of Ammonia, HF
4. Evaluation of Formulated Cleaning Chemical
   - Defect on blanket wafer
   - Properties of Formulated Chemical
   - Results of Tungsten Surface Analysis
   - Results of Application to Pattern Wafer

SAMSUNG THE NEXT CREATION STARTS HERE

Typical Tungsten CMP Process

Ⅰ. Polishing

Ⅱ. Cleaning

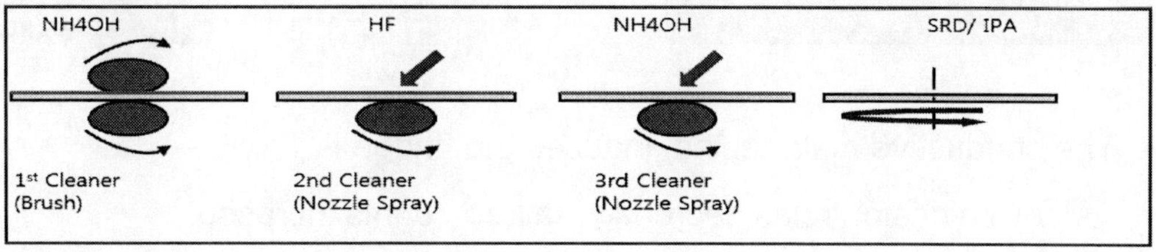

- Tungsten CMP process consists of two process, a main polishing process and a post-CMP cleaning process.
- In the post-CMP cleaning step, all of the wafers were treated with wet chemicals to remove the remaining chemicals and abrasives on the wafer.

SAMSUNG THE NEXT CREATION STARTS HERE

Defects Induced W CMP Process

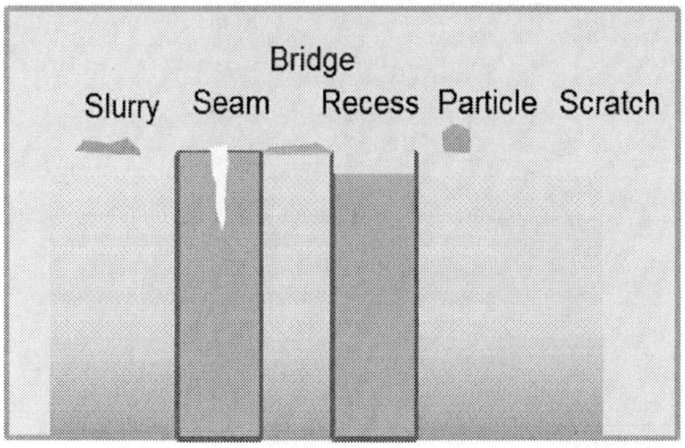

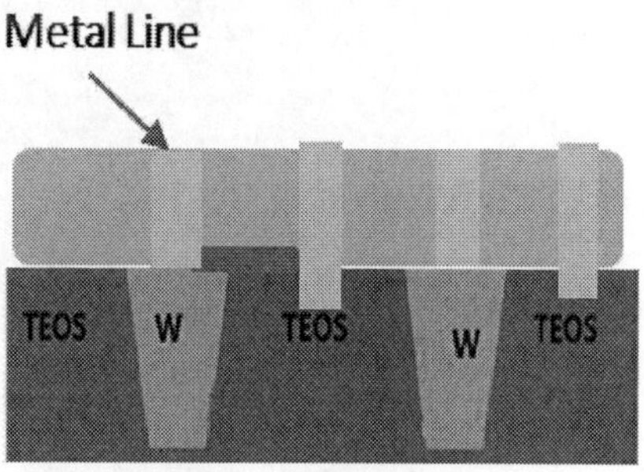

Metal Line Bridge

- Defects to be removed from W CMP process include slurry residues, organic particles, scratches and bridge.
- Bridge : Materials on the surface after W CMP causes the subsequent Metal line connection
- Conducting materials can result in the product yield loss.

SAMSUNG THE NEXT CREATION STARTS HERE

Problem Statement

- Bridge Fail on W CMP

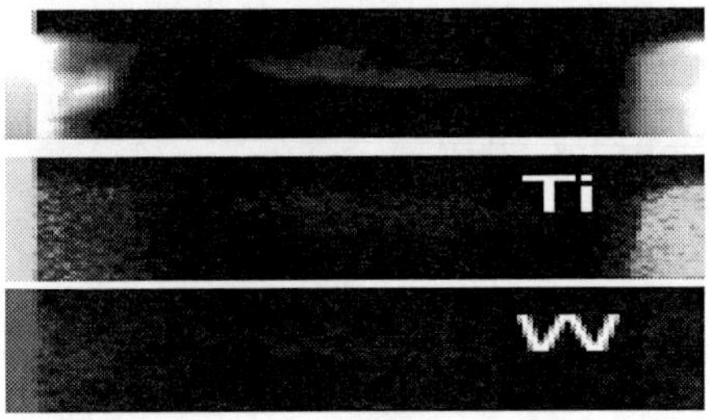

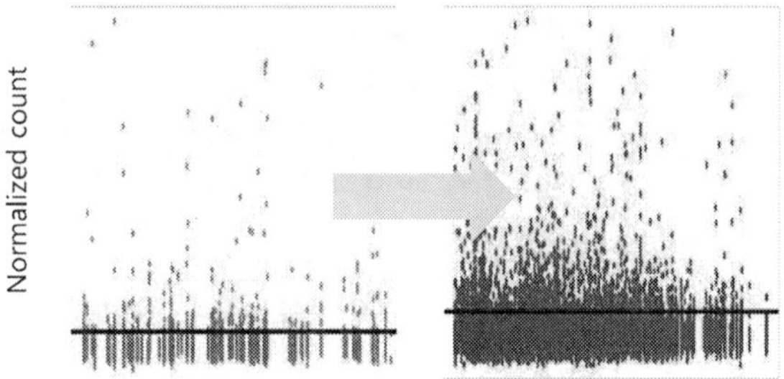

- The conductive material that induces the bridge is Ti, W.
- As device design rules decrease , failure counts increase.
- Residues that contribute to this issue are undetectable and do not correlate to total defects on the wafer.

SAMSUNG THE NEXT CREATION STARTS HERE

Cross Contamination Source

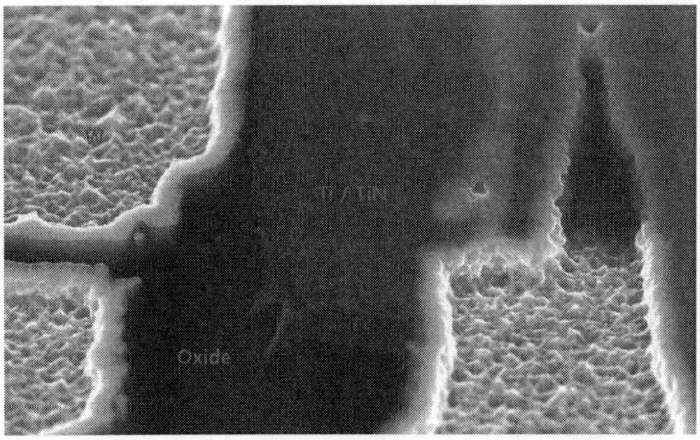

V_SEM Image

- The wafer surface has multiple films such as W, Ti, TiN, Oxide.
- These can cause brush contamination.

SAMSUNG THE NEXT CREATION STARTS HERE

Evaluation of Brush contamination

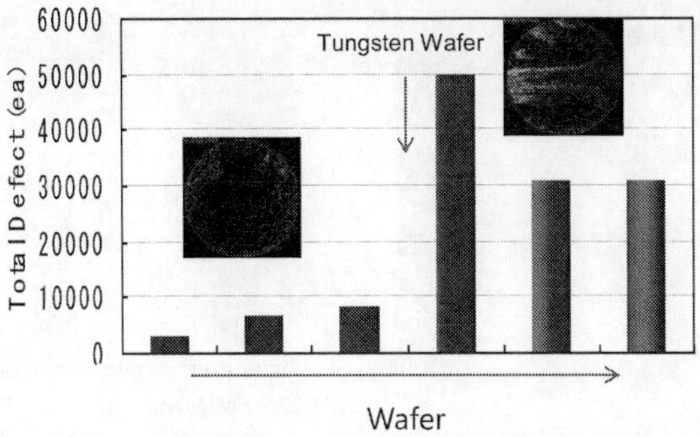

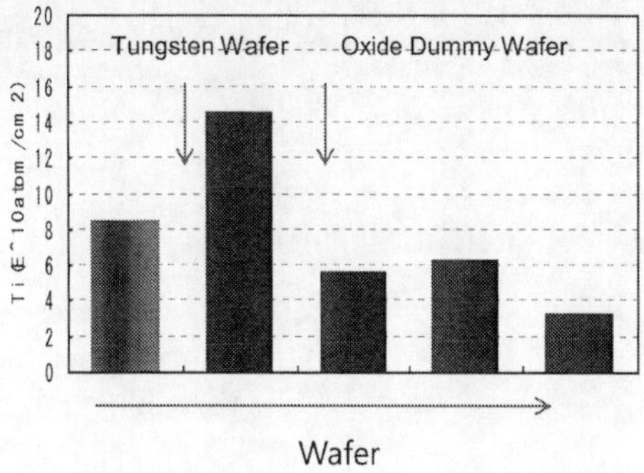

- Defects do not increase significantly during oxide film wafer polishing.
- Defect and metal contamination increase after polishing tungsten wafer.

Brush Cleaning can contaminate wafers.
→ Bridge fail Increase according to Wafer Polishing.

Evaluation of Brush contamination

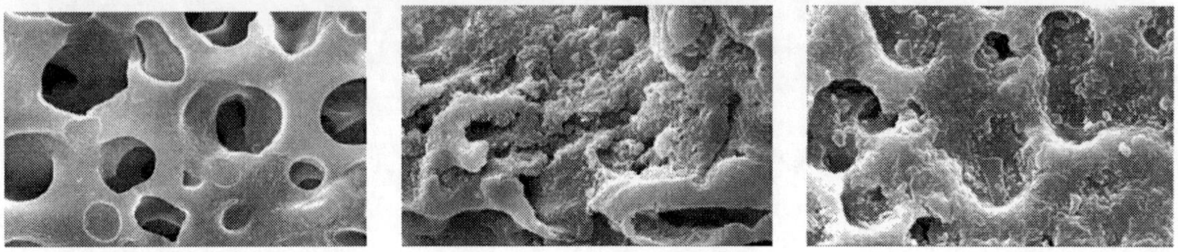

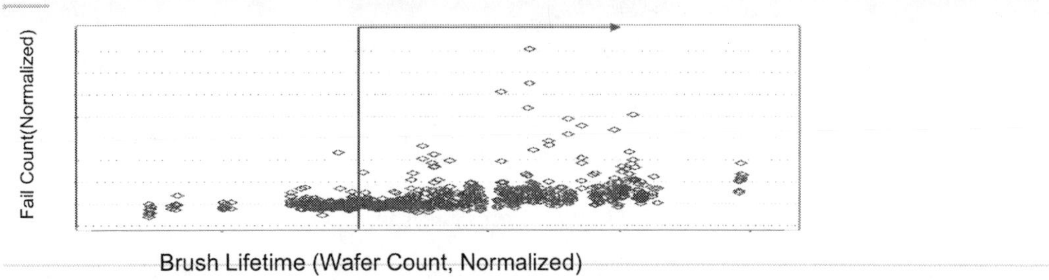

- Brush contamination is associated with an increase of defect and affects an increase in the number of failures due to invisible defects.
- As a result, the defects or failures were controlled by limiting brush usage time.

Effect of Cleaning Chemicals

	Brush	Spray1	Spray2	SRD
1	NH_4OH	X	X	o
2	NH_4OH	HF	X	o
3	NH_4OH	HF	NH_4OH	o

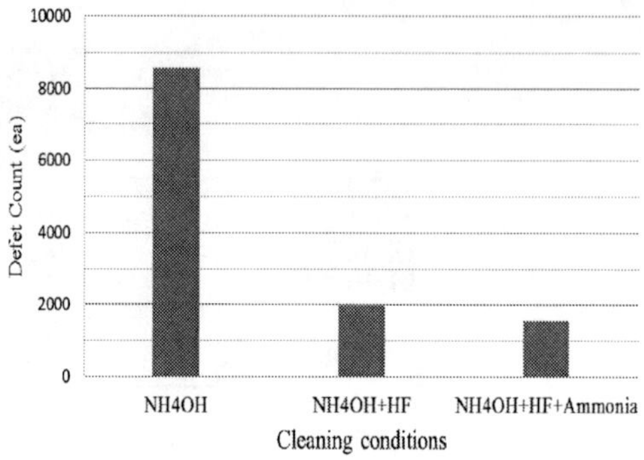

- HF Cleaning is most effective in removing particles
- Is it possible to increase HF cleaning time?
  → W protrusion, pattern damage, etc
- Is it possible to increase NH_4OH cleaning time?

Effect of NH_4OH on Devices

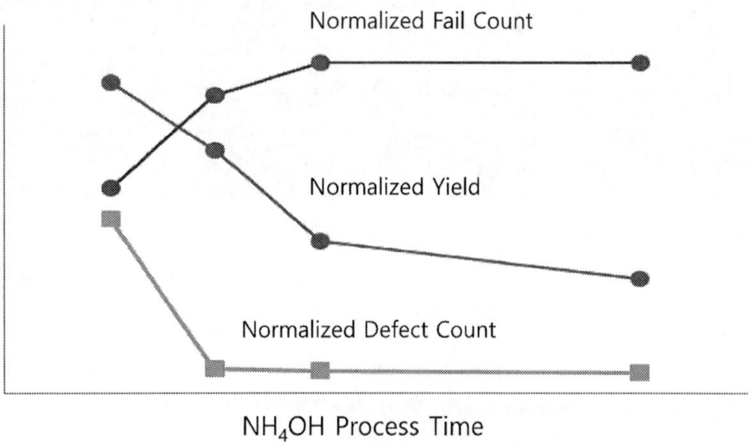

- As the NH_4OH cleaning time increases, visible or detectable defects are reduced.
- The number of failures count due to invisible defect is increased regardless of defect reduction.

Target W CMP Cleaning

- NH4OH
  1) remove particle.
  2) Low W compatibility.
- HF
  1) Improve clean performance.
  2) Cause the loss of TiN barrier.
  3) Make negative impact on the device performance.
  4) Damage the tungsten plug and the underlying layer.
- Brush :
  1) Become loaded with slurry particles.
  2) The efficiency was reduced gradually.

	NH4OH	HF	Target Formulated Chemical
Tungsten	X	O	O
TiN	O	X	O
Clean Efficiency	△	O	O
Brush Lifetime Dependency	O	O	?

Therefore, the use of a formulated chemistry is quite critical in Tungsten Post CMP clean process.

SAMSUNG THE NEXT CREATION STARTS HERE

Target Defect Removal

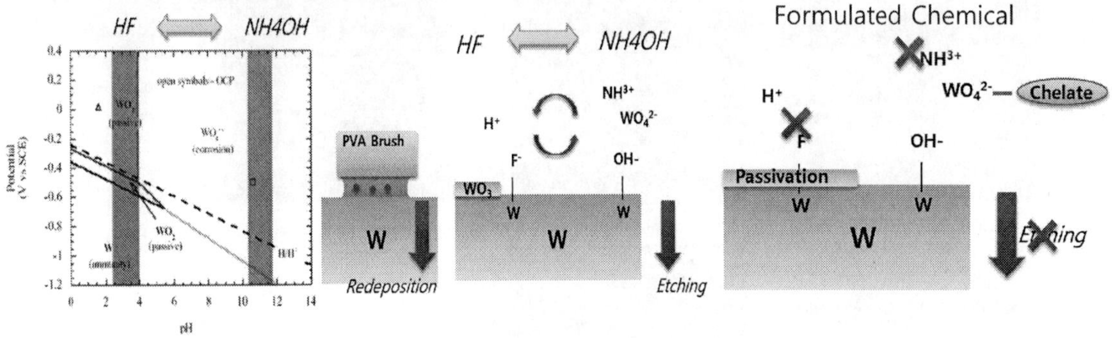

- Surfactant : Prevention of re-adsorption of particle
- Chelating Agent : Metal decontamination
  → Metallic contamination removed from the surface is stabilized in the solution by chelating agents
- pH: Acid (W etch prevention)
  → Re-adsorption side effect is overcome with surfactant
- Additive
  → Silica particle removal and organic residue removal

SAMSUNG THE NEXT CREATION STARTS HERE

Corrosion Potential of W/Ti Cleaning Chemical NH_4OH

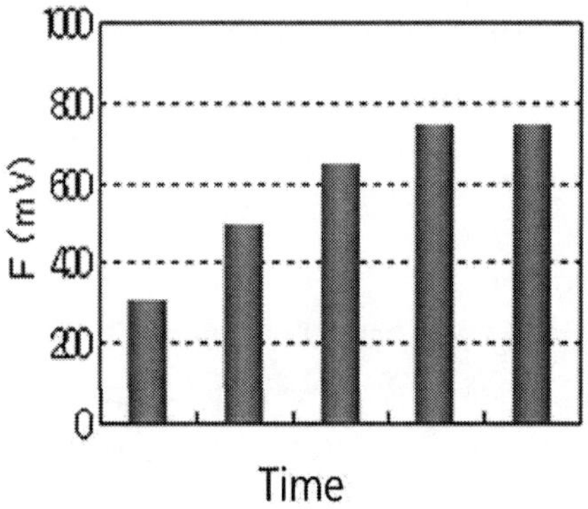

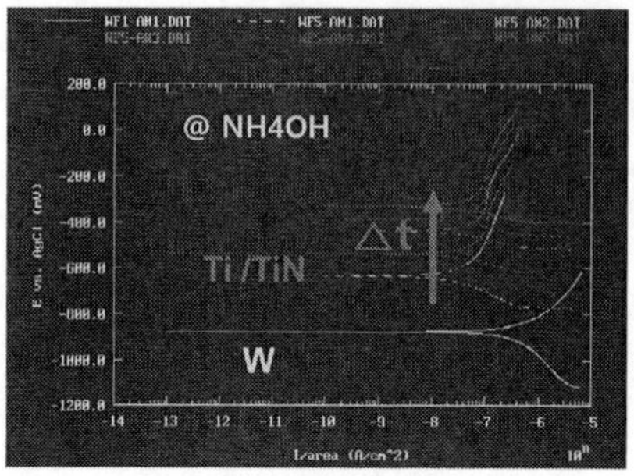

Evaluation of potential change according to ammonia cleaning time

1. The potential shifts to positive depending on the NH4OH exposure time.
2. The barrier metal increases the possibility of being reduced or precipitated.

Particle Removal

Ammonia Clean

Ammonia + dHF Clean

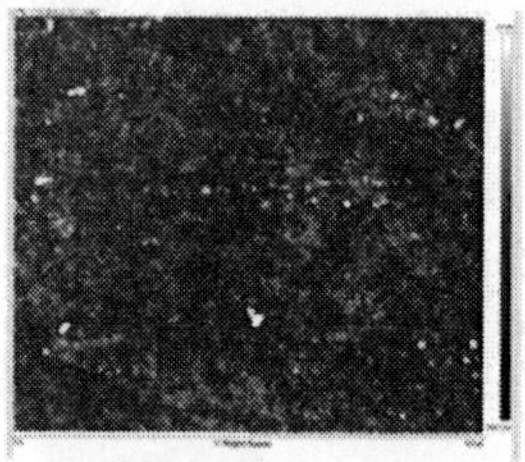

Ammonia + dHF Clean+ Ammonia

DIW Clean

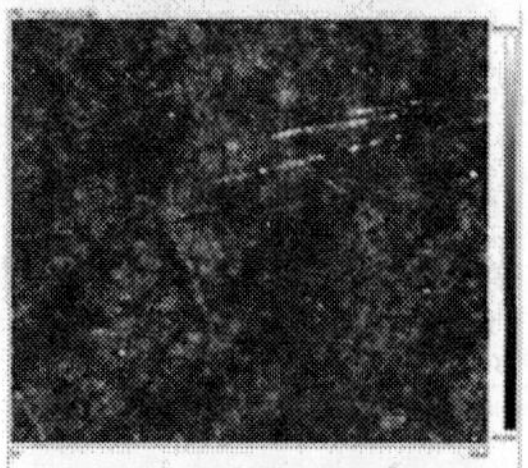

Formulated Chemical

Image courtesy of VERSUM

- Formulated chemicals show the smallest number of defects.
- There is no correlation between the reduction of the defect and the yield.
- After selecting the best formulation for particle removal on blanket wafers, pattern wafers were cleaned with new chemical.

Particle Removal

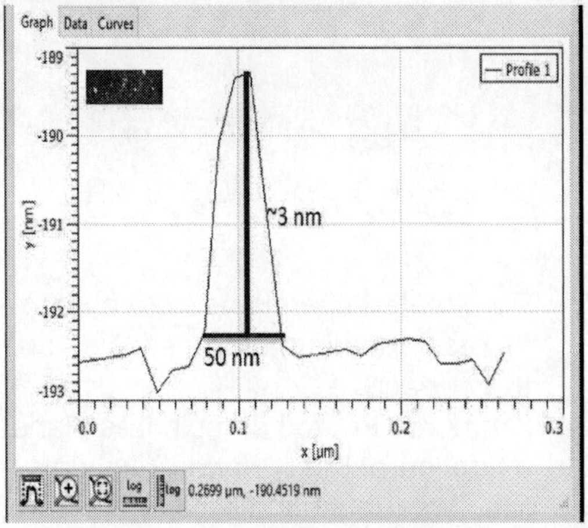

Typical defect after NH_4OH/HF

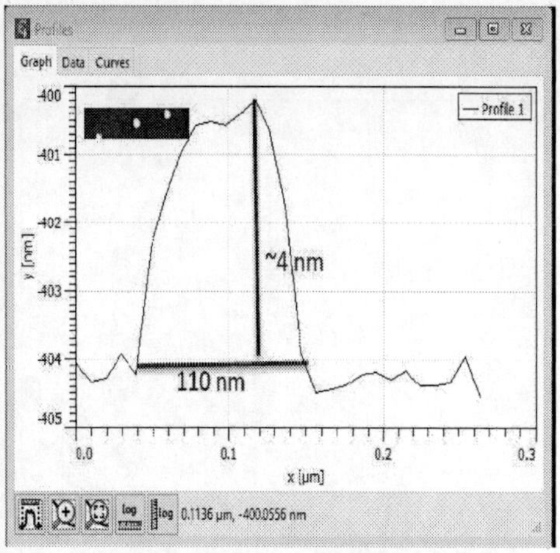

Typical defect after DIW Clean

Image courtesy of VERSUM

- Cleaning processes like NH_4OH/ HF appeared more successful in removing large particles however appeared insufficient for removing small particles.

Properties of Formulated Chemical

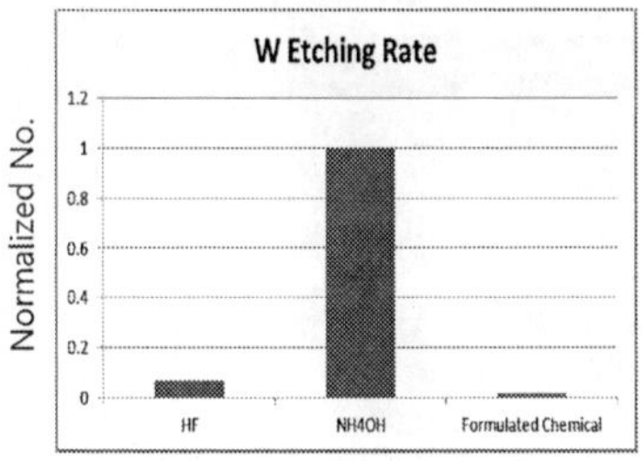

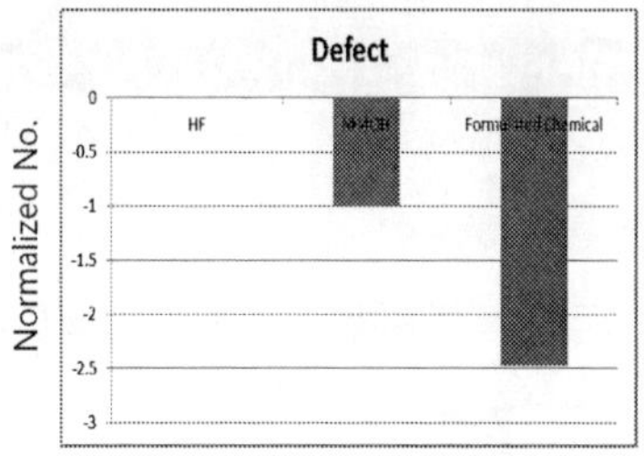

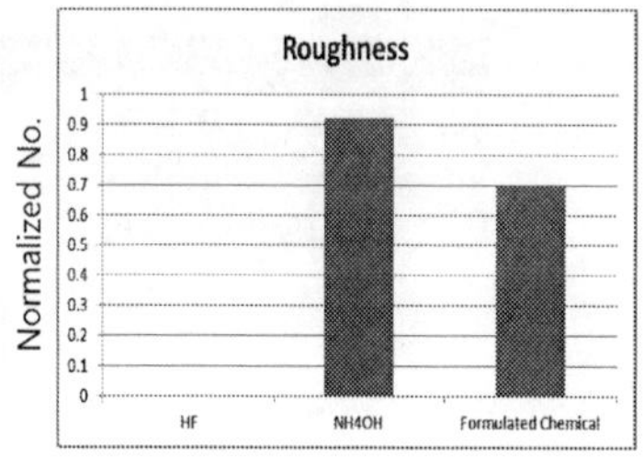

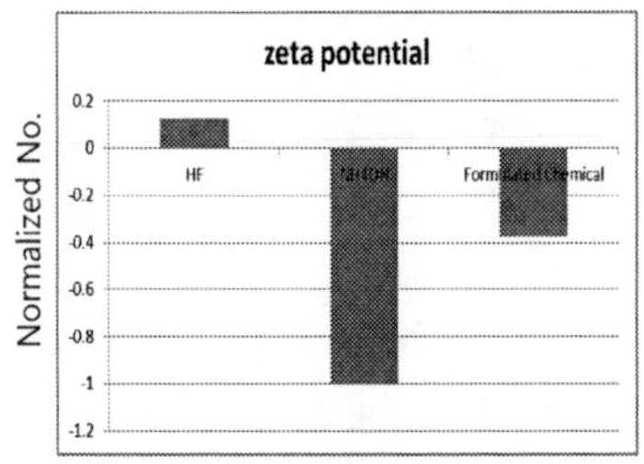

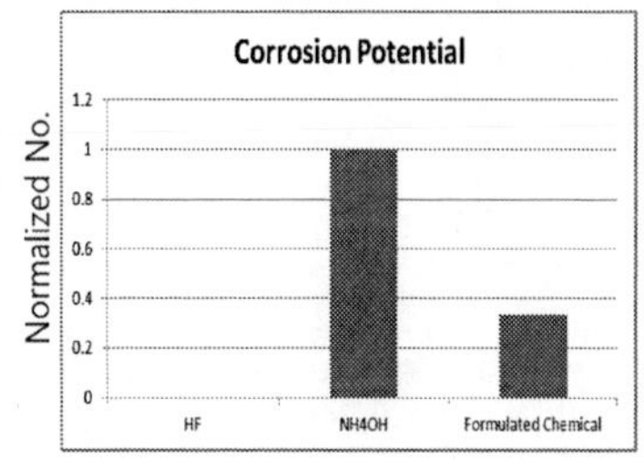

Tungsten surface analysis result

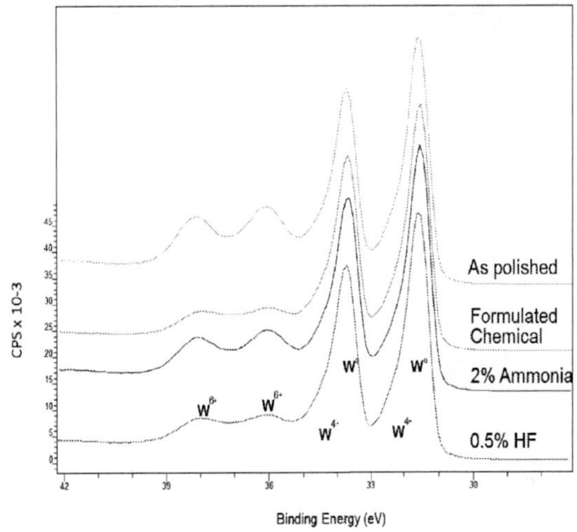

Image courtesy of VERSUM

$W^{+} + 3H_2O \rightarrow WO_3 + 6H^{+} + 6e^{-}$

$WO_3 + H_2O \rightarrow WO_4^{2-} + 2H^{+}$

- Formulated chemistries are able to increase stable metallic species on the surface.
- Ammonia treated sample has higher oxidation of the surfaces, likely a result of tungsten dissolving to form WO_4^{2-} and re-depositing back.
- Re-deposition may be a concern for electrical leakage.

SAMSUNG THE NEXT CREATION STARTS HERE

Evaluation Results on Pattern Wafer

	Brush	Spray1	Spray2
A	NH_4OH	HF	NH_4OH
B	Chemical	Chemical	Chemical

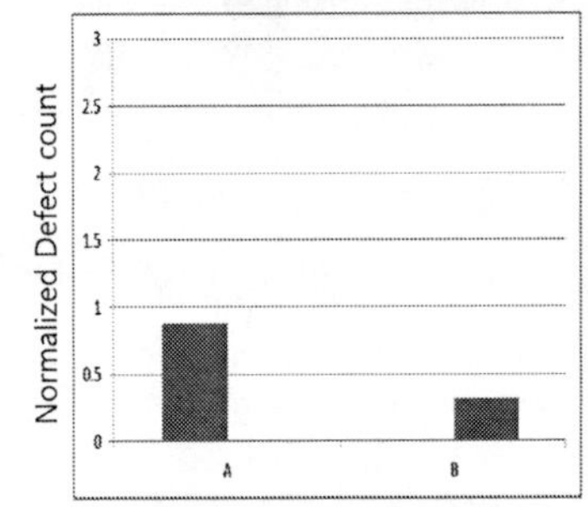

Defect after CMP

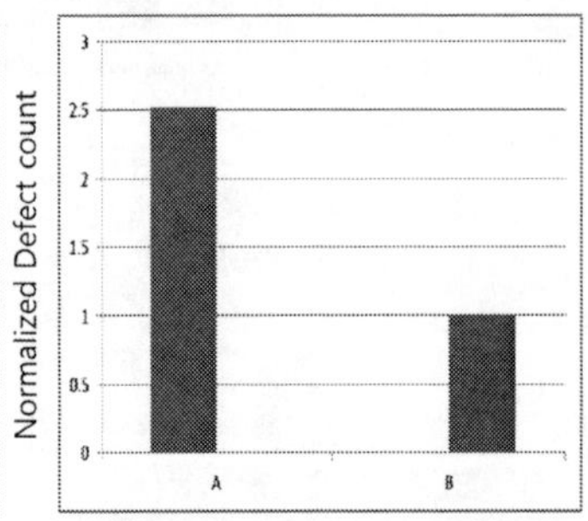

Defect after Film Dep. Followed by CMP

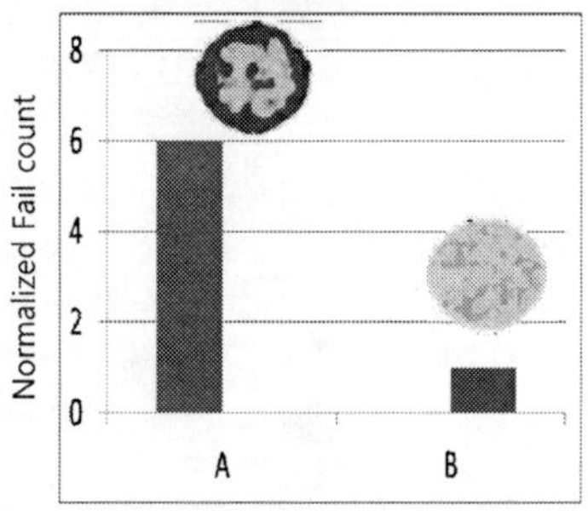

- Defects decreased under the B condition.
- It has been confirmed that it has decreased even after subsequent film deposition to see smaller defects.
- Formulated Chemical is effective for particle removal and bridge failure reduction.

SAMSUNG THE NEXT CREATION STARTS HERE

Summary

1. The CMP process has been mainly improved with tool, slurry, and pad for planarity, uniformity and defect.

2. As the device design rule decreases, bridge failure must be improved.

3. Improvements with existing chemicals (NH_4OH, HF) are limited.

4. In the memory devices, post CMP cleaning chemicals have been applied to the Cu processes, but there is also a need for new chemicals for W CMP.

5. It has been confirmed that the bridge generation by environment, chemicals can be reduced and the process margins can be increased using formulated chemicals.

SAMSUNG THE NEXT CREATION STARTS HERE

Post Etch Residue Removal on Cobalt Contact

Wilson Yeh*, Wade Lee, Carol Cheng, Alan Chen, Cindy Chiang, Paul Wang, Lester Lee, Sang-Min Wong, and Chi Yen
EKC Technology, Semiconductor Technologies, DuPont Electronics & Imaging
Contact info: Wilson.Yeh@dupont.com; Sang-Min.Wong@dupont.com

Introduction

In advanced technology nodes 7nm and beyond, cobalt has been introduced as contacts in the middle-of-the-line (MOL) of IC fabrication. Cobalt offers major advantages such as greater volume of the contacts and does not require thicker barrier layers like tungsten requires. Although replacing tungsten contacts with cobalt contacts is gaining attraction, customers would still maintain existing processes utilizing tungsten contacts. Therefore, there is a need to develop new chemistries which are able to clean both cobalt and tungsten contacts after dry etching. One key attribute of the chemical is the ability to clean dry etching residues without damaging the underlying cobalt or tungsten metals. In response to the industry need, we have developed cobalt post-etch cleaning solutions with superior cobalt and tungsten compatibility.

Results and Discussion

Co Residues Cleaning Approaches

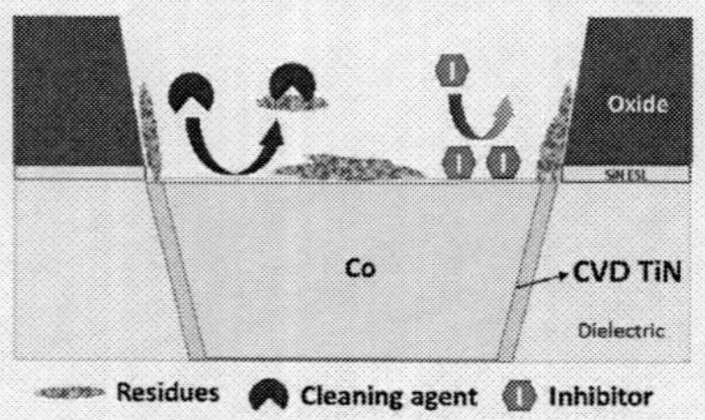

- ❑ Introduce an effective cleaning agent to remove the Co post-etch residues.
- ❑ Employ Co and W dual inhibitors with high passivation efficiency.
- ❑ Adopt suitable pH regime for the better Co and W compatibility and clean capability.

Co and W Selective Protection by Dual Inhibitors

	EC-series With only Co inhibitor	EC-series With only W inhibitor	EC-series With both Co and W inhibitors
N signal of XPS on Co	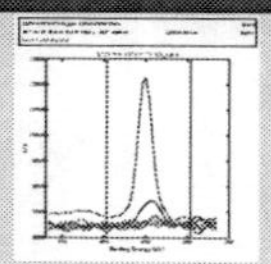	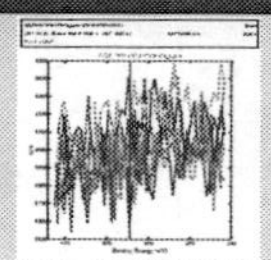	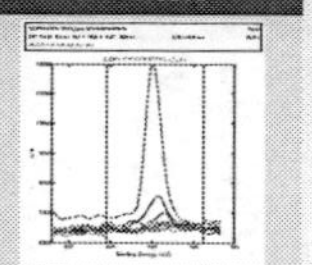
N signal of XPS on W	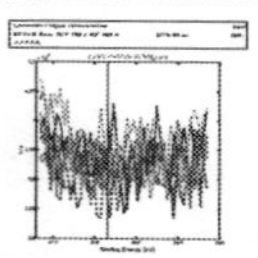	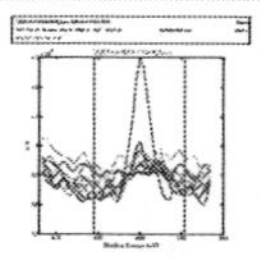	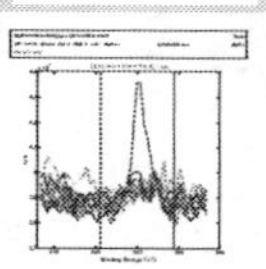

Selective Passivation

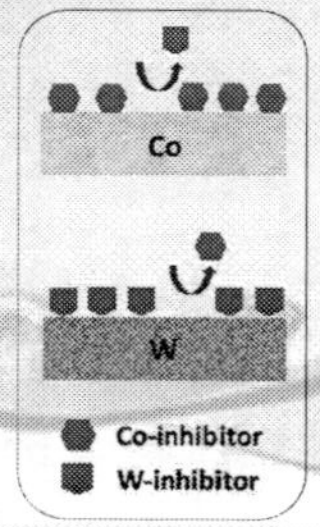

Metal Compatibility Performance

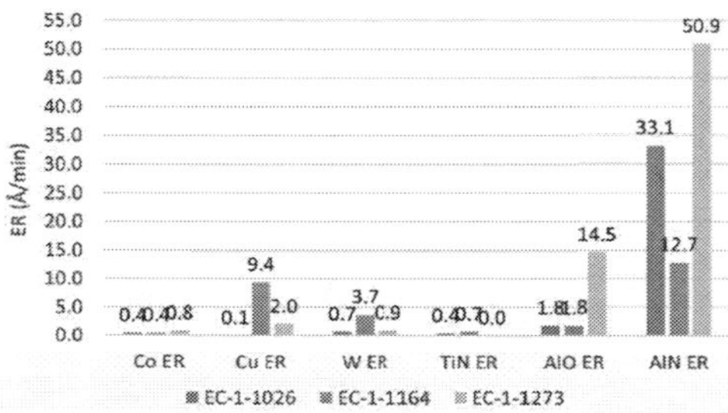

Chemical	Co ER	Cu ER	W ER	TiN ER	AlO ER	AlN ER
EC-1-1026	0.4	0.1	0.7	0.4	1.8	33.1
EC-1-1164	0.4	9.4	3.7	0.7	1.8	12.7
EC-1-1273	0.8	2.0	0.9	0.0	14.5	50.9

(a) The unit of ER (etch rate) is Å/min. (b) Testing condition: Co: 50°C 10mins; Cu: 50°C 5mins; W: 50°C 5mins; TiN: 50°C 10mins; AlN: 50°C 5mins; AlO: 50°C 10mins.

Clean Capability Evaluation

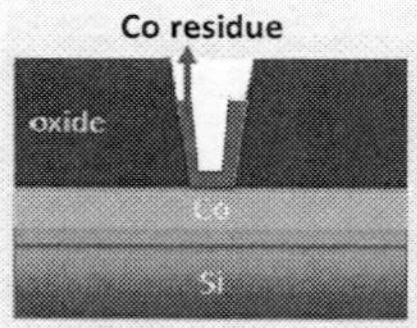

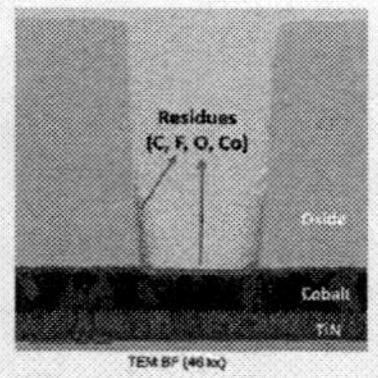

Chemical	Before clean	EC-1-1164 clean
Rating (1 to 5)[a]	1	5
SEM X-section	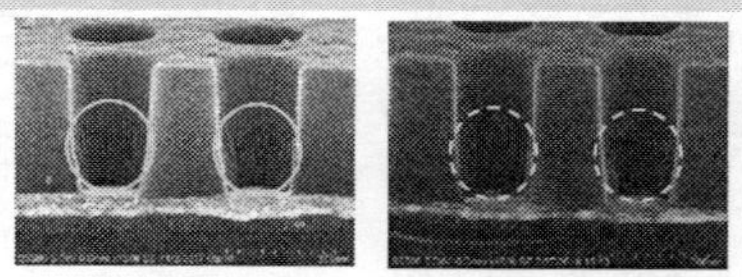	
TEM BF	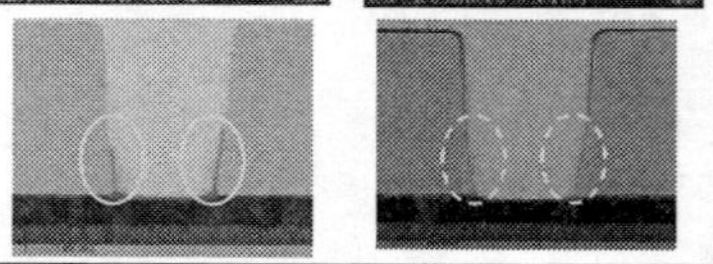	

a. Rating the clean performance from 1 to 5, the higher the better.

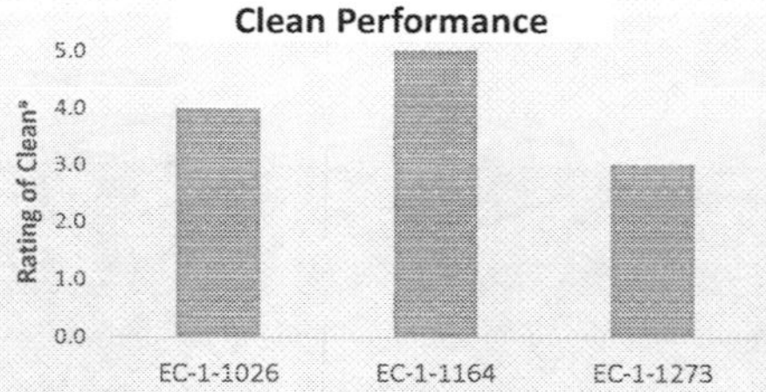

a. The higher the better for clean

Before clean EDX

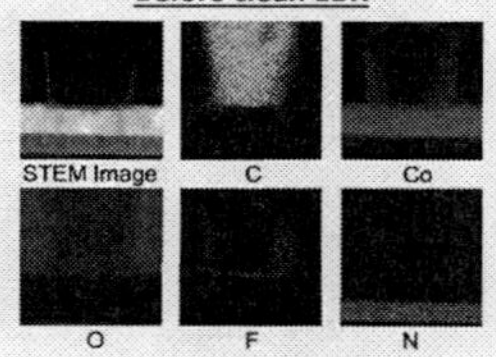

After EC-1-1164 clean EDX

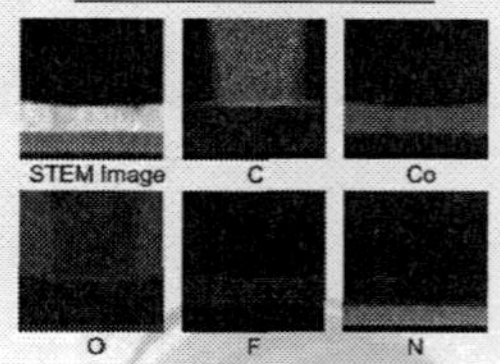

Conclusions

In this study, we report series of novel wet cleaning formulations for post-etch residue removal on cobalt contact. Our formulations possess good capability to remove post-etch cobalt residues, high compatibility with Co, W and TiN BM, as well as tunable AlN and AlO etch rate. For the extreme Co and W compatibility requirements, the selective protection of the dual inhibitors toward Co and W were demonstrated by X-ray photoelectron spectroscopy (XPS). We also establish the in-house capability for clean performance evaluation by customized Co pattern wafer, so as to understand the tuning knob of Co residue cleaning.

Improved Post Etch Wet Clean Method for Defectivity Reduction in 14nm BEOL Process

Srishti Agrawal, Asha Sharma, Sunil Singh, Teck Jung Tang and, Sherjang Singh
GLOBALFOUNDRIES Inc., Malta, NY

Abstract

Wet cleaning, rinse and dry steps are important parts of the Back End of Line (BEOL) process integration in advanced semiconductor technology nodes. This ensures the removal of post etch residues from vias and trenches. The processing time, concentration, temperature, and flowrate of the chemical are critical factors influencing particle removal efficiency. In this study, process time for DHF clean was optimized to evaluate the effects to reduce the defect density (D_0). By tuning the parameter, D_0 reduction was achieved by 23% when compared to POR improving single and multiple line open defects resulting in significant device yield improvement.

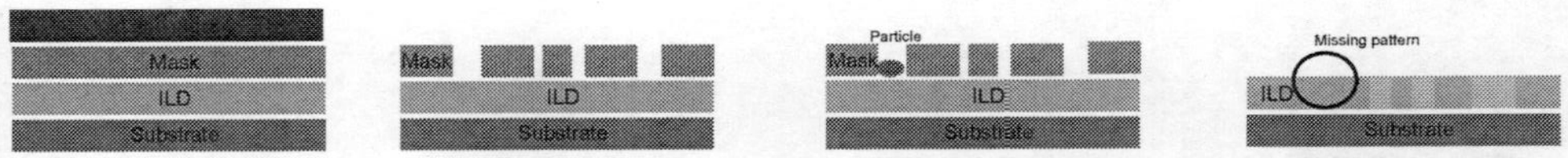

Conventional Stack | Hard Mask Etch during Cleans | Post clean particle left in stack | Post clean particle resulting in blocked ILD etch

Particle free stack is necessary for high quality patterning, therefore particle removal efficiency is important. Particles left in the stack can cause blocked etch which lead to missing metal pattern

Post clean residue particles can result in various patterning issues depending on their shape, size, location and properties

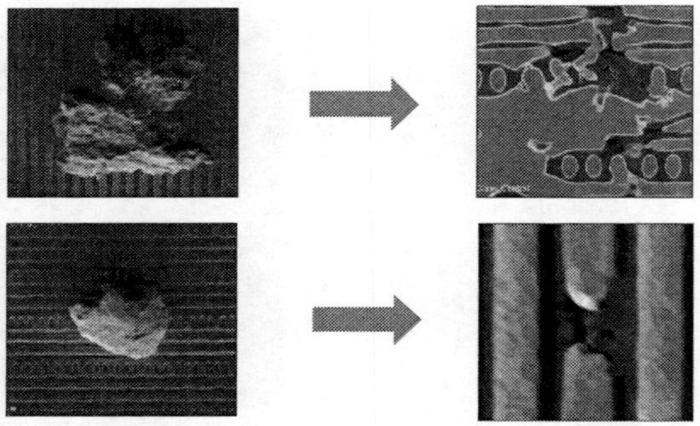

Post etch residue not cleaned | Post CMP Final Pattern

Particle Removal Comparison

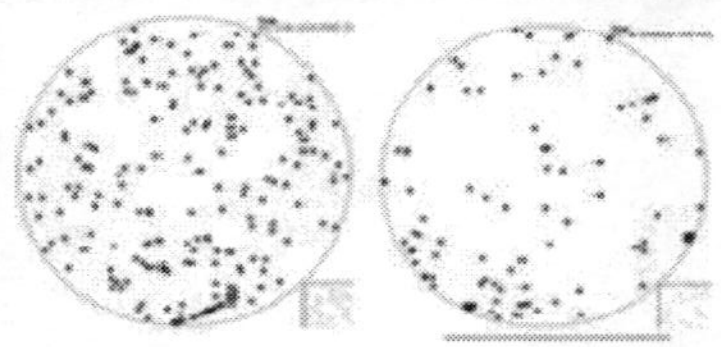

POR | Rcp 1

Total Defects

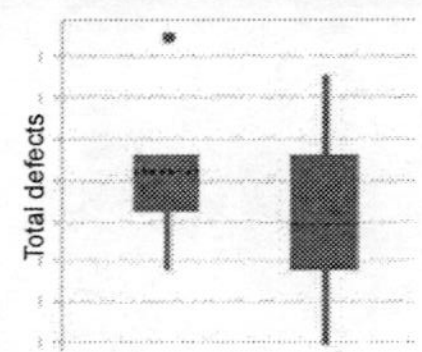

D0 Cumulative Percentage

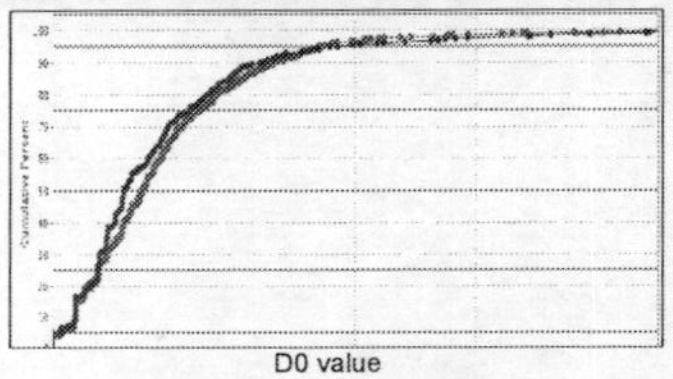

Yield Percentage

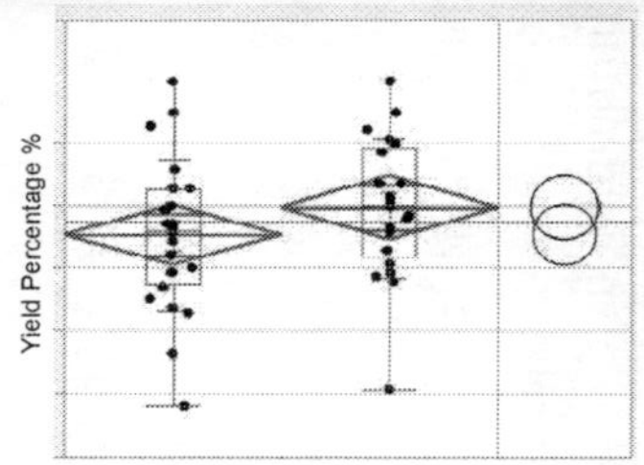

Summary

- BEOL cleaning process helps to improve metal patterning and enhance device properties.
- Particle removal efficiency was improved by tuning the process time for DHF clean.
- ~23% D_0 reduction observed with the new CIP recipe resulting in yield improvement.

Acknowledgements

14 nm BEOL Process Integration team and,
14 nm BEOL CFM team

Srishti Agrawal
GLOBALFOUNDRIES INC., 400 Stone Break Extension, Malta, NY 12020
Email. [illegible]
Phone. 585-880-1722

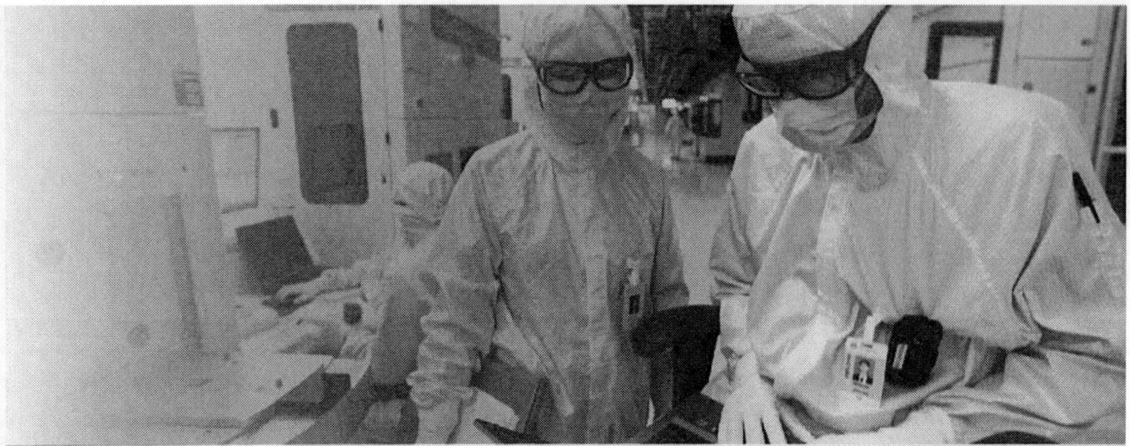

Greener SC1 Process For Single Wafer Cleans

Dhiman Bhattacharyya, Ph.D.
GLOBALFOUNDRIES
400 Stonebreak Rd. Ext., Malta, NY - 12020
Mobile: 518-948-3123
Email: dhiman.bhattacharyya@globalfoundries.com

April 10, 2018

GLOBALFOUNDRIES

Background

- Standard Clean-1 (SC1), also known as RCA-1 clean and APM (ammonia peroxide mixture), is extensively used as one of the major components of the semiconductor wet cleans processes.
- It is a mixture of ammonium hydroxide and hydrogen peroxide in deionized water. The ratio of the two chemicals may vary for different processes.
- Due to high volume usage of SC1, majority of the waste going down through IWD (industrial waste drain) consists of this alkaline solution.
- Requires pretreatment to an acceptable condition for discharge to the sewers.
- Requires control over the quantities and rates of discharge.

GLOBALFOUNDRIES 2

Background

- SC1 flow rate also varies by tool configuration, vendors and finally by process windows.
- Usual flow rates range from 1000 ml/min to 2500 ml/min
- Reduction in the SC1 flow rate helps in multiple ways:
  1. Reduces chemical consumption and associated chemical cost,
  2. Reduces SC1 waste volume and associated cost to neutralize the waste.
- Reduction in the SC1 flow rate risks the wafers in multiple ways as well:
  1. Possible more particles
  2. Lower etch rate
  3. Within wafer uniformity
  4. Finally, lower yield

GLOBALFOUNDRIES 3

Background

- Considering following assumptions, calculations show that reduction in SC1 flow rate reduces the effective process cost significantly.

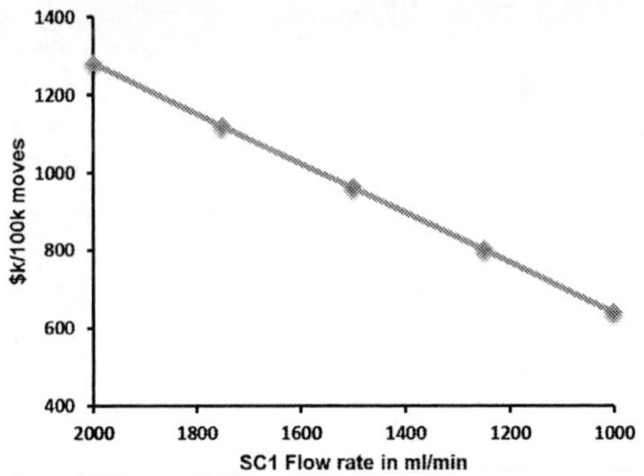

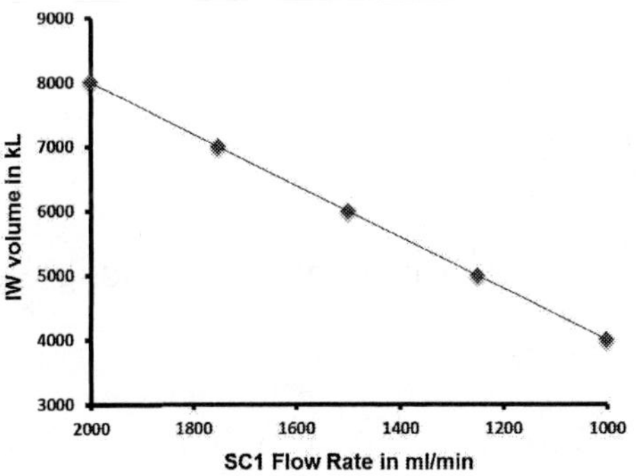

Assumptions:

- SC1 ratio 1:1:50 ($NH_4OH:H_2O_2:H_2O$)
- SC1 Flow time: 1 min
- Total process steps using 1 min SC1 = 40
- Moves @ each of 40 process steps = 100k

GLOBALFOUNDRIES 4

Approach

- Evaluated the effect of SC1 flow rates on etch rate on blanket wafer
- Evaluated the surface particle adders from various SC1 flow rates
- Compared the PRE (particle removal efficiency) on product wafer
- Compared the inline thickness on product
- Compared sort yield between low and high SC1 flow rates

GLOBALFOUNDRIES 5

Etch rate on blanket wafer

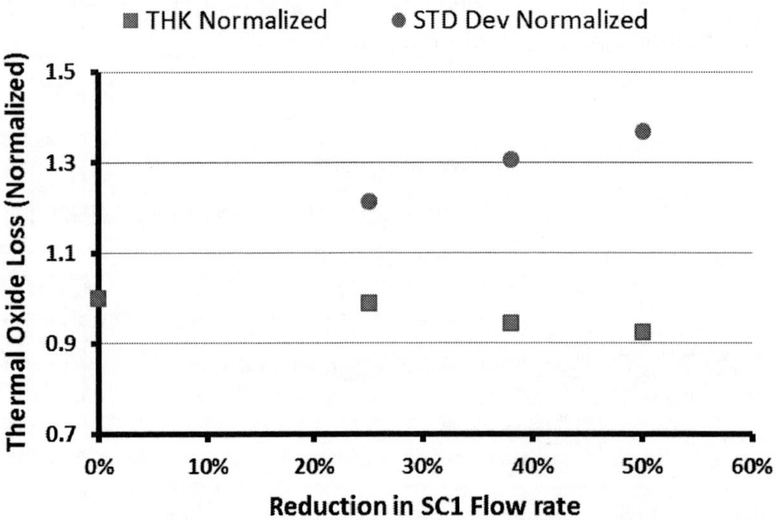

- A minor difference in thickness etched and sigma were observed among various SC1 flow rates for SC1 only process.

GLOBALFOUNDRIES 6

Particle Performance on blanket wafer

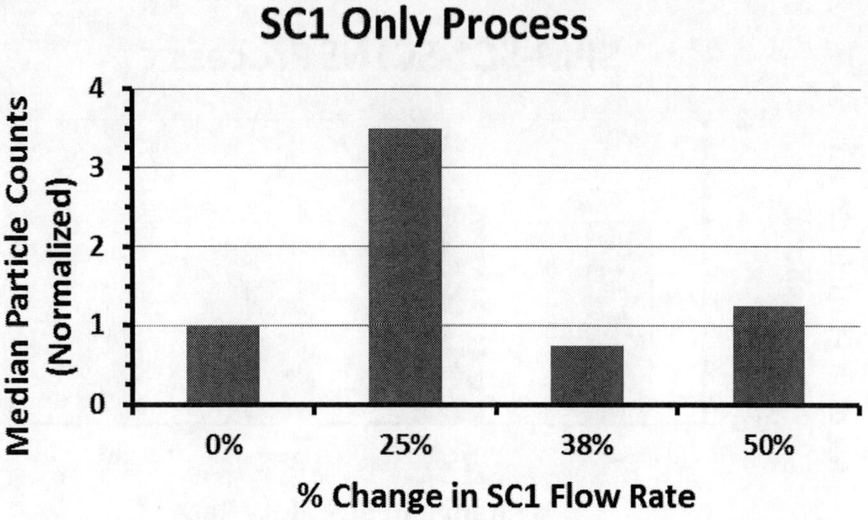

- No significant difference in particle performance was observed among various SC1 flow rates for SC1 only process.

GLOBALFOUNDRIES 7

Particle Performance on blanket wafer

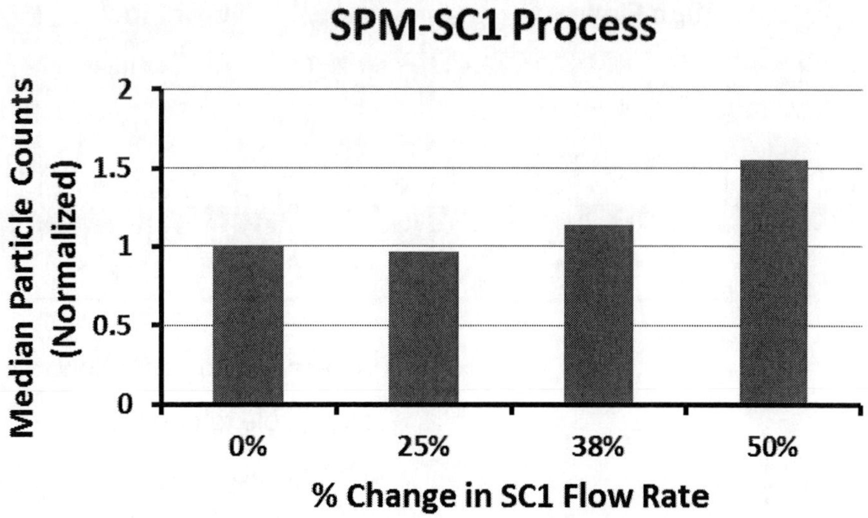

- No significant difference in particle performance was observed among various SC1 flow rates for SPM-SC1 only process.

GLOBALFOUNDRIES 8

Particle Performance on blanket wafer

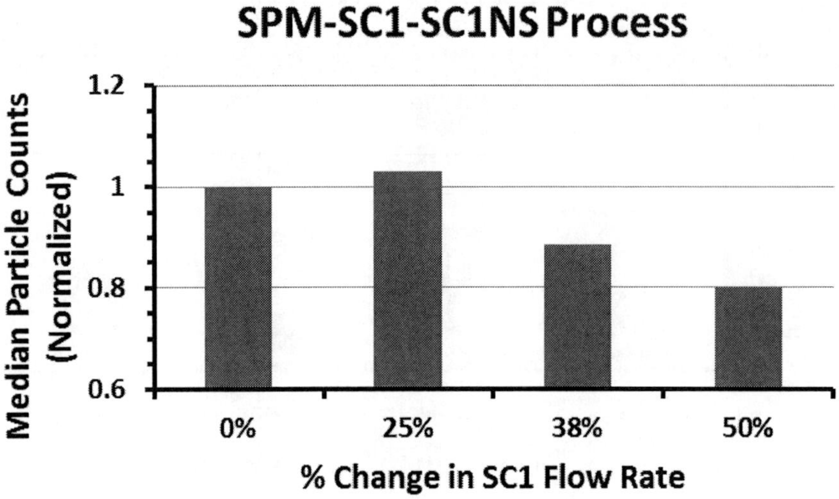

- No significant difference in particle performance was observed among various SC1 flow rates for SPM-SC1-SC1NS only process.

GLOBALFOUNDRIES 9

Particle Trend on Product

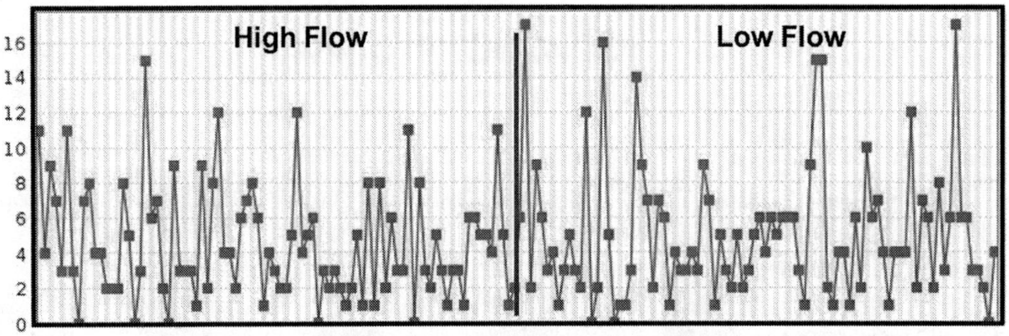

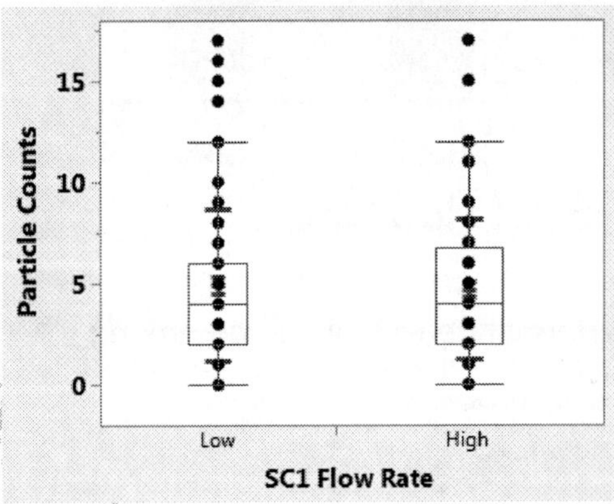

- Process with low SC1 flow rate showed comparable to or better than the process with high SC1 flow rate in particle performance.

GLOBALFOUNDRIES 10

Inline Trend on Product

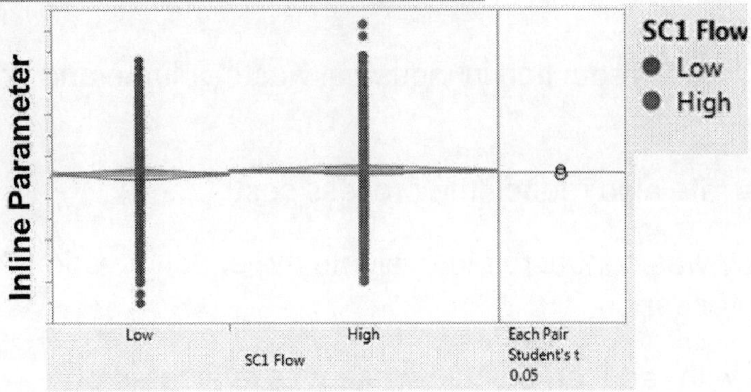

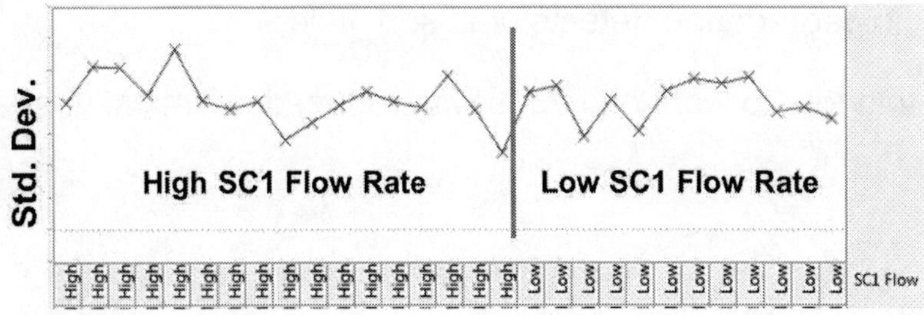

- Process with low SC1 flow rate showed comparable data to the process with high SC1 flow rate.

GLOBALFOUNDRIES® 11

Yield Analysis on product

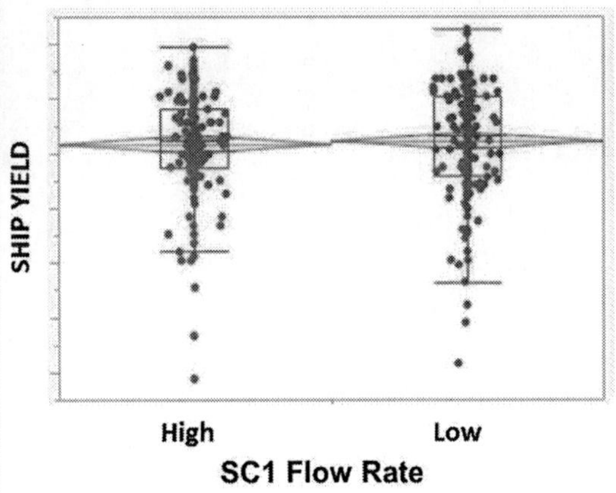

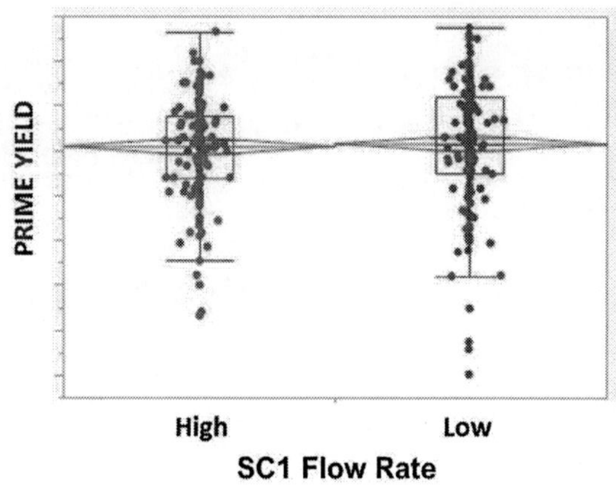

- Process with low SC1 flow rate showed comparable to or better than the process with high SC1 flow rate for both Ship Yield and Prime Yield.

GLOBALFOUNDRIES® 12

Summary

- Reduced SC1 usage means reduction in industrial waste volume and pre-treatment cost.
- Reduction in SC1 flow rate also reduce the process cost.
- A comprehensive study was conducted to evaluate the effect of various flow rates of SC1 on process.
- An optimized process with reduced SC1 flow rate was evaluated on product showing no degradation defects or in sort yield.
- Encourages tool vendors to work on BKMs with reduced chemical usage.

GLOBALFOUNDRIES 13

Acknowledgment:

All colleagues in GLOBALFOUNDRIES

GLOBALFOUNDRIES 14

From Fundamental to Application: One Universal Solution Toward Highly Efficient PCMP Ceria Cleaning in Advanced FEOL Processing

Jhih-Fong Lin*, Paul Bernatis, Pei-Yu Tai, Chia-Hui Bai, Eric Lee, Wen-Hsuan Chang, Ling Chang, Akira Kuroda and Chi Yen

EKC Technology, DuPont

1 Introduction

With the semiconductor manufacturing pitches advance node, ceria based slurries have been widely applied to achieve rapid removal rates of silicon oxide (TEOS) in front-end-of-line (FEOL) chemical mechanical planarization (CMP) processing. In addition to the relatively strong chemical bond with the oxide surface, opposite charge and tiny size of ceria abrasive makes ceria post-CMP cleaning a greater challenge than before. As a result, formulating chemistry starts to emerge in market and gradually replaces commodity chemistry such as diluted HF(dHF), ammonia SPM and SC1, in both logics and memory applications.

2 Fundamental

Two templates EKC-CE072 and EKC-CE150 have been developed and demonstrates their capabili ty on SiN and TEOS cleaning, respectively. Consi dering to replace commodity chemistry such as SC1 as well as acidic POR in customer side. One universal solution toward varying cleaning issue is urgently needed.

(a)

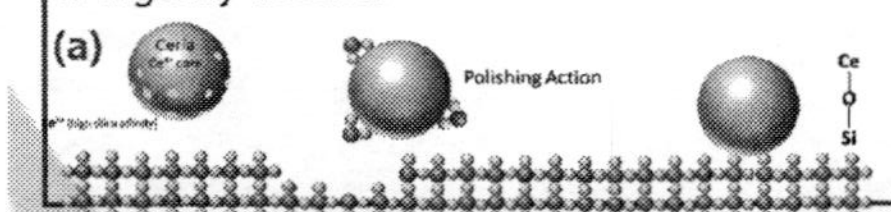

(b)

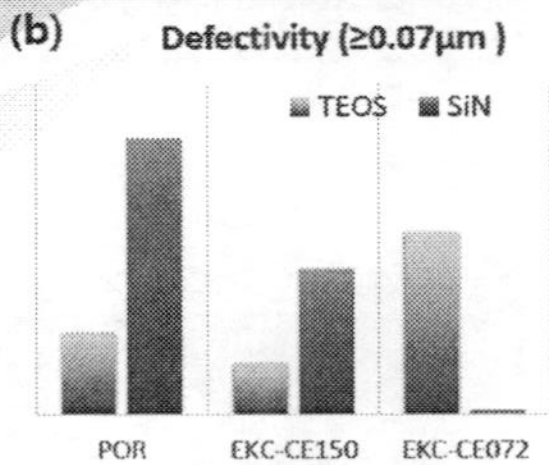

Figure 1. (a) Ceria CMP Processing and **(b)** cleaning performance of EKC-CE150, as well as EKC-CE072 versus POR

3 Inspection Methods

(a)

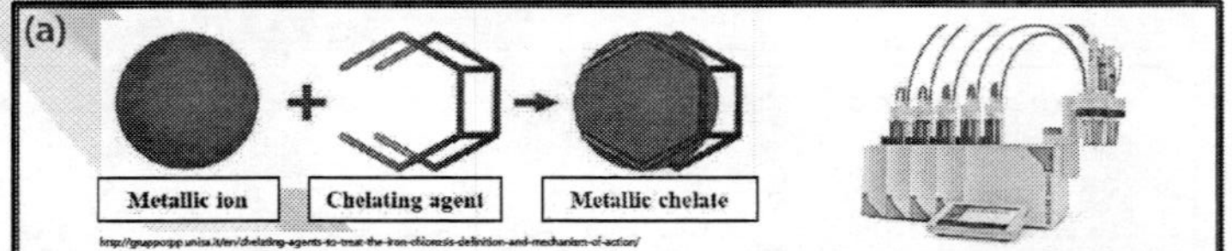

(b)

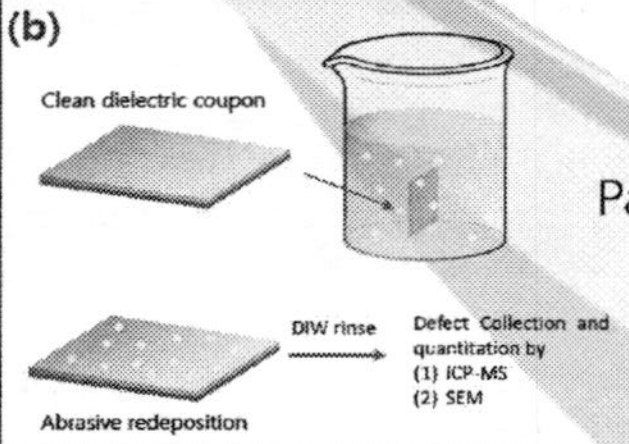

$$\text{Passivation\%} = \frac{[Ce]_{remained\ on\ surface}}{[Ce]_{total}}$$

Figure 2. (a) Titration-based technique as the chelator screening methodology and **(b)** methodology for abrasive-diel ectric passivation inspection.

4 Results and Discussion

Component Screening

(a)

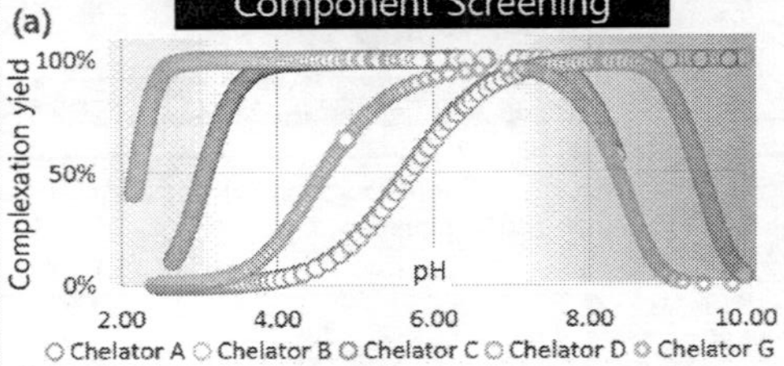

(b)

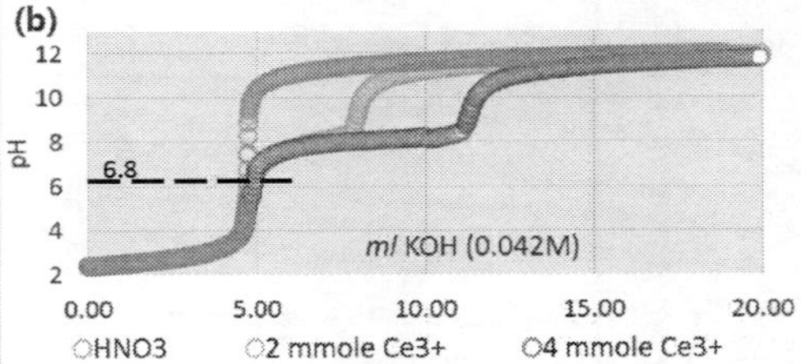

(c)

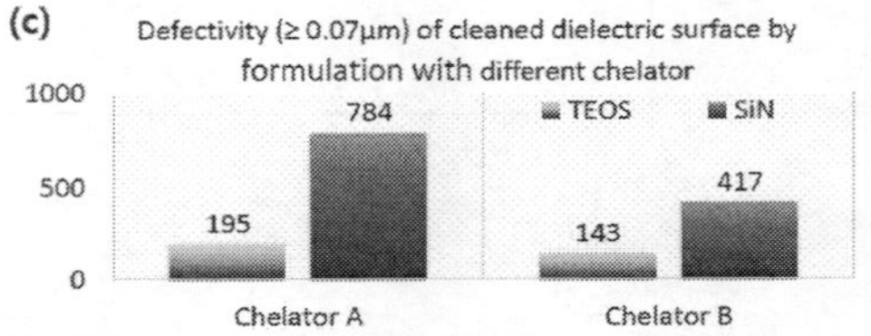

Figure 3. (a) Complexation behavior of Ce^{3+} versus varying chelators cross pH 2-10 ;**(b)** hydrolysis of Ce^{3+}, which explains the dropping of complexation yield for some chelators at pH above 7; **(c)** impact of chelation agent switching in cleaning perfor mance, in which the POU of formulation is acidic.

Formulation Optimization

(a)

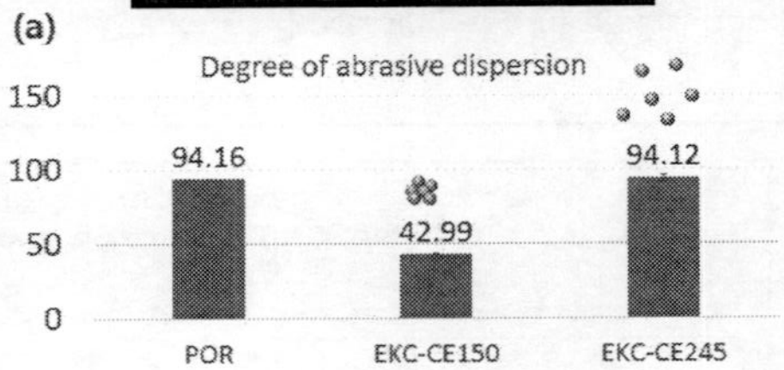

(b)

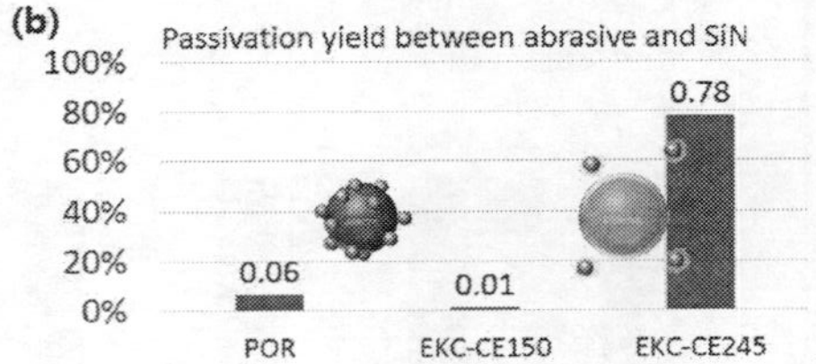

(c)

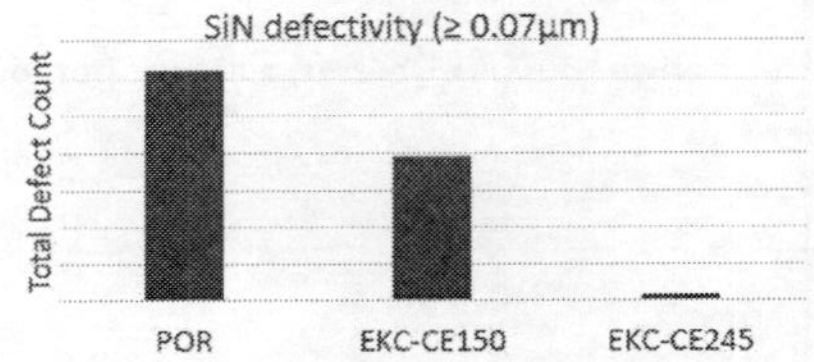

Figure 4. (a) Dispersion and **(b)** passivation capability of formulation. **(c)** Cleani ng performance of corresponding formulation on SiN surface in practice.

5 Conclusions or Future Application

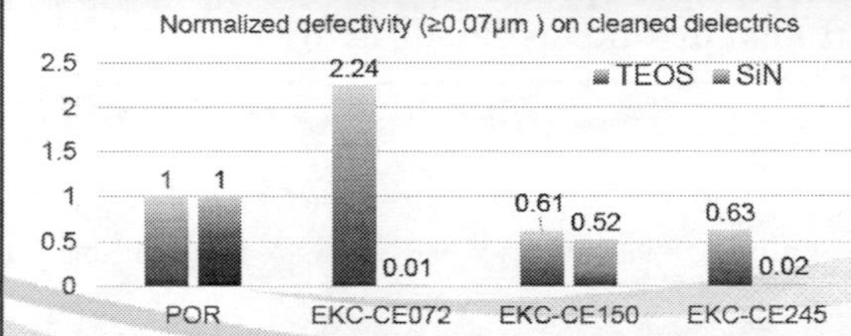

- Voltammetric titration method and filtration-based methodology provides more significant insight for PCMP processing such as the working window of cleaner formulation and abrasive behavior in formulating chemistry
- The ceria cleaner solution EKC-CE245 is developed by synergizing strength of EKC-CE072 as well as EKC-CE150.
- Compared to commercial POR, EKC-CE245 can achieive 37% and 98% reduction in defectivity on TEOS and SiN surface, respectively.

jhih-fong.lin@dupont.com

The Last Puzzle toward Cost-Effective FEOL CMP Processing: New-generation Cleaner for Positive-charged Silica Slurry

Jhih-Fong Lin*, Paul Bernatis, Pei-Yu Tai, Chia-Hui Bai, Eric Lee, Wen-Hsuan Chang, Ling Chang, Akira Kuroda and Chi Yen

EKC Technology, DuPont

1 Introduction

Oxide(silica) removal is the most time/cost-consuming application in FEOL CMP (chemical mechanical planarization) processing. Recently, positively-charged silica with lower cost of ownership indicates another direction in CMP processing. With tailored surface engineering on abrasives, this kind of slurry can greatly enhance the interaction between abrasive and desired polished surface, as a result enables elimination of solid content at point of use. On the other hand, the enhanced affinity (also removal rate) of positive-charged silica to polished surface brings out the additional cleaning issue in practice which commodity chemistry couldn't help.

2 Technical Issue

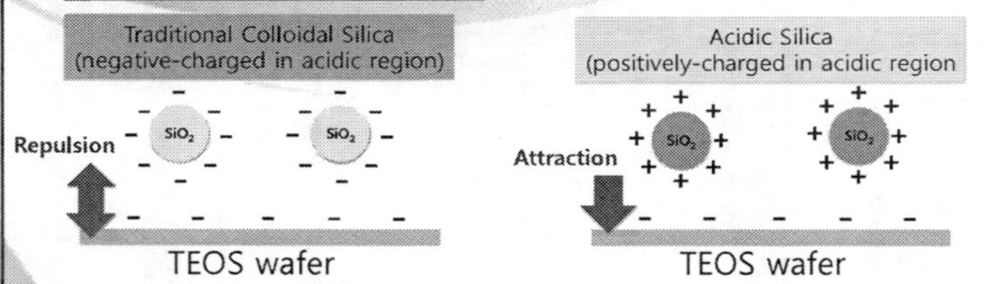

Figure 1. interaction of different silica abrasives to TEOS surface

3 Cleaning Principle

Strong repulsion force results in excellent performance

EKC-5710 would give strong electrostatic repulsion force between TEOS/SiN surface and slurry particle

Surface charge Z.P. on various substrates [mV]

Substrate	TEOS	SiN
Surface charge Z.P. [mV]	+16	+34

Surface charge Z.P. on abrasive [mV]

Particle	(Positively-charged colloidal SiO2 abrasive)*
Surface charge Z.P. [mV]	+3.5

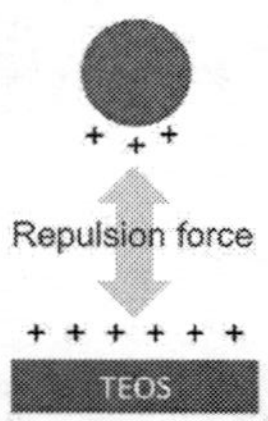

4 PCMP Cleaning Evaluation

(i) Great material compatibility to dielectric surface
(ii) No organic residue/contamination
(iii) Broaden cleaning window versus commodity chemistry

EKC-5710 provides a total-solution to acidic/positive charged silica slurry PCMP processing

(a)

	Cleaner	TEOS	SiN	PolySilicon
Compatibility (Å/min)	dHF (0.5%)	63	3	<1
	EKC-5710	<1	<1	<1

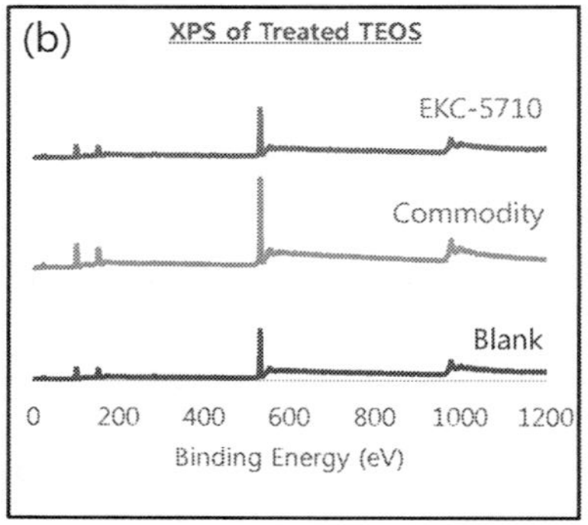

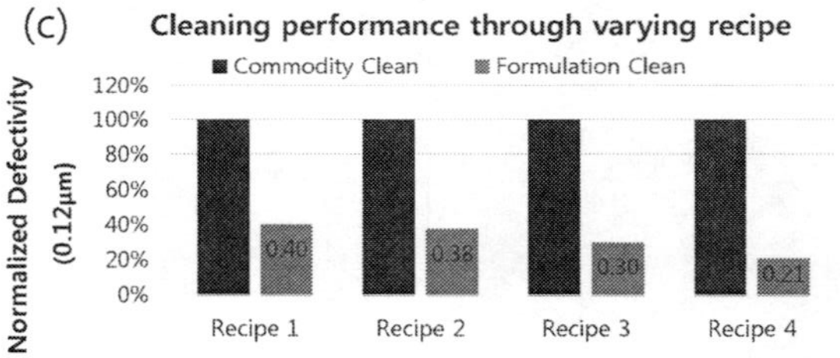

Clean Configuration*	Mega	Brush 1	Brush 2
Commodity Clean	Alkaline	dHF	Alkaline
Formulation Clean	Alkaline	EKC-5710	Alkaline

(i)Tool: AMAT; (ii) Inspection: KLA-SP2 (iii) Target: TEOS/HDP; (iv) Cabot D9228 slurry

Figure 2. (a) Material compatibility of commodity chemistry and EKC 5710; **(b)** Surface analysis of wet-clean ed TEOS wafer; **(c)** PCMP cleaning performance upon positively-charged silica CMP.

5 Conclusions

Positively-charged silica CMP has been widely applied for ILD and STI Processing, the reverse surfac e charge makes acidic silica abrasive with higher TEOS removal efficiency. Meanwhile, the electrostat ic attraction force also pose another technical challenge in PCMP cleaning. With tailored surface en gineering on positively-charged silica and TEOS surface, EKC-5710 enables strong repulsion force an d prevents the particle redeposition during the PCMP cleaning process. In addition, EKC-5710 also d emonstrates promising dielectric compatibility without the organic residue/contamination left on cle aned surface. On the other hand, EKC-5710 also shown much broaden working window than comm odity chemistry. The enhancement of defect removal indicates EKC-5710 reduces significant amount of particular defect that could be last puzzle toward the low-cost ownship FEOLCMP process.

jhih-fong.lin@dupont.com

Polysilicon PostCMP Cleaning: Material Solution beyond the Emerging Technical Hurdle in FEOL Processing

Jhih-Fong Lin*, Paul Bernatis, Pei-Yu Tai, Chia-Hui Bai, Eric Lee, Wen-Hsuan Chang, Ling Chang, Akira Kuroda and Chi Yen

EKC Technology, DuPont

1 Introduction

Polycrystalline silicon, as one of important materials employed in varying applications, for example, polysilicon contacts, polysilicon capacitors, poly silicon gate and polysilicon floating gate. In these applications, chemical mechanical planarization (CMP) is utilized to achieve surface smoothing or global planarization for following manufacturing. Generally speaking, the surface of polysilicon is predominately hydrogen-terminated and hydrophobic after the CMP processing. This incompatible surface characteristics makes normal aqueous cleaner hard to remove abrasive pa rticle from polished surface in post-CMP (PCMP) cleaning process.

2 Fundamental

(i) Wetting Capability; (ii) Particle-polysilicon interaction
Both approaches are employed for polySi cleaner development

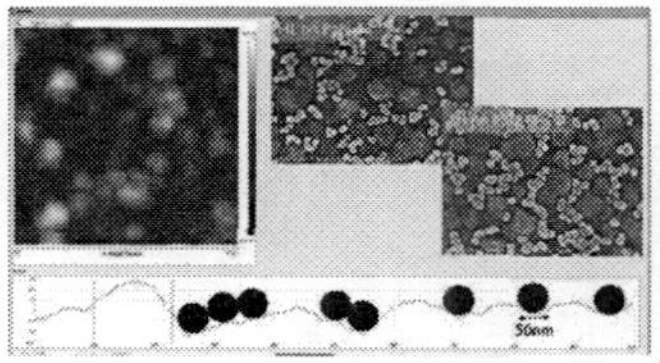

(i) Excellent wetting capability enables characteristics:

- "Flowability" to penetrate into the interface between polySi surface and abrasive
- Improved organic residue removal

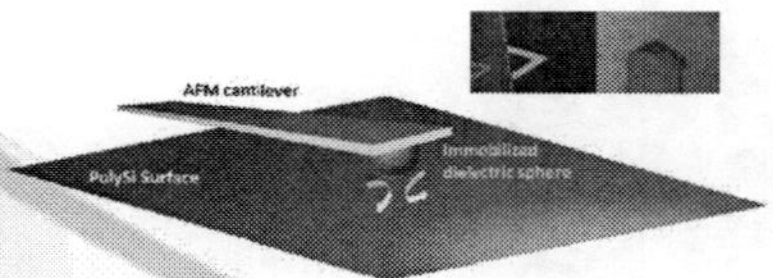

(ii) Impacts of particle-polySi interaction can be derived by:

- Adhesion force of abrasive to polySi surface
- Electrostatic passivation/attraction (Surface charge)

3 Results and Discussion

Formulating Chemistry Makes Synergy Possible

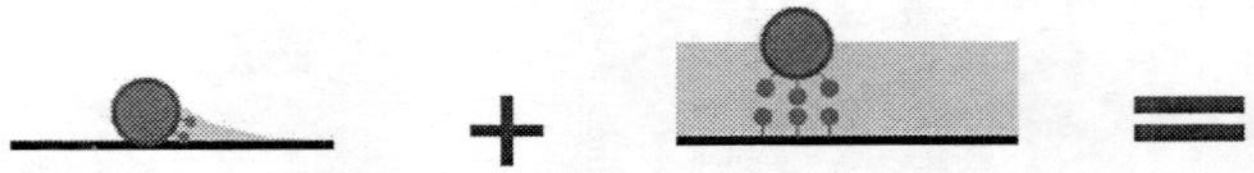

(i) Wetting Capability + (ii) SiO_2-PolySi interaction = EKC-142 (alkaline), EKC-447 (acidic)

Static contact angle of polySi		
EKC-477	EKC-142	DIW
44°	36°	75°

Adhesion force		
EKC-477	EKC-142	DIW
1.5nN	0.1nN	11.6nN

Compatibility		
EKC-477	EKC-142	NH_4OH
<1 Å/min	<1 Å/min	42 Å/min

Both EKC-142 and EKC-447 show promising cleaning Performance

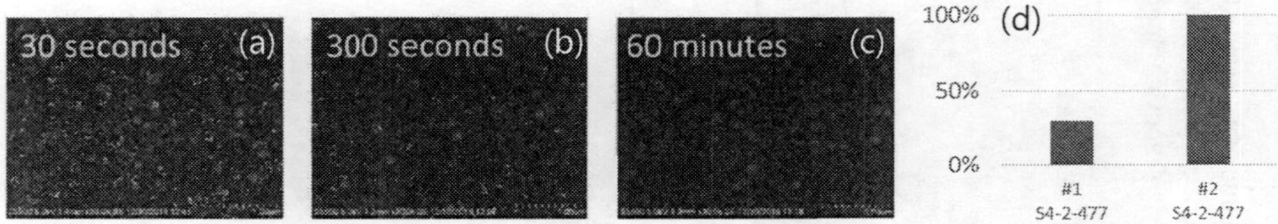

Figure 1. (a-c) Abrasive-removal versus cleaning time in EKC-142; **(d)** normalized defect count (v.s. NH_4OH), characterized by KLA in SP5 resolution.

5 Conclusions or Future Application

Currently, the cleaning process for polysilicon CMP is mainly done by alkaline commodity such as ammonia or SC1. Through our inhouse study, these chemicals can only enable electrostatic passivation (repulsion force) between abrasive particle and polysilicon surface. Instead of that, through the formulation design, our EKC formulating cleaner can easily reach the criteria listed above and outperform the commodity chemistry in practical PCMP cleaning.

EKC DuPont RESTRICTED *jhih-fong.lin@dupont.com*

Post-CMP W Cleaners with Excellent Tungsten Compatibility for both Buffing and Brush Cleans

Zoey Chiang, Max Tsai, Ruby Chen, Chuan-An Kao, Cami Shih, and Chi Yen*
EKC Technology, Electronics and Imaging, DuPont Taiwan Limited
zoey.chiang@dupont.com, chi.yen@dupont.com*

1 Introduction

Tungsten chemical and mechanical planarization (WCMP) is an important procedure to create a flat and smooth surface in advanced semiconductor manufacturing process for a better depth of focus in the photolithography step. With integrated circuits becoming more complex, the requirement in W compatibility and clean performance becomes more critical. Fig. 1 shows common defectivities after W Post-CMP clean. How to develop test methods that simulate the real operating process becomes an important issue for us. This study focuses on in-house method development to reduce device-dependent W plug recess and to improve W compatibility at a high temperature for buffing clean.

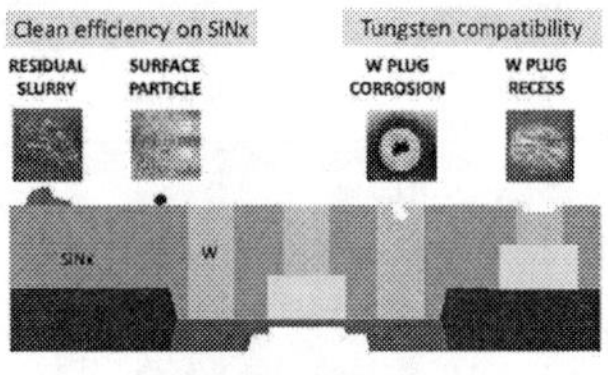

Figure 1: Schematic of four types of common defectivities after W Post-CMP clean.

2 W plug recess

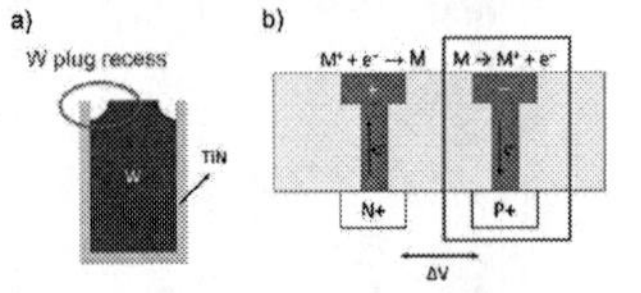

Figure 2: a) Schematic of W plug recess after Post-CMP clean. b) The proposed mechanism of electrochemical effects for the device-dependent W corrosion.

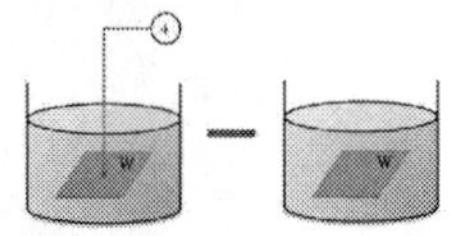

Figure 3: The device-dependent W plug recess is simulated by applying positive charges on W surface.

It has been shown that W plugs suffer from corrosion in the stacked metal/via interface. The layout dependent failure happens only on the source/drain contact output nodes, suggesting that it is the electrochemical reaction that causes corrosion when the wafer undergoes post-CMP clean processes. We developed a method to mimic W plug recess by applying positive charges on W surface to simulate the electrochemical reaction. The extraction of W is determined by measuring W concentration in solution using ICP-MS. W loss purely due to charge effect is obtained by subtracting W loss without charge from W loss with positive charges applied, as shown in Fig. 3. With our hypothesis of inhibiting W surface to prevent electron transfer from W to P+, we develop a co-inhibitor system that helps to minimize W plug recess.

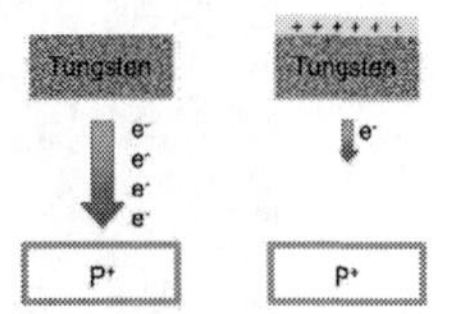

Figure 4: Hypothesis – we can build up a pile of positive charges on top of W surface to reduce the tendency of electron transfer from W to P+ to minimize W loss due to electrochemical effects.

3 W loss at a high T

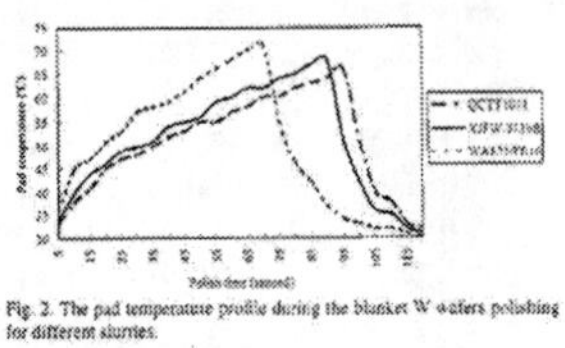

Fig. 2. The pad temperature profile during the blanket W wafers polishing for different slurries.

Y.-L. Wang et al. *Materials Chemistry and Physics* 52 (1998) 17-22

Figure 5: Pad temperature increases with time during the polishing process due to heat accumulation. Similarly, the surface temperature is expected to increase during the buffing process.

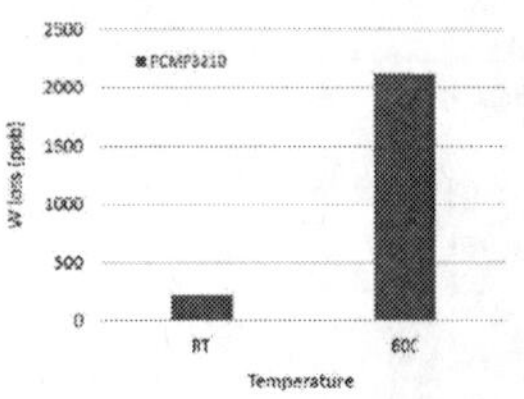

Figure 6: W loss with our commercialized product PCMP3210 becomes much higher at 60C than at RT. This suggests that a stronger inhibitor is required for better protection of W at a high T.

W compatibility test for buffing process

- Cut 3cm x 3cm square tungsten coupons
- Immerse the coupon in formulation at **60C**
- Measure the concentration of tungsten dissolved in the formulation with ICP-MS

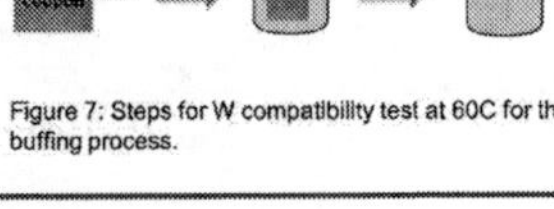

Figure 7: Steps for W compatibility test at 60C for the buffing process.

A mechanical force is applied during buffing clean similar to the CMP process. It has been shown in the literature that there is heat accumulated on pad surface during the polishing process, resulting in an increased temperature (T). The high surface temperature is also expected in the buffing process. Fig. 6 shows the W loss with our commercialized PCMP3210 at various temperatures. W loss increases significantly with an increasing temperature from room temperature (RT) to 60C. This suggests that a stronger inhibitor system is required at a high T for better W inhibition. Thus, screening out inhibitors which provide better W protection at a high temperature becomes our next challenge.

4 Results and Discussions

- **W plug recess is minimized by our co-inhibitor system**

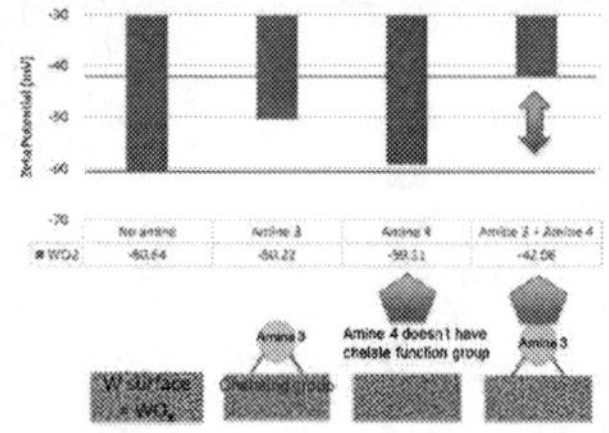

Figure 8: Zeta potential on W surface. A significant increase in WOx surface zeta potential is observed with the combination of Amine 3 + Amine 4, with Amine 3 chelating on W surface and Amine 4 stacking on Amine 3 by Van der Waals forces. This suggests that our co-inhibitor system provides positive charges on W to modify its surface charge.

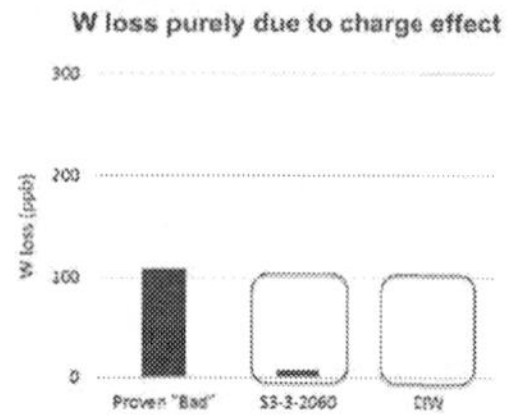

Figure 9: W loss purely due to positive charge effect. The result shows the same trend as that observed on customer site. DIW shows the lowest W loss probably due to the low conductivity. Our product S3-3-2060 shows a comparable W loss to that of DIW.

- **W loss at 60C is minimized with a more effective inhibitor**

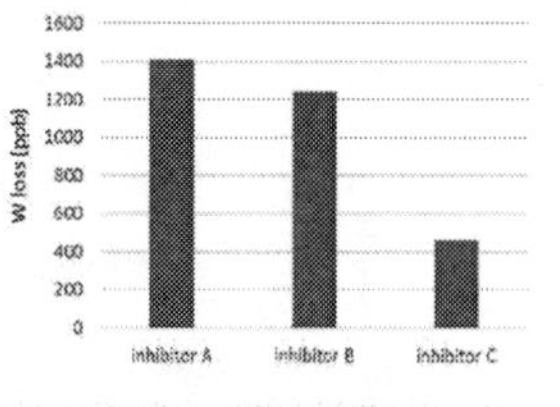

Figure 10: W loss at 60C for inhibitor screening. Having the most functional groups on a molecule, inhibitor C shows the best inhibition of W compared to inhibitors A and B.

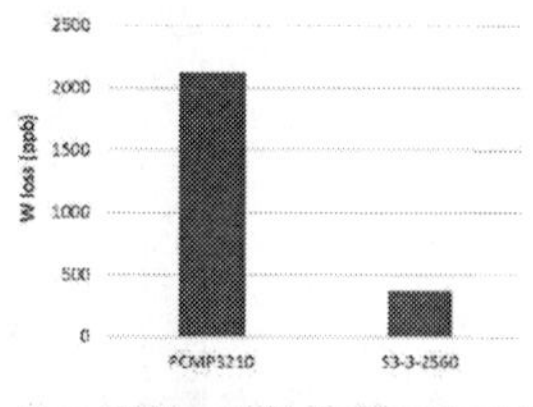

Figure 11: W loss at 60C with different formulations. EKC S3-3-2560 with the most effective inhibitor C shows a much lower W loss at 60C than PCMP3210.

- **EKC cleaners outperform ammonia in clean capability**

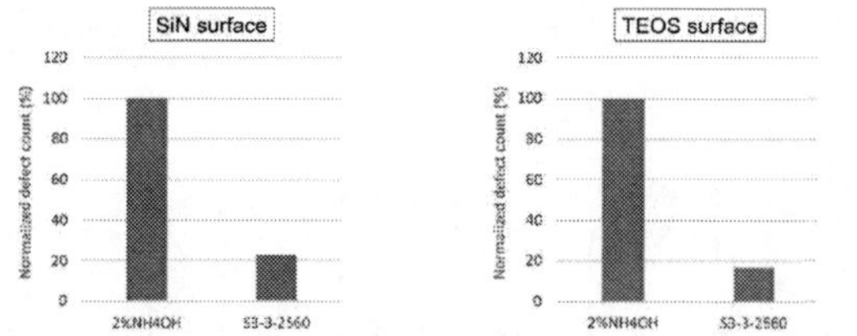

Figure 12: Normalized defect count on blanket SiN and TEOS surfaces with full-wafer (12") demonstration via Ebara FREX300L with inspections done by KLA SP1 0.15μm for SiN and KLA SP1 0.16μm for TEOS. EKC S3-3-2560 decreases defect counts on both SiN and TEOS significantly compared to 2% NH_4OH.

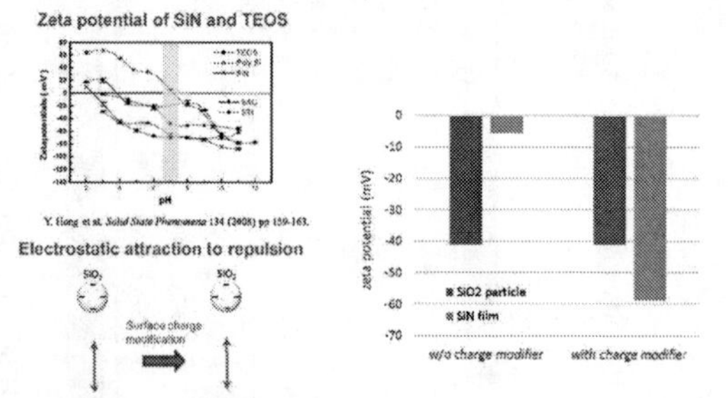

Figure 13: Zeta potential of SiN and TEOS at various pH (upper left) shows that the two materials carry opposite charges at neutral regime where S3-3-2560 falls in. The schematic showing the idea of surface charge modification to turn electrostatic attraction to repulsion between SiN and SiO_2 (lower left). Zeta potential of SiN film and SiO_2 particle with and without surface charge modifier (right). With the surface charge modifier, zeta potential of SiN film becomes more negative, suggesting that it provides a stronger electrostatic repulsion between SiN and SiO_2 and thus better particle removal capability.

5 Conclusions

In this study, we developed methods to simulate the device-dependent W plug recess and W loss in buffing clean. We figured out key components that help to reduce W loss in both cases. In addition, we are able to improve clean performance on SiN significantly by our surface charge modification technique. Our products are able to fulfill multiple requirements and are applicable for both brush and buffing cleans.

Characterization of Ultrasonic Ultra-High Purity Final Cleaning of Metal Oxide Plasma Coated Parts

Osama Khalil*, John Treviño, John Deem, David Zuck

QUANTUM GLOBAL® TECHNOLOGIES, LLC

Global Headquarters: Quakertown, PA, USA ✧ Global Center for R&D: Phoenix, AZ, USA

Introduction

art and sub-component cleanliness for ultra-high urity semiconductor manufacturing tools is ecoming increasingly critical for maximizing tool roductivity and product yield.

igh volume manufacturers continue to redefine ritical particle characteristics and contamination oncentrations as technologies advance.

ltrasonic particle removal from metal oxide lasma coated parts is a focus area for **improving hamber performance.**

ptimal ultrasonic parameters **minimize substrate r coating damage,** while removing process-ffecting **loosely** or **partially-adhered particles,** eading to ***Atomically Clean Surface™*** through the uantumClean ***Final Surface Finish™*** process.

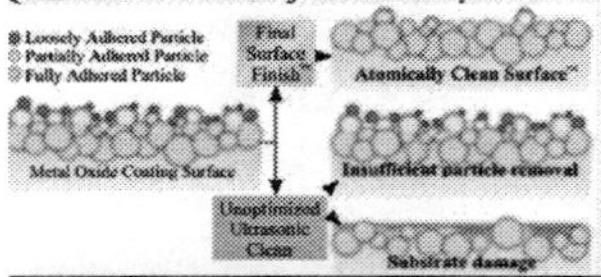

Experiment Overview

he purpose of this study was to characterize the mpacts of **ultrasonic cleaning parameters** such as: upplied power | Part orientation | Immersion time

amples for this experiment were prepared using arious common substrates with **QuantumClean® oatings** including:

- C-Coat™
- M-Coat™
- Z-Coat™
- White Yttria

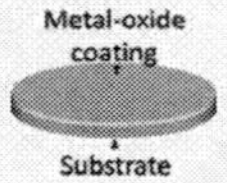

article behavior was traced with an in-line **liquid article counter (LPC).**

article behavior can be modeled and controlled y applying experimental data to a model derived to escribe particle behavior (y) in a flowing rinse tank s a function of ultrasonic power (u)[1,2]:

$$T_D^2 \frac{dy}{dt^2} + 2\zeta T_D \frac{dy}{dt} + y = K\left(\alpha \frac{du}{dt} + u\right)$$

 microscope was used to demonstrate the change f **surface morphology** as the result of ultrasonic rocessing.

Results

Particle Generation by Power Input

Finely tuning ultrasonic power input to maximize loose particle removal and minimize substrate impact is critical to attaining improved chamber performance by reducing risk of particle shedding in the chamber.

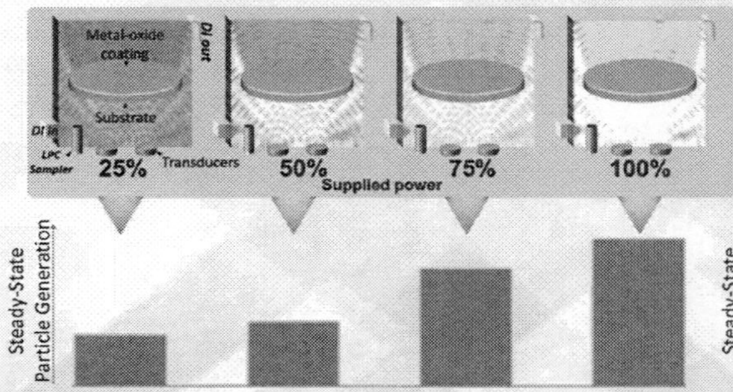

Particle Generation by Part Orientation

Orienting samples at varied angles relative to the transducers affects steady-state particle generation.

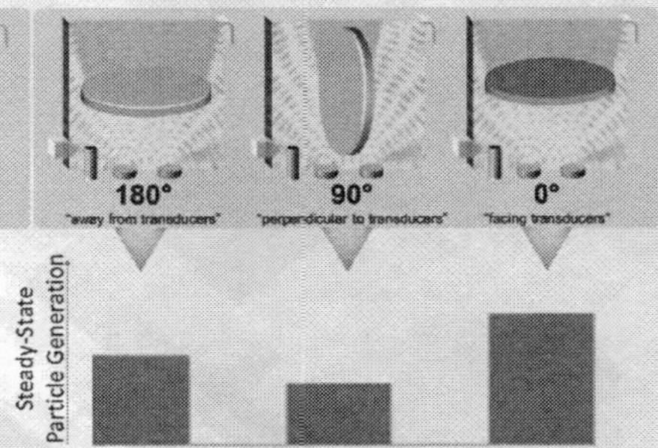

Surface Damage by Power

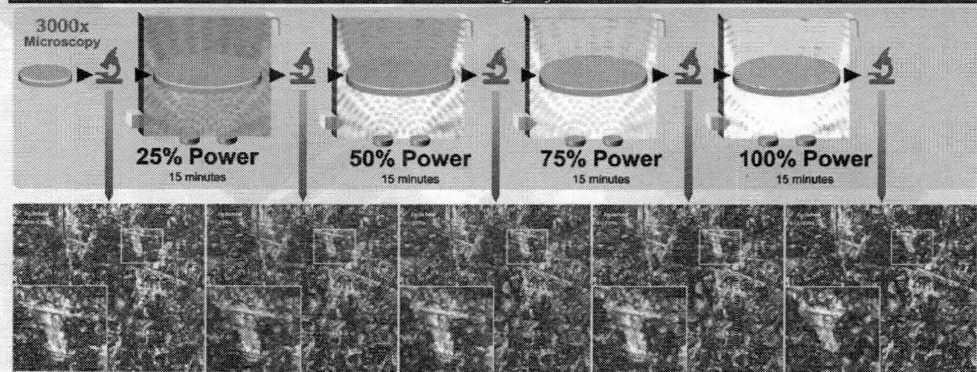

References:
[1] Bequette, B.W. (2002). Process Control: Modeling, Design and Simulation.
[2] Khalil, O.M. (2017). Process Model Identification for Optimizing Particle-to-Part Selectivity during Ultrasonic Parts Cleaning.

Model-Based Design

- A full characterization of ultrasonication for final parts cleanir goes beyond particle removal dynamics.
- The figures below show two examples of QuantumClean's use model-based characterization and design.
- Experiment and model driven design enable QuantumClean develop optimized ultrasonic equipment for robust partic removal

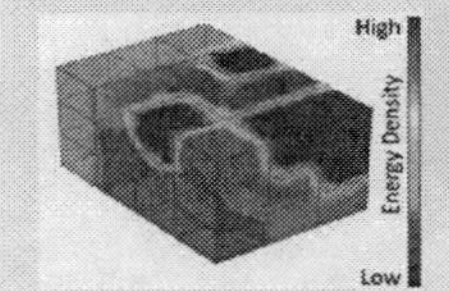

Acoustic energy density throughout the volume of an ultrasonic rins tank

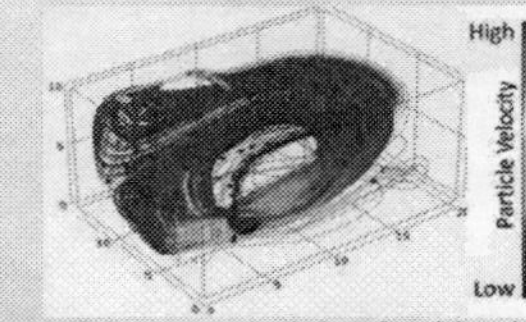

A time-variant finite element simulation of particle trajectories withi the flowing fluid space of an ultrasonic rinse tank.

Conclusion

Final Surface Finish™ (FSF™) is the process most critical in par cleaning for achieving ***Atomically Clean Surfaces™ (ACS™)*** sub-l nm manufacturing. The ultrasonic parameters that can remove tl process-affecting loosely adhered and surface particles witho damaging the substrate have been quantitatively determined. Throug experiments such as this, and use of computer simulation-base design, QuantumClean® continues to broaden its technolog leadership through full characterization and optimization of par cleaning processes so critical to yield.

Contact

Author: Osama Khalil
osama.khalil@quantumclean.com
3925 E. Watkins St • Ste 100 • Phoenix AZ

QUANTUM GLOBAL® TECHNOLOGIES, LLC
QGTCommercialGroup@quantumclean.com
1900 AM Drive • Quakertown PA USA

QUANTUMCLEAN®
ADVANCED TECHNOLOGY CLEANING CENTE

UNITED STATES	ASIA & EUROP
ARIZONA E. Phoenix • W. Phoenix	SINGAPORE
CALIFORNIA Fremont • San Jose	TAIWAN Tainan • Hsin
COLORADO Colorado Springs	REP OF KOREA Pyeon
OREGON Hillsboro	ISRAEL Qiryat Gat
MAINE Scarborough	SCOTLAND UK Glen
TEXAS Carrollton	IRELAND Celbridg

QUANTUMCLEAN®
Validated Ultra-High Purity. Maximum Productivity.™

SPCC 2018
Surface Preparation and Cleaning Confere
April 10-11 Cambridge, MA

BALAZS™ NanoAnalysis

Single Nanoparticle ICP-MS Analysis of Process Chemicals and UPW Used in Surface Cleaning and Preparation

Lisa M. Mey-Ami*, Jinjin Wang, Hugh Gotts and Fuhe Li

Air Liquide – Balazs NanoAnalysis
46409 Landing parkway, Fremont, CA 94538, USA

Introduction

Nanoparticles are ultrafine particles having size ranges from 1 to 100 nm. Nanoparticles are often present in the cleaning chemicals and UPW used in surface cleaning and preparation. Nanoparticles are very difficult to be removed and they can cause random device defects and lower the manufacturing yield, especially of current and next generation semiconductor devices.

As the leading chip makers have already begun the 10 nm processes in 2017, the control of the process/cleaning chemicals and UPW for nanoparticles is becoming crucial to the success and profitability of the semiconductor manufacturing. However, monitoring nanoparticles is a rather challenging task.

This paper introduces a non-traditional ICP-MS analysis for nanoparticles. Nanoparticle results obtained from UPW and the chemicals commonly used in the surface cleaning and preparation will be presented.

Why Nanoparticle Analysis is needed

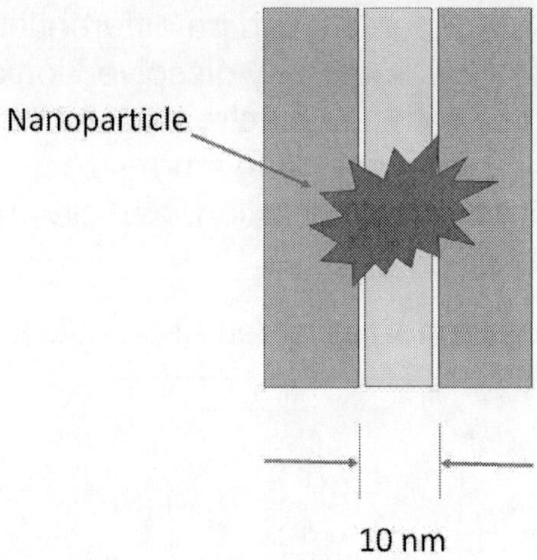

Semiconductor manufacturing processes

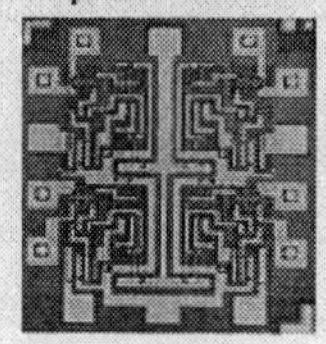

10 µm – 1971
6 µm – 1974
3 µm – 1977
1.5 µm – 1982
1 µm – 1985
800 nm – 1989
600 nm – 1994
350 nm – 1995
250 nm – 1997
180 nm – 1999
130 nm – 2001
90 nm – 2004
65 nm – 2006
45 nm – 2008
32 nm – 2010
22 nm – 2012
14 nm – 2014
10 nm – 2017
7 nm – *~2019*
5 nm – *~2021*

10-100 nm particles are now critical. Monitoring & controlling single nm and near 1 nm particles will soon be required.

Nanoparticle Analysis Fundamentals

ICP-MS has been used regularly for the determination of trace metals in UPW, process chemicals and extraction solutions from OEM tool parts. The detection limits are often as low as sub parts per trillion (ppt). Since trace metals are completely dissolved and aspirated into the ICP plasma continuously, producing a constant signal intensity, only total elemental concentration is determined.

Continuous signal from dissolved ionic analyte

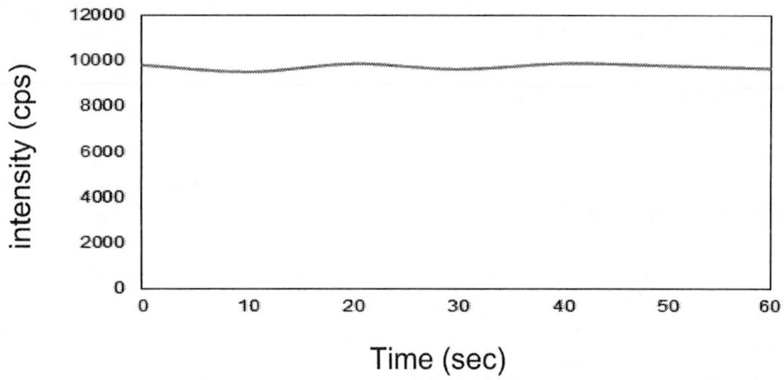

Nanoparticle Analysis Fundamentals

Using time resolved data acquisitions with short dwell times, the transient signal of each individual nanoparticle can be differentiated from the steady-state ICP-MS signal generated by dissolved ionic species. This unique feature coupled with fast electronics enables ICP-MS to characterize nanoparticles down to single nm sizes, producing results in the form of particle concentration, particle median size, and size distribution information.

Each nanoparticle gives a transient signal, collected at point/0.1 ms

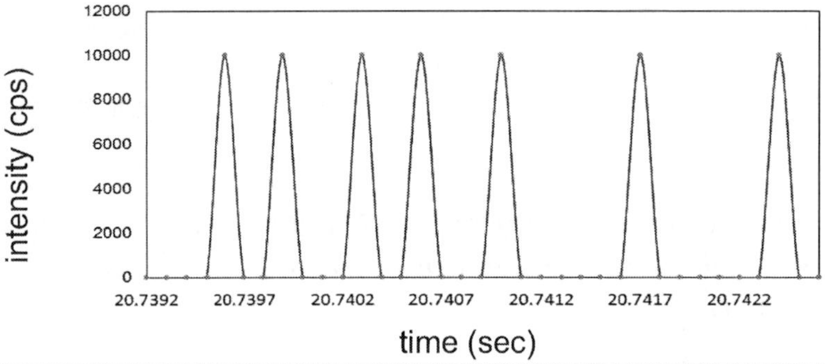

Typical Nanoparticle Analysis Figures of Merit

- Particle Concentration (number of particles/mL)
- Ionic Concentration (dissolved concentration; ng/mL)
- Particle Size Distribution
- Median Particle Size; nm
- Elemental Composition of Particles
- BED (background equivalent particle diameter; nm)

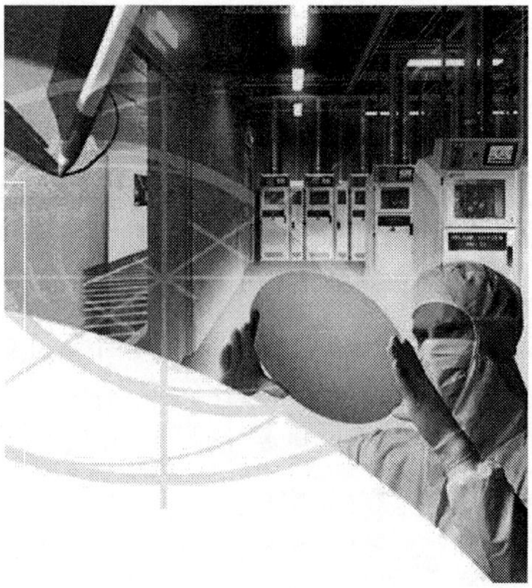

AIR LIQUIDE, THE WORLD LEADER IN GASES, TECHNOLOGIES AND SERVICES FOR INDUSTRY AND HEALTH

Analysis of a UPW Sample

Element	Particle conc (particles/mL)	Ionic conc (ppb or ng/mL)	BED (nm)	Median Size (nm)
Li	34175	0.008	21	40
Mg	4143	<0.001	8	21
Sn	668	<0.001	5	21
Al	567613	<0.001	10	18
Na	4679	<0.001	24	59
K	45950	<0.001	20	35
Fe	75953	0.021	9	16
Pb	0	<0.001	4	NA
Ni	501	0.005	10	33
Mn	52413	0.014	9	16

Analysis of a UPW Sample - continued

Element	Particle conc (particles/mL)	Ionic conc (ppb or ng/mL)	BED (nm)	Median Size (nm)
Ca	1865164	0.004	15	32
B	8286	<0.001	12	29
Cr	99053	<0.001	8	15
V	268	0.003	4	15
Zn	6629	0.007	12	29
Ti	0	<0.001	11	NA
Cu	63801	0.004	6	12
Sb	0	<0.001	4	0
Si	62420	0.110	28	52
Ba	0	<0.001	3	NA

Particle Size Distribution: Fe vs. Si

(similar concentration of particles but different median particle size)

Normalized frequency

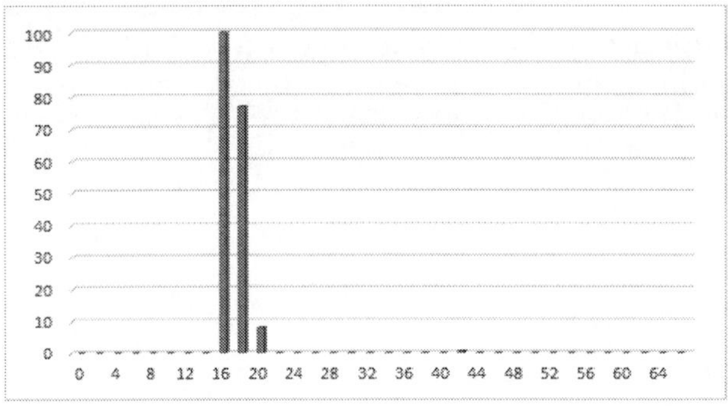

Fe

(75953 particles/mL)

Normalized frequency

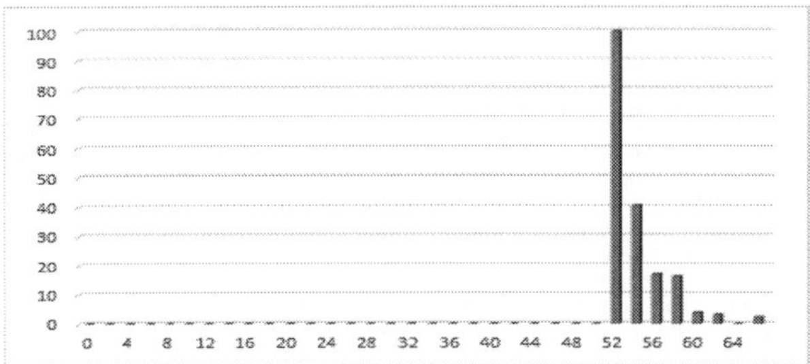

Si

(62420 particles/mL)

particle size (nm)

Analysis of a 35% H_2O_2 Sample

Element	Particle conc (particles/mL)	Ionic conc (ppb or ng/mL)	BED (nm)	Median Size (nm)
Li	18546	<0.001	17	36
Mg	6048	<0.001	9	23
Sn	0	<0.001	5	NA
Al	73080	0.007	11	21
Na	2714	0.012	22	59
K	54695	<0.001	20	36
Fe	12683	<0.001	11	18
Pb	0	<0.001	4	NA
Ca	1053523	<0.001	15	30
B	7308	0.016	13	31

Analysis of a 35% H_2O_2 Sample - continued

Element	Particle conc (particles/mL)	Ionic conc (ppb or ng/mL)	BED (nm)	Median Size (nm)
V	3276	0.012	6	19
Cr	73332	0.019	9	17
Zn	1454	<0.001	9	27
Ti	0	<0.001	11	NA
Cu	29573	<0.001	5	11
Sb	0	<0.001	4	NA
Ba	0	<0.001	4	NA
Ni	0	<0.001	10	NA
Si	68655	2.2	44	84
Mn	22632	<0.001	8	16

Analysis of a 25% HF Sample

Element	Particle conc (particles/mL)	Ionic conc (ppb or ng/mL)	BED (nm)	Median Size (nm)
Li	55705	<0.001	21	38
Mg	20430	0.004	10	21
Sn	3400	0.057	8	22
Al	35753	<0.001	15	30
Na	232106	<0.001	22	72
K	70792	<0.001	17	33
Fe	6218	<0.001	9	21
Pb	0	<0.001	3	NA
Ca	36673	<0.001	15	30
B	673	<0.001	10	30

Analysis of a 25% HF Sample - continued

Element	Particle conc (particles/mL)	Ionic conc (ppb or ng/mL)	BED (nm)	Median Size (nm)
Cr	34604	<0.001	15	30
V	26648	0.034	9	18
As	0	<0.001	8	NA
Zn	448	<0.001	9	29
Ti	897	<0.001	17	52
Cu	7859	0.001	5	13
Sb	0	<0.001	5	NA
Ba	680	<0.001	6	20
Ni	227	<0.001	8	27
Mn	696	<0.001	6	16

Two 35% HNO3 Samples: Grade A vs. Grade B

Element	Nitric Acid Grade	Particle conc (particles/mL)	Ionic conc (ppb or ng/mL)	Median Size (nm)
Mg	A	722	<0.001	29
Mg	B	1254522	0.212	45
Sn	A	947	<0.001	20
Sn	B	1284869	1.012	43
Al	A	23453	<0.001	27
Al	B	36180	22.676	118
Na	A	0	<0.001	NA
Na	B	1715238	1.632	96
K	A	31712	<0.001	33
K	B	1025138	0.133	62

Two 35% HNO3 Samples: Grade A vs. Grade B - continued

Element	Nitric Acid Grade	Particle conc (particles/mL)	Ionic conc (ppb or ng/mL)	Median Size (nm)
Fe	A	12508	<0.001	39
Fe	B	1710222	0.362	42
Ca	A	42067	<0.001	30
Ca	B	107447	5.992	102
Ti	A	3486	<0.001	82
Ti	B	25842	1.521	65
Pb	A	0	0	NA
Pb	B	34923	0.036	15
Ni	A	0	<0.001	NA
Ni	B	3459	0.176	34

Conclusion

The single nanoparticle ICP-MS enables direct measurement of liquid-borne nanoparticles in UPW and cleaning chemicals, providing not only nanoparticle concentration but also particle median size and size distribution information. In addition, elemental composition of nanoparticles can also be identified.

ICP-MS' ability to distinguish between ionic and nanoparticle concentration results gives more comprehensive understanding of elemental impurities in a sample solution. There were many examples in which the ionic concentrations were low for a particular element but particle concentrations high for that same element.

The nanoparticle results have been utilized to assist engineers with their process optimization. The databases and the trending charts built using the nanoparticle results can also be used to control the quality of UPW and cleaning chemicals used in various surface preparation and cleaning steps, e.g. final rinsing.

PALL Pall Corporation

Filtration. Separation. Solution.SM

Filter Material Cleanliness Characterization by Electrophoretic Method

Makonnen Payne and Rao Varanasi, *Scientific and Laboratory Services - Microelectronics*

Pall Corporation (makonnen_payne@pall.com) Port Washington, NY, USA

Glen W. Wildermuth and Arthur J. Ackermann

Microfier, Inc. (info@microfier.com) St. Louis, MO, USA

Introduction

- The cleanliness and particle removal requirements for filtration for the semiconductor industry continue to increase in order to keep pace with the development of sub-10nm devices.
- As a part of the filter development process, new materials are constantly being evaluated to improve the cleanliness of the final product
- The lack of analytical methods to directly evaluate materials in a way that is meaningful to the microelectronics industry can hinder development
- To attempt to address this issue, we use an electrophoretic method to collect and concentrate contaminants from prototypical filters to determine their level of cleanliness.

PALL Pall Corporation

Problem Statement

- Two new materials of construction (MOC) are being evaluated for use in a 2nm filter prototype
- Standard analytical methods (extractions and/or particle count) show differences in cleanliness, but:
  - Testing efficiency can be improved to decrease prototype cycle time since standard testing can take multiple weeks
  - End-user observes contaminants on wafer that were not previously detected

PALL Pall Corporation

Experimental

- Flow system
  - Two 10-inch filters in series
    - Housing # 1 - Upstream Filter
      - 10 nm rated filter
    - Housing # 2 - Downstream Filter
      - 2 x 2nm rated prototype (hereafter, labeled Prototype-V and Prototype-D)
      - Filtration medium is the same in both cases; MOC different
  - Electrophoretic particle capture device
- Test fluid is water
  - Flow is single pass for at least 24 hours and a maximum of 96 hours
- Where reported, liquid particle count (LPC) measured using Liquid-Borne Particle Sensor KS-19F from Rion Co., Ltd.

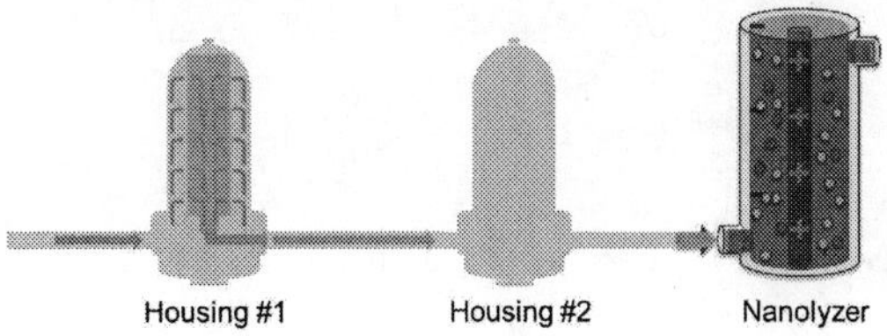

PALL Pall Corporation

Method – Electrophoretic Analysis

- The electrophoretic analysis is performed using the Nanolyzer from Microfier, Inc.

- The technology takes advantage of the fact that
  - An electromagnetic field can be established in UPW because it is non-conductive
  - Particles in UPW carry surface charge and are mobile in an electrical field
  - As particle size decreases, charge/volume ratio increases, dramatically improving removal.

- The main advantage of the technology is that it captures and agglomerates particles on an electrodes while liquid is under laminar flow
  - Allows for analysis of the chemical nature of the particles and captured contaminants

Pall Corporation

Method - Particle Capture

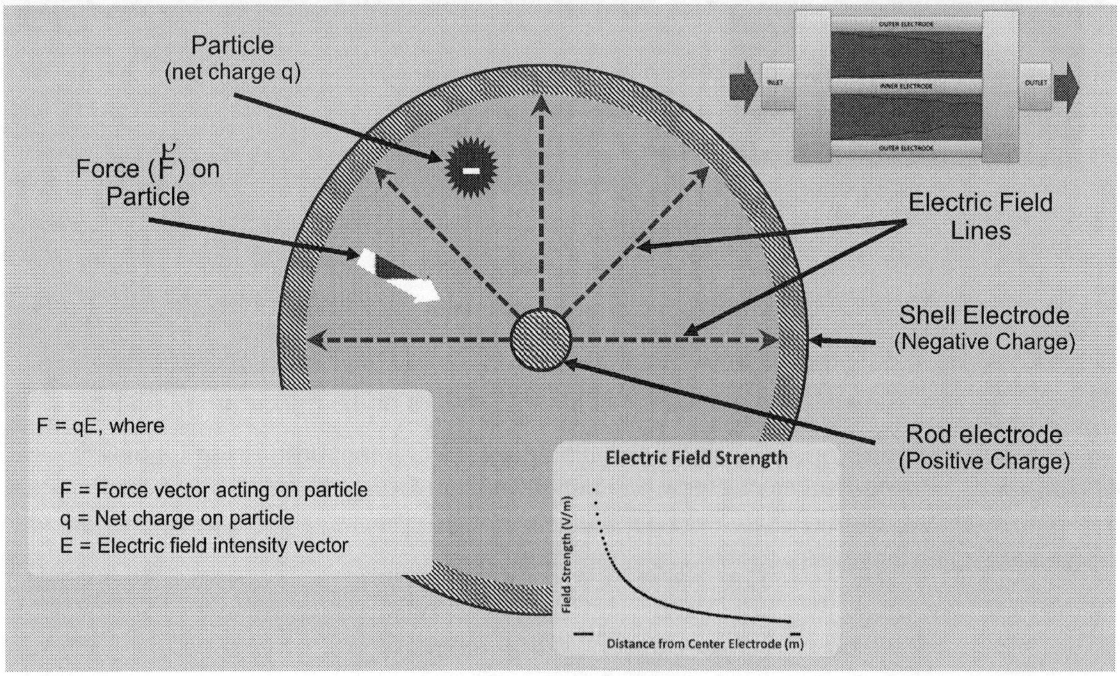

Pall Corporation

Method - Particle Release

- In order to release the captured particles, an oscillating voltage is **applied** to the chamber in long then short cycles. The release voltage is chosen such that is high enough to drive off the captured particles

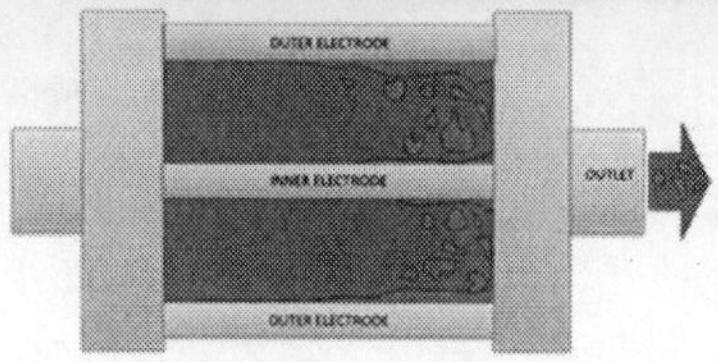

- The **measured** inter-electrode current (IEC) that occurs as a result of the movement of the released particles is the response to:
  - # of released particles
  - Charge density of the released particles
  - Particle composition

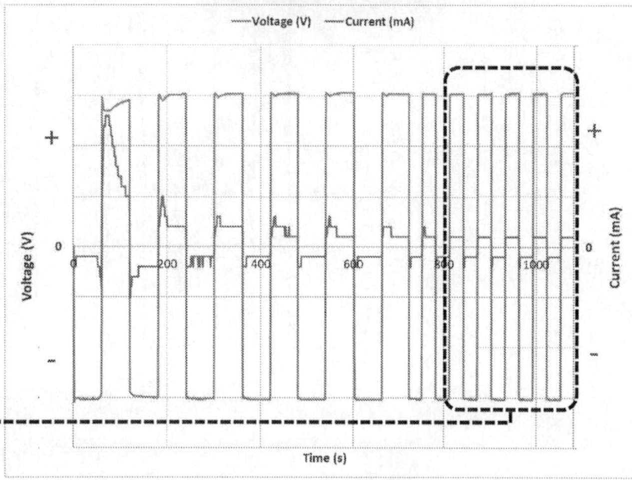

- Clean system will be seen as:
  - Flat voltage profile
  - Low and flat current response

PALL Pall Corporation

Results - UPW Baseline

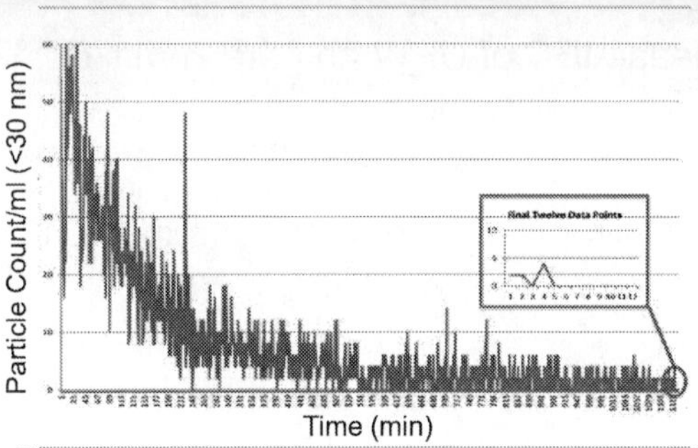

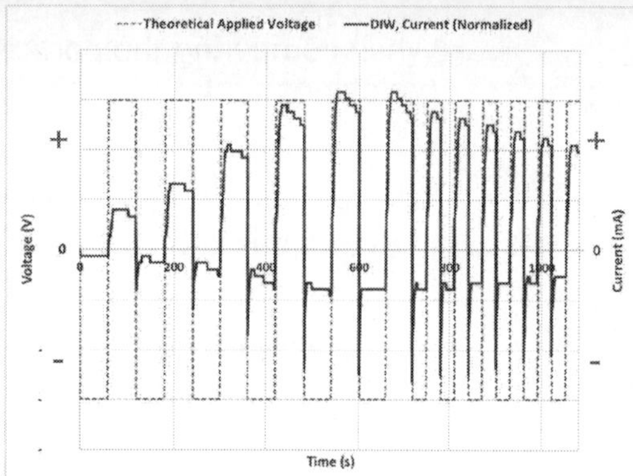

- Particle counts are below detection limit on 30 nm channel of Rion KS-19F after the system has stabilized.
- However, particles are measureable after they have been aggregated by the Nanolyzer for at least 24 hours
- There is a measurable IEC after a particle capture and subsequent particle release cycle

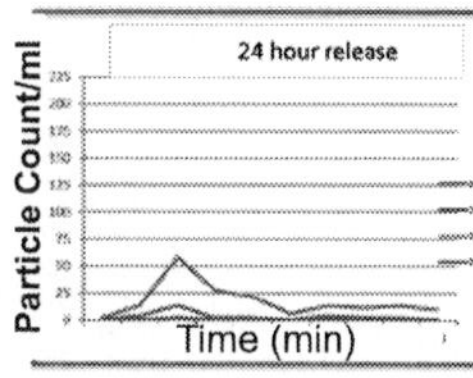

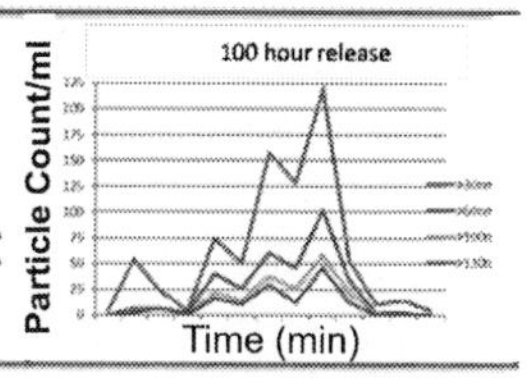

PALL Pall Corporation

Results – IEC of Prototype-D versus Prototype-V

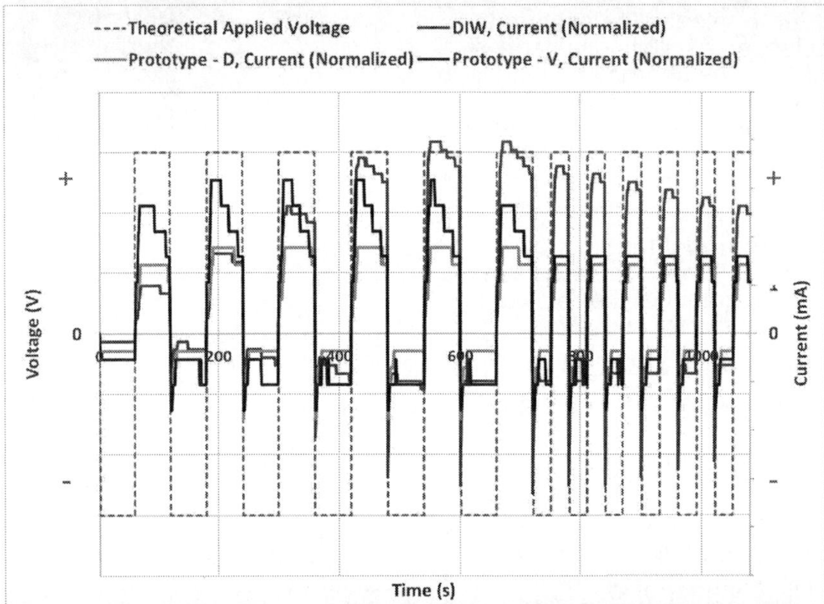

IEC plots indicate difference in cleanliness between filters

PALL Pall Corporation

Results – IEC of Prototype-D versus Prototype-V

- When comparing the particle release IEC of UPW and the 2nm filter prototypes prototype:
  - Overall, the water from the 2nm filter is cleaner than from the 10 nm filter alone
  - Test material in Prototype-D cleaner than the material used in Prototype-V
  - The contaminant species from Prototype-V different than that from the water or Prototype-D

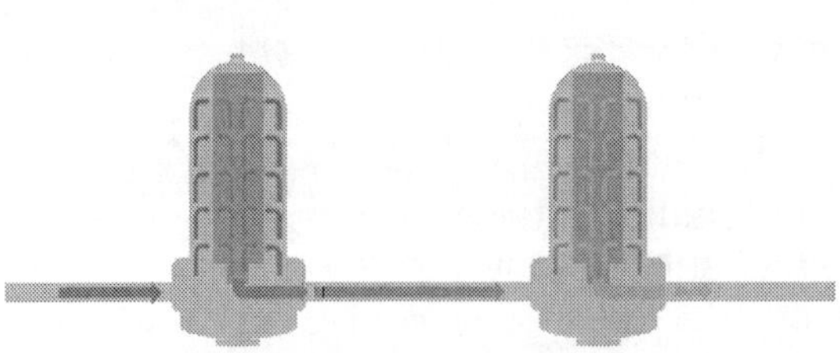

PALL Pall Corporation

Method - SEM Sample Collection

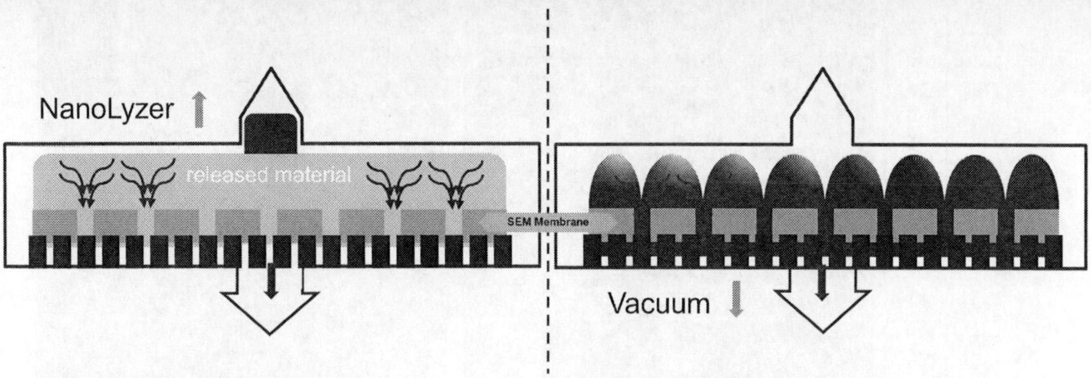

- During particle release the agglomerated particles were sent to an SEM membrane for analysis
- 30 nm track etch membrane used in this case
- Vacuum draw-down needed after release cycle complete
- The majority of the material deposited in layer on the membrane whose structure is dependent on the type of aggregates captured

PALL Pall Corporation

Results - SEM

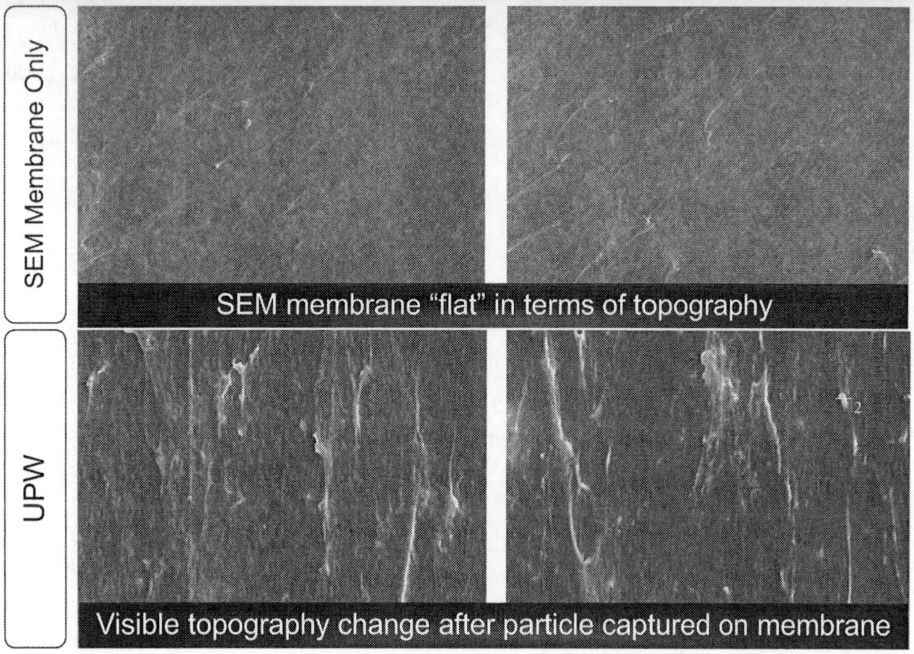

PALL Pall Corporation

Results - SEM

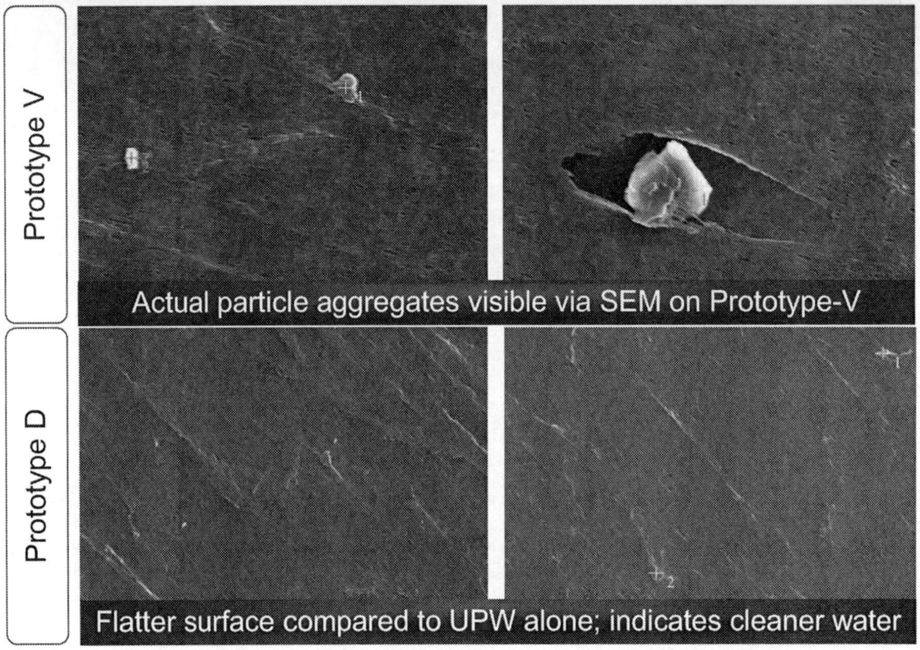

Pall Corporation

Results - EDX

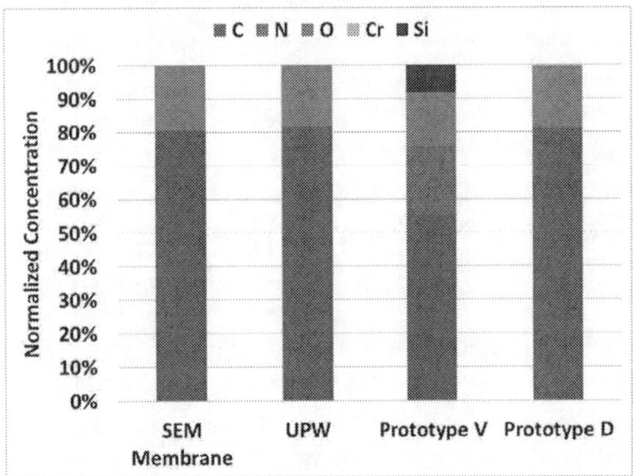

Trace Contaminants			
SEM Membrane	UPW	Prototype V	Prototype D
Na, Si, Al	Si, Al	Cr, Na, Al, Mg, Si, S	P, Si, Al, Na

- Al and Si are common contaminants across the samples
- Prototype-V has the largest array of species in its trace contamination profile

EDX of released particles confirms the interpretation of the IEC results

Pall Corporation

Summary

- **Smaller pore size leads to cleaner water**
  - Lower current in the IEC curves from the 2nm prototypes
- **Prototype-D cleaner than Prototype-V**
  - No difference in media used, contaminants from MOC
- **Electrophoretic method is a viable option for evaluating filter cleanliness and performance**
  - IEC is sensitive to the species that are released from the filters
  - The release (elution) time of the particles gives insight to contaminants species
  - Results verified by standard techniques

Pall Corporation

Next Steps

- Test performance against 20nm liquid particle counter
  - Where reported, 30 nm LPC used
- Verify performance of 2nm product independent of the upstream filter
- Determine filter performance in other application fluids
  - All tests presented performed in UPW
- Generate on-wafer data to validate electrophoretic method
  - Determine significance of bulk contamination vs trace contamination
- Test materials as single components
  - All data reported based on fully constructed filters

Pall Corporation

Acknowledgements

- Scientific and Laboratory Services
  - Ibrahim Mohamed-Ali
  - Kevin Alfonso
  - Ed Marino
  - Mike Gofkowski
  - Shawn Hubbard
- Device Research and Development
  - Jian Tan

PALL Pall Corporation

Metrologies of Amine and Ammonium in Post Etch Residue Removal

Jingjing Wang, Chuannan Bai, Guang Liang, Helen Lu and Eugene Shalyt
ECI Technology, 60 Gordon Drive, Totowa, New Jersey 07512, USA

Introduction

Post Etch Residue Removal (PERR) Challenges in the semiconductor industry driven to sub-10 nm node

- Removing residues completely from vias and metal lines with diminished feature sizes and increasing number of interconnect layers
- Not corroding metals or affecting low-κ dimensions and properties

Most modern PERR formulations include multiple subcomponents for various purposes such as corrosion inhibitors and chelators, while ammonium and amines remain the most common and basic active ingredients. The monitoring and control of these species become challenging because traditional metrologies may not be satisfactory due to the complexity of the solution.

Metrology of Ammonium for online Process Control

Method	Typical Spec	Pros	Cons
NIR	3% acc., 1% RSD	Real time analysis (<1 min) non-contact	Method establishment requires set of "learning" Standards and can be limited by wide pH range.
Electrochemical	5% acc., 3% RSD	Fast (<5 min) Minimal interference Low cost	May require chemical use (buffer)
Reagent Photo-metric	3% acc., 1% RSD	No need for full Standard	Requires multiple reagents Certain amines can interfere
IC	3% acc., 1% RSD	Free matrix interference	High cost >15 min analysis time

References

1. G.W. Gale, R. Small, K.A. Reinhardt, W. Kern (Eds.), "Handbook of Silicon Wafer Cleaning Technology", 2nd ed., William Andrew Inc., New York, pp. 201-265, 2008.
2. E. Shalyt, G. Liang, P. Bratin, C. Lin, "Real Time Monitoring for Control of Cleaning and Etching Solution" (Proceed. of SPWCC Conference, USA, 2007).
3. Solarzano, L *Limnol. Oceanogr.*, 14:799-800, 1969

• Measuring Ammonium by Electrochemical Method

Ammonium can be measured by Ion Selective Electrode (ISE) method. Under ideal conditions, the potential (E) of an ISE is given by the Nernst equation:

$$E=E_0-(2.303\frac{RT}{nF})\ \log[NH_4^+]$$

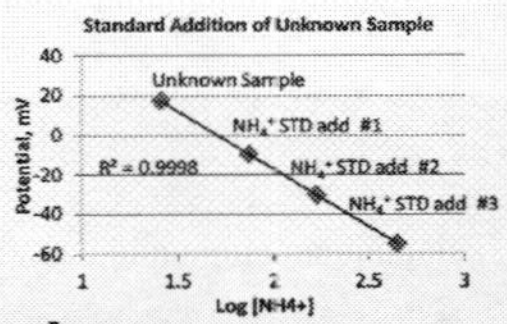

• Measuring Ammonium by Photometric Method

Ammonium can be converted to indophenol dye by the reaction:

2 ⬡–O⁻ + NH_3 + $3ClO^-$ → -O-⬡-N=⬡

Potential interference Test:

- Hydroxylamine (HA)
- Monoethanolamine (MEA)
- Diethanolamine (DEA)
- Triethanolamine (TEA)

Note: 0.18 μmol amines spiked into 0.18 μmol of NH_4^+

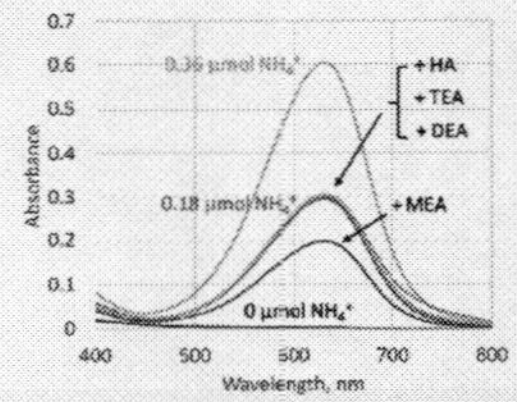

• Measuring Ammonium by IC

Ammonium can be be analyzed by ion chromatography, free from potential interferences of foreign ions.

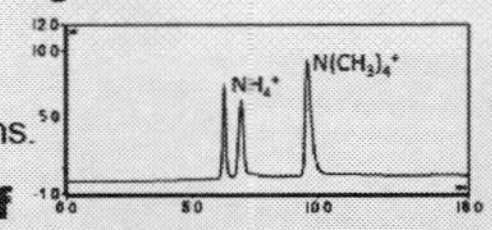

• Method Comparison of Ammonium Analysis in PERR Samples

Ammoniums in PERR samples were measured by photometric, electrochemical, and NIR methods. The results obtained by the three methods are highly comparable.

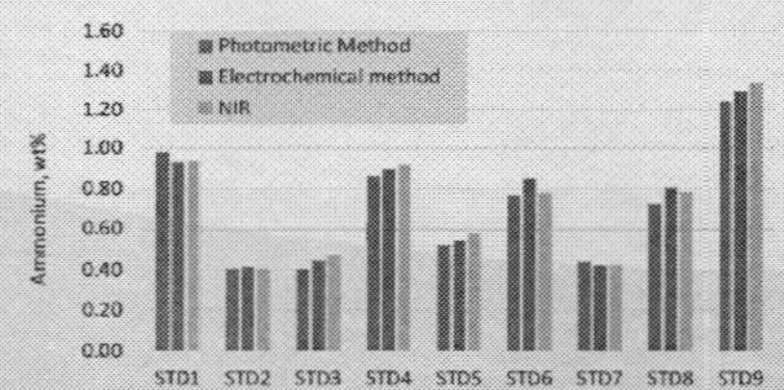

Metrology of Amines for Online Process Control

Method	Typical Spec	Pros	Cons
NIR	3% acc., 1% RSD	Real time analysis (<1 min) non-contact	Method establishment requires set of "learning" Standards
Acid Base Titration	3% acc., 2% RSD	No need for full Standard Low cost	Requires chemical use (reagents and titrants) Potential interference from bases with similar pKa as the analyte amine

• Measuring Amine by NIR

Amines can be measured by Near-infrared (NIR) spectroscopy combined with multivariate calibration.

$$c_{analyte}=c_{\lambda_1}\times A_{\lambda_1}+\ldots c_{\lambda_i}\times A_{\lambda_i}+\ldots c_{\lambda_n}\times A_{\lambda_n}+c_0$$

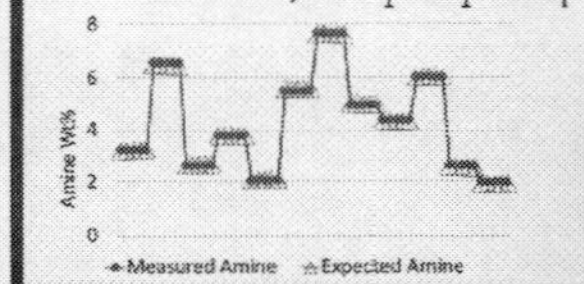

Samples	Expected Amine wt%	Measured Amine wt%	Relative Error
STD3	2.61	2.65	1.55%
STD3+Ammonia	2.59	2.55	2.47%
STD5	2.03	2.08	0.00%
STD5+TMAH	1.97	1.91	-1.72%
STD6	5.29	5.29	-2.81%
STD6+Solvent	5.14	5.02	-2.45%
STD6+Amine	9.29	9.25	-0.47%
STD6+Chelator	5.25	5.27	0.32%
STD6+DIW	4.80	4.80	-0.14%

• Measuring Amine by Titration

As an organic base, amines can be titrated by acid forming conjugated acids. Direct titration tends to have large errors due to other sources of acids/bases.

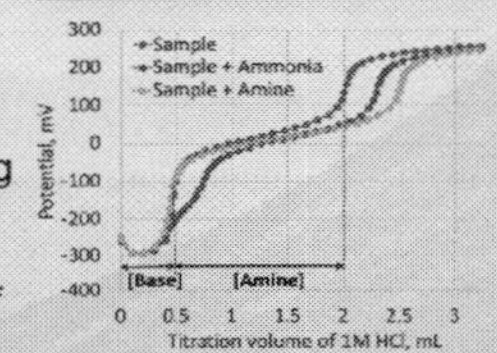

The method is developed to avoid interference from the alkalinity of the sample. Common sources of alkalinity such as ammonia does not impact the analysis of amine.

Conclusion

Near Infrared (NIR) technology has been an attractive metrology of ammonium and amine in PERR due to its real-time, non-contact, and non-reagent monitoring capabilities. Alternative metrologies (titrimetric, electro-chemical, photometric, IC, etc) can be preferable depending on solution complexity and process control requirements.

TECHNOLOGY

SPCC 2018

We Keep The Chemistry Right